A FIRST COURSE IN PROBABILITY AND STATISTICS

A FIRST COURSE IN PROBABILITY AND STATISTICS

B. L. S. Prakasa Rao

University of Hyderabad, India

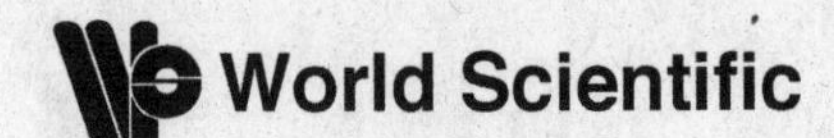

NEW JERSEY • LONDON • SINGAPORE • BEIJING • SHANGHAI • HONG KONG • TAIPEI • CHENNAI

Published by

World Scientific Publishing Co. Pte. Ltd.
5 Toh Tuck Link, Singapore 596224
USA office: 27 Warren Street, Suite 401-402, Hackensack, NJ 07601
UK office: 57 Shelton Street, Covent Garden, London WC2H 9HE

British Library Cataloguing-in-Publication Data
A catalogue record for this book is available from the British Library.

A FIRST COURSE IN PROBABILITY AND STATISTICS

ISBN-13 978-981-283-653-3
ISBN-10 981-283-653-5
ISBN-13 978-981-283-654-0 (pbk)
ISBN-10 981-283-654-3 (pbk)

Printed in Singapore.

In memory of my father

Bhagavatula Rama Murthy
(1906-1966)

for his love for Mathematics

Preface

As soon as one sees [illegible] book [illegible] question arises. When there are so many books [illegible] Introduction to Probability and Statistics, what is the need [illegible] another one? I have been teaching [illegible] courses in [illegible] at Master's level [illegible] at the Indian Statistical Institute and at undergraduate level at various universities [illegible]. During the last [illegible] years I have used a variety of textbooks for teaching undergraduate courses. Most of these books at the undergraduate level are too [illegible] [illegible] [illegible] [illegible] [illegible] Calculus is a [illegible] Probability and Statistics. Most of the [illegible] contains [illegible] statistical methods [illegible] [illegible] data analysis without understanding [illegible] basic concepts [illegible]. The student [illegible] knowledge to [illegible] when a method can be applied [illegible] normal probability density function [illegible] not enough [illegible] symmetric [illegible] would still need to know [illegible] probabilities of various events. Keeping these [illegible] in mind, [illegible] basic concepts [illegible] and Statistics in this book.

Thanks are due to Dr. R. [illegible] University of [illegible] [illegible] for their help in preparing the [illegible] [illegible] [illegible] [illegible] [illegible] support [illegible] of this project.

[illegible]

[illegible], [illegible] 2009

Preface

As soon as one sees this book, an immediate question arises. When there are so many books in print on the introduction to Probability and Statistics, what is the need or use of another one? I have been teaching courses in Statistics at Master's level in India at the Indian Statistical Institute and at undergraduate, graduate level at various universities in U.S.A. during the last forty two years. I have used a variety of text books for teaching undergraduate courses. Most of these books at the undergraduate level are too simplistic and problem oriented. Explanation of basic concepts and methods of Statistics need a reasonably good mathematical background. In my opinion, a good course in Calculus is a prerequisite for a First Course in Probability and Statistics. Most of the pocket calculators and computers contain Statistical packages for applying statistical methods. Overemphasis on the usage of these packages for data analysis without understanding the basic concepts is dangerous. The students must acquire the knowledge to differentiate when a method can be applied and when it cannot be. For instance, to understand what a normal probability density function is, it is not enough if I explain to them that it is unimodal, symmetric and bell-shaped. They would still need to know its functional form to compute the probabilities of various events. Keeping these ideas in mind, I tried to cover the basic concepts of Probability and Statistics in this book.

Thanks are due to Dr. R. Radha of University of Hyderabad and my children Vamsikrishna and Venumadhav for their help in preparing the Figures and Tables and Camera-ready copy for the book. As with all of my other books, my wife Vasanta and my son Ramagopal were always supportive of this project.

B.L.S.Prakasa Rao
Hyderabad, August 15, 2008

Contents

Chapter 1

WHY STATISTICS?

1.1 Introduction

Rational decision making at any stage needs information. If the government of a country wants to find out how many schools have to be started in a particular region, how many hospitals have to be built in a particular district or whether how many highways have to be extended or improved, it needs to collect the information on the present state of affairs and make an assessment on the needs and finances available prior to taking action. If an industry wants to expand its base in manufacturing, it needs to have information on its capacity to produce the product and on the demand for the product. If the government of a region wants to find out the average income of a household in a particular area or percentage employed in a particular state, a scientific decision can be made only after it obtains observations or data on such issues. If a manufacturer of drugs wants to introduce a new drug into the market for some disease, it has to study the performance of the new drug it discovered through clinical trials and possibly compare its performance to other drugs presently used possibly in the market for the same disease so that it can arrive at a rational conclusion. Such a procedure needs data on the recovery aspects of the new drug besides its side effects as compared to the old drug. If a meteorologist wants to predict or forecast the weather during a particular period in a year, say, in the month of November in a particular year, he or she can make a rational decision only if the information on the weather patterns during the month of November and possibly earlier months in the previous years is available. If a stock broker wants to know how the share prices of a company are likely to fluctuate in a specified time of the year, information on the volatility in the share prices of the company during the previous trading seasons is needed besides the general patterns of volatility of share prices in the stock market. If a commodity salesman wants to find how the prices of a particular commodity is likely to fluctuate in a given time of the year, he or she has to have information on the prices during the corresponding previous trading season. There is a common link that is apparent in all these situations. Any rational decision making process involves data and information. The subject of Statistics deals with developing the methods and the techniques for such a decision making process. Modelling of an observed data is an important statistical problem as it possibly brings out or indicates the causes leading to or underlying the phenomenon giving rise to this data. If a particular model does not explain the data or does not help in forecasting or predicting the data reasonably accurately or the actual observations do not follow the pattern indicated by the model, it is necessary to take a fresh look at the data and try to refine or improve the model. Of course it is important not to clutter the model to be constructed with too much information for the simple reason that too many restrictions lead to a model which is problem specific and will not be applicable to similar but slightly different problems. A model should be such that it encompasses or unifies several similar problems taking into account the common features of these problems for unified study or discussion. A basic requirement for such a study of the model building is to know the distribution or shape of the data and other relevant features of the data.

1.2 Representation of Data

We will now describe a method of representing a data or information through an example.

Suppose we are interested in estimating the average income of all the people employed in a particular city. A first stage in obtaining the estimate is to prepare a list of all people, choose a representative sample of the people employed in the city and then collect the information from the sample chosen. It is clear that if our sample consists of people only from the low income group or if all the people sampled come from the high income strata , the sample chosen is not representative. How to choose a representative sample from the city? We will not deal with this question here. The subject of study of methods of choosing samples is known as *Survey sampling.* Suppose we have obtained such a set of observations. Let us denote the sample of observations by $X_i, 1 \leq k \leq n$. We choose an origin x_0 and a positive real number h, called binwidth and divide the whole real line into intervals $[x_0 + mh, x_0 + (m+1)h)$ of length h, called bins. On every such bin of length h, erect a rectangle with its height equal to $\frac{1}{nh} \times$ (number of observations falling in that bin). The graph so obtained from the data is called a *histogram* for the data. The shape of a histogram for a data depends on the choice of the origin x_0 as well as the bin width h of the rectangles erected. There is no unique representation for any data using a histogram because the choice of the origin and the choice of the bin-width h determine the groups into which the data is divided. For practical reasons, it is convenient to choose h so that the number of groups is neither too large nor too small. This can be achieved by finding the *range* r of the observations, that is, the difference between largest and the smallest of the observations. If we want to divide the data into k classes , we can choose h to be $\frac{r}{k}$. An important use of a histogram is that it will indicate the shape of the data , for instance, whether it is unimodal or multimodal, whether it is bell-shaped or whether it is skewed from the left or from the right and so on. This is especially useful in modelling the data as we shall see later in this book. In order to develop statistical methods useful for analysis any data, we will first introduce the notion of probability of an event and study related concepts in the next chapter.

[illegible]

Chapter 2

PROBABILITY ON DISCRETE SAMPLE SPACES

2.1 Introduction

Let us consider an experiment where the outcome is unpredictable. A simple example of such an experiment is the tossing of a coin with the head on one side and the tail on the reverse side. If such an experiment is performed, the outcome of the experiment might be either a head or a tail or the coin might possibly even roll away from the place where you have tossed. We call such experiments as *random experiments* as opposed to the experiments which you perform in a physics laboratory or a chemistry laboratory where the outcome is predictable from past information. Suppose a random experiment consists of tossing a coin 10 times and recording the number of heads observed. Let us repeat the experiment 100 times and record the number of heads observed and then 1000 times, 10000 times and so on and collect the information obtained in a tabular form as given below.

Table 2.1

Number of tosses	Number of heads observed	Proportion
10	4	0.4
100	52	0.52
1000	492	0.492
10000	5196	0.5196

The last column in the above table indicates the proportion of heads which is the ratio of the number of heads observed to the total number of tosses. It is called the *relative frequency* of the outcome H where we indicate by the letter H that head has been observed on tossing the coin. As you can see from the last column of the above table , the proportion of heads is stabilizing around 0.5 as the number of tosses is increasing. We notice an interesting phenomenon in this observation. Even though we are not able to predict the outcome of a single toss of the coin, we see a pattern in the proportion of heads as the number of tosses is increased or as the random experiment of tossing a coin is repeated. This leads us to the concept of *probability* through what is called the *frequency interpretation* of the concept of probability. It is of course true that not all random experiments are repeatable in nature. Even in such cases, we do talk about the chances for a particular outcome of the experiment. For instance, if you ask me what are the chances that man or woman will step foot on the planet Mars before the end of Year 2020, I might say there is a chance of about 5%. Note that the quantification of 5% for reaching Mars did not come about by repeating the experiment of reaching Mars! It indicates my belief in such a statement. It is subjective and leads to the *subjective interpretation* of probability. As we shall see later in this chapter, the mathematics behind both the interpretations of the concept of probability is the same. Without much further ado, we will now introduce the concept of probability.

2.2 Probability

Sample space

Let us consider at first random experiments $\mathcal{E}$ where the number of possible outcomes of the experiment $\mathcal{E}$ are finite in number. For instance, if the experiment $\mathcal{E}$ consists of tossing a coin once, the possible number of outcomes is two, namely "head" or "tail". It is possible of course that the coin might roll away or it might disappear and so on. But we ignore such unlikely outcomes as we are in the process of building realistic models for further use. The list of outcomes of the experiment can be listed as a set $S = \{H, T\}$ where H stands for the outcome that a head has appeared on tossing and T for the outcome that a tail has appeared on tossing the coin once. Suppose we toss the coin twice. Let us record the outcome of the first toss and the outcome of the second toss in that order. Then the list of all possible outcomes is

$$S_1 = \{(H, H), (H, T), (T, H), (T, T)\}.$$

Here the element (H, T) denotes the event that the outcome of the first toss is a head and the outcome on the second toss is a tail. The set S_1 is called a *sample space* for this random experiment $\mathcal{E}$. A *sample space* for a random experiment $\mathcal{E}$ is a list of all the possible outcomes. Before designating a set as a sample space for an experiment, it is necessary to make sure that every element of the sample space corresponds to *one and only one* outcome of the experiment and every outcome of the experiment corresponds to *one and only one* element of the sample space. A listing of the elements in a sample space depends on the type of recording of the observations of the random experiment. For instance, if, in the experiment $\mathcal{E}$ of tossing a coin twice, we are informed or we can observe only the number of heads that have appeared, then a sample space can be specified in the form $S_2 = \{0, 1, 2\}$ where '0' denotes that ' there is no head on any of the two tosses', '1' denotes that 'exactly one head has appeared' and '2' denotes that 'two heads have appeared'. Hence the sample space of the same random experiment $\mathcal{E}$ can be represented in two different ways S_1 and S_2 depending on the type of recording of the outcome of the random experiment. However note that the sample space S_1 gives more detailed information than the sample space S_2. We will now discuss few more examples of random experiments.

Example 2.1. Suppose that a random experiment consists of tossing a coin three times and then recording the outcome of each toss. The sample space for the experiment is

$$S = \{HHH, HHT, HTH, HTT, THH, THT, TTH, TTT\}.$$

Example 2.2. Suppose that a random experiment consists of throwing a die once and then recording the face that appeared. A sample space for the experiment is $S = \{1, 2, 3, 4, 5, 6\}$ where the element "1" denotes the number of dots on the face that appeared is one and so on. If the recording of the outcome of the experiment indicates only whether an odd number or an even number has appeared, then the sample space can be formed in the form $S = \{Odd, Even\}$ where "Odd" indicates the subset of elements $\{1, 3, 5\}$ and "Even" indicates the subset of elements $\{2, 4, 6\}$.

Example 2.3. Suppose that the random experiment consists of throwing a die twice and then recording the faces that appeared at the first and at the second throw in that order. The sample space for the experiment is

$$\begin{aligned} S = \{&(1,1), (1,2), (1,3), (1,4), (1,5), (1,6), (2,1), (2,2), (2,3), (2,4), \\ &(2,5), (2,6), (3,1), (3,2), (3,3), (3,4), (3,5), (3,6), (4,1), (4,2), \\ &(4,3), (4,4), (4,5), (4,6), (5,1), (5,2), (5,3), (5,4), (5,5), (5,6), \\ &(6,1), (6,2), (6,3), (6,4), (6,5), (6,6)\}. \end{aligned}$$

As you can see from the sample space described above, a listing of all the elements in a sample space can be quite cumbersome at times. Even though the number of elements of a sample space is finite, it might be a very large number.

An element of a sample space is called a *sample point* or a *simple event* and a subset of a sample space is called an *event*. We say that an event A of a sample space S has occurred if the outcome of the experiment corresponds to a sample point in S belonging to the subset A.

There are many situations where the number of possible outcomes of a random experiment is infinite, either countably infinite or uncountably infinite. Recall that a set is countably infinite if its elements can be arranged in a sequence.

Let us consider the random experiment of recording the number of telephone calls received by you in a given period of time , say between 8.00 a.m and 9.00 a.m. on a particular day. The number of calls received may be nil, one or two or three and so on. Even though, it is impossible to get a very large number of calls due to physical constraints, it is clear that no specific bound can be placed on the total number of calls received. A sample space for this experiment can be written in the form

$$S = \{0, 1, 2, \ldots, k, \ldots\}$$

where the element 'k' of the set denotes that the number of calls received were k in number. Here '0' denotes the event that there were no calls. Note that the sample space here consists of a countably infinite number of sample points.

Let us consider another random experiment of selecting a real number from the closed interval $[0, 1]$. The number of possibilities for such a choice is uncountable. A sample space for this experiment is $S = \{x : 0 \leq x \leq 1\}$. Note that the sample space here consists of an uncountable number of elements.

Sample spaces which have either finite or countably infinite number of sample points will be called *discrete sample spaces*. Any subset of a discrete sample space is called an *event*. If a sample space has an uncountable number of sample points, then a specified collection of subsets only will be called events.

Before we conclude this section, we consider a couple of more examples leading to discrete sample spaces.

Example 2.4. Let us consider a random experiment where the result of the outcome can be categorized as either a success S or a failure F. If the experiment is repeated n times and we record the outcome each time, then the sample space S consists of 2^n elements (why?). For instance, the subset corresponding to the event $A =$ "All Successes" is the set consisting of a single element $\{(S, S, \ldots, S)\}$. A listing of all elements of the sample space is too lengthy for large n.

Example 2.5. Let us consider the experiment discussed in Example 2.4 and suppose we repeat the experiment until we obtain a success. Then the sample space corresponding to the experiment can be written in the form $\Omega = \{S, FS, FFS, FFFS, \ldots; E\}$ where, for instance, FS denotes a failure on the first trial and a success on the second trial of the experiment and E denotes the event that there was no success in the repeated trials.

Operations on events

Let S be a sample space corresponding to a random experiment $\mathcal{E}$. If A and B are events in a sample space S, then the subset $A \cup B$ stands for the occurrence of either the event A or the event B or both and the subset $A \cap B$ stands for the the event that both A and B have occurred. In a similar manner, all the set-theoretic operations on the subsets of the set S can be interpreted in the language of events of the sample space corresponding to a random experiment $\mathcal{E}$. In the following table, capital letters indicate the subsets of a universal set S or the corresponding events as subsets of a sample space S. The set Φ denotes the empty set or the null set.

Table 2.2

Operation	Set-theoretic	Probabilistic
Union	$A \cup B$	Either the event A or the event B or both
Intersection	$A \cap B$	Both the events A and B
Complementation	$\bar{A}$	Negation of the event A
Universal set	S	Sure event
Null set	Φ	Null event
Null intersection of the sets A and B	$A \cap B = \Phi$	Events A and B are disjoint

These operations can be explained through the shaded areas in the Venn diagrams given below.

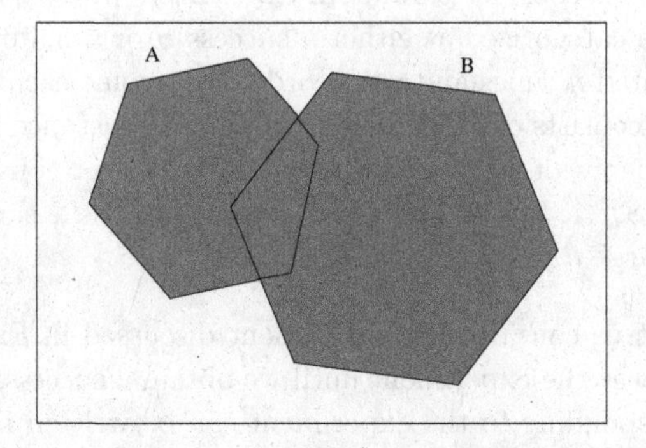

Fig. 2.1 Union of two events A and B

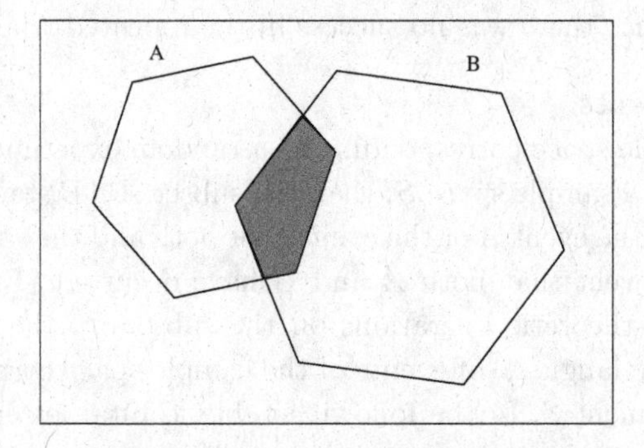

Fig. 2.2 Intersection of two events A and B

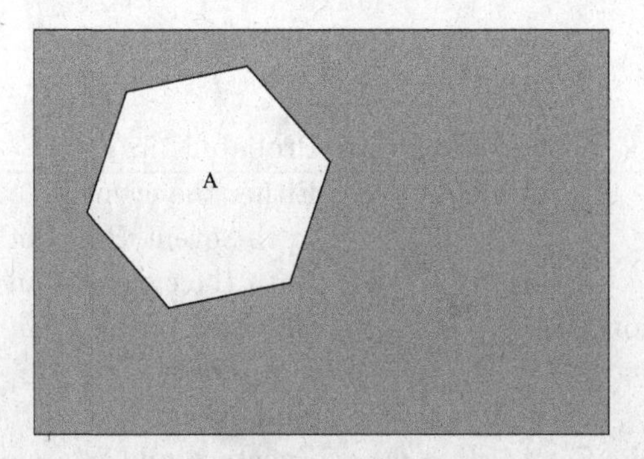

Fig. 2.3 Complement of an event A

The following rules governing the set-theoretic operations are useful in the study of events in a sample space corresponding to a random experiment.

1) $A \cup (B \cup C) = (A \cup B) \cup C = A \cup B \cup C.$
2) $A \cap (B \cap C) = (A \cap B) \cap C = A \cap B \cap C.$
3) $\overline{(A \cup B)} = \bar{A} \cap \bar{B}.$
4) $\overline{(A \cap B)} = \bar{A} \cup \bar{B}.$
5) $\bar{A} \cup A = S.$
6) $A \cup (B \cap C) = (A \cup B) \cap (A \cup C).$
7) $A \cap (B \cup C) = (A \cap B) \cup (A \cap C).$

Example 2.6. Let us consider the experiment of throwing a die twice. Let A be the event that an odd number appeared on the first throw and B be the event that an odd number appeared on the second throw. Then the subset of the sample space S corresponding to the event A is given by

$$A = \{(1,1),(1,2),(1,3),(1,4),(1,5),(1,6),(3,1),(3,2),(3,3),(3,4),(3,5),(3,6),(5,1),(5,2),(5,3),(5,4),(5,5),(5,6)\}$$

and the subset corresponding to the event B is given by

$$B = \{(1,1),(2,1),(3,1),(4,1),(5,1),(6,1),(1,3),(2,3),(3,3),(4,3),(5,3),(6,3),(1,5),(2,5),(3,5),(4,5),(5,5),(6,5)\}.$$

Then $A \cap B$ corresponds to the event that both the throws lead to odd numbers and $A \cup B$ corresponds to the event that at least one of the throws shows an odd number. Check out the corresponding subsets of S for these events.

Permutations and combinations

Before we go into the discussion on the concept of probability, we will study a fundamental principle or basic rule in counting which is useful in enumerating the number of sample points in a subset of a sample space or in the whole sample space.

Suppose that two jobs J_1 and J_2 have to be performed. Further suppose that the job J_1 can be done in m ways and the job J_2 can be done in n ways. How many way can both the jobs J_1 and J_2 be done? It is clear that for every one way of performing J_1 , there are n possible ways of doing J_2. Hence the total number of ways of doing both the jobs J_1 and J_2 is the product of m and n which is mn. This is the basic principle underlying the theory of permutations and combinations. This principle can be extended obviously to more than two jobs.

A *permutation* is an ordered arrangement of a collection of objects where as a *combination* ignores the order but recognizes the collection of objects. For instance, let us consider n objects numbered $1, 2, \ldots, n$.. The ordered arrangement $(1, 2, \ldots, n)$ and $(2, 1, \ldots, n)$ are two different permutations of the same combination $(1, 2, \ldots, n)$. However the number of combinations or unordered arrangements of these n objects is one.

Rule #1: The number of permutations of n objects is $n(n-1)\ldots 1$ which we denote by $n!$ (called "n factorial").

We can explain this by a simple idea. Let us think of filling up a set of n empty boxes with the n objects such that there is *one and only one* object out of the n in each box. Each such arrangement gives a permutation of the n objects. Note that the first box can be filled by any one of the n objects in n ways, the second box in $(n-1)$ ways (as one object is already in the first box), the third box in $(n-2)$ ways and so on. There will be only one box by the time we reach for the n-th box which is the last box and there will be only one object left. The last object can be placed in the last box in one and only one way. Applying the basic principle of counting discussed above, all the n jobs can be performed in

$$n(n-1)\ldots 1 = n!$$

ways. As a convention, we define $0!=1$.

Let us now find out the number of permutations of r objects out of n. The method of counting is the same as above. We now suppose that there are r empty boxes to be filled up by r objects out of n objects such that each box contains one and only one of the n objects. Note that the first box can be filled by any one of the n objects in n ways, the second box in $(n-1)$ ways (as one object is already in the first box), the third box in $(n-2)$ ways and so on. There will be only one box left by the time we reach for the r-th box and there will be only $(n-r+1)$ objects left out of the n objects. The last box can be filled in $(n-r+1)$ ways by any one of the $(n-r+1)$ objects which were left over. Applying the basic principle we discussed above, we get that all the r boxes can be filled up in

$n(n-1)\ldots(n-r+1)$ ways. Observe that

$$n(n-1)\ldots(n-r+1) = \frac{n(n-1)\ldots 1}{(n-r)(n-r-1)\ldots 1}$$
$$= \frac{n!}{(n-r)!}.$$

Here after we denote the above number by the symbol n_r^P. It denotes the number of permutations of r objects out of n objects and we have the following rule.

Rule # 2: The number of permutations of r objects out of n objects is $n_r^P = \frac{n!}{(n-r)!}$.

Let us now turn to the problem finding the number of combinations of r objects out of the n objects. Let n_r^C denote such a number. A little reflection shows that every permutation of r objects has to come out of a permutation of some combination of r objects chosen out of n. Since the total number of permutations of r out of n is n_r^P and since the number of permutations of r out of r is $r!$ by Rule# 1, it follows that $n_r^P = n_r^C r!$. Hence

$$n_r^C = \frac{n_r^P}{r!} = \frac{n!}{(n-r)!r!}$$

and we have the following rule.

Rule # 3: The number of combinations of r objects out of n objects is $n_r^C = \frac{n!}{(n-r)!r!}$.

Note that every combination of r objects out of n can be interpreted as a selection of r objects out of n and leaving out the rest. In other words , we are dividing the complete set of n objects into two classes, one class consists of r and the other contains the remaining $(n-r)$. Hence the number of combinations of r out of n is the same as the number of combinations of $(n-r)$ out of n, that is,

$$n_r^C = n_{n-r}^C.$$

This can also be checked by the Rule # 3.

Now we ask the following question: In how many ways can one partition a set of n objects into k groups such that the first group contains n_1 number of objects, the second group contains n_2 and so on till the k-th group contains n_k objects? Let us denote such a number by N. It is clear that there are $n!$ permutations in total of the n objects by Rule # 1. Further more every such permutation of n objects has to be obtained by choosing one of the N possible partitions of n objects into k groups and then by permuting the objects in each group of the partition. It follows therefore by Rule # 1 and the fundamental principle of counting that

$$n! = N(n_1!)\dots(n_k!).$$

Therefore

$$N = \frac{n!}{n_1!\dots n_k!}.$$

We will discuss applications of these different ways of counting later in this section.

Probability measure

Let us consider a random experiment $\mathcal{E}$ with a sample space S. For the present discussion, we will assume that the number of sample points in S is either finite or countably infinite. Such sample spaces are called *discrete sample spaces.* The case when the sample space has possibly uncountable number of elements will be discussed later in this book.

Definition. A *probability measure* $P(.)$ defined on a discrete sample space S is a real valued function defined on the class $\mathcal{F}$ of all subsets A of S satisfying the following conditions:

(i) $0 \leq P(A) \leq 1$ for all subsets A contained in S;
(ii) $P(S)$=1; and
(iii) if $A_1, A_2, \dots$ are disjoint events, then $P(\cup_{i=1}^{\infty} A_i) = \sum_{i=1}^{\infty} P(A_i)$.

Remarks. Here the class $\mathcal{F}$ is the class of all subsets of the sample space S. However, for sample spaces which possibly have uncountable number of sample points, this class is too big for defining a probability measure. We will not go into the reasons here.

It is clear from the definition of the probability measure that if A and B are disjoint events in S, then

$$P(A \cup B) = P(A) + P(B)$$

and in general if $A_1, A_2, \dots, A_n$ are finite number of disjoint events in S, then

$$P(\cup_{i=1}^{n} A_i) = \sum_{i=1}^{n} P(A_i). \tag{2.1}$$

In particular, for any event A in S,

$$P(\bar{A}) + P(A) = 1 \tag{2.2}$$

since A and $\bar{A}$ are disjoint events. Hence

$$P(\bar{A}) = 1 - P(A)$$

for all events A in S.

Suppose A and B are two events such that the occurrence of the event A implies the occurrence of the event B. Then the set A is contained in the set B as a subset of the sample space S. Check that $B = A \cup (B - A)$ where $B - A$ denotes the set of all points in B not contained in A. Note that the sets A and $B - A$ are disjoint. Hence, by the property (2.1) of a probability measure, $P(B) = P(A) + P(B - A)$ and we have the property

$$P(B - A) = P(B) - P(A) \tag{2.3}$$

whenever an event A implies another event B or equivalently the set A is contained in the set B as subsets of the sample space S. Since the probability of the set $B - A$ has to be nonnegative, it follows that $P(B) - P(A) \geq 0$ or equivalently

$$P(B) \geq P(A) \tag{2.4}$$

whenever an event A implies another event B.

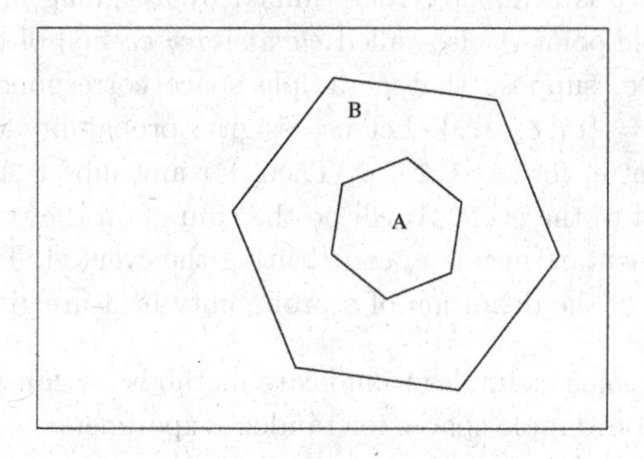

Fig. 2.4 Event A contained in Event B

Let us now consider any two events A and B which need not be disjoint. It can be checked that the set $A \cup B$ can be represented as $A \cup (B - (A \cap B))$ and the sets A and $B - (A \cap B)$ are disjoint.

Hence

$$P(A \cup B) = P(A) + P(B - (A \cap B))$$

by the property (2.1). Since the set $A \cap B$ is contained in the set B,

$$P(B - (A \cap B)) = P(B) - P(A \cap B)$$

by the property (2.3). As a consequence of these relations, we obtain that

$$P(A \cup B) = P(A) + P(B) - P(A \cap B) \tag{2.5}$$

for any two events A and B. This is known as the *Additive law* of probability.

Check that, for three events A, B and C,

$$P(A \cup B \cup C) = P(A) + P(B) + P(C) - P(A \cap B) - P(B \cap C) - P(C \cap A) + P(A \cap B \cap C)$$

and, in general, for any set of n events $A_i, 1 \leq i \leq n$ in S,

$$P(\cup_{i=1}^n A_i) = \sum_{i=1}^n P(A_i) - \sum_{1 \leq i \neq j \leq n} P(A_i \cap A_j) + \cdots + (-1)^{n-1} P(A_1 \cap A_2 \cap \cdots \cap A_n).$$

As a consequence of (2.1), we note that the probability measure on a discrete sample space is completely determined by assigning probabilities for individual sample points (also called *elementary events*) of the sample space. For instance, suppose that a sample space corresponding to an experiment $\mathcal{E}$ is $E = \{e_1, e_2, \ldots\}$. Let us assign a probability value p_i to the elementary event e_i for $i = 1, 2, \ldots$. Then, for any subset $A \subset S$, the probability assigned to the event A will be the sum of all the probabilities assigned to the elementary events e_i constituting the event A. This follows from condition (iii) of the definition of a probability measure given above.

We now discuss some examples to indicate methods for construction of probability models on sample spaces for random experiments.

Example 2.7. Consider again the example of tossing a coin once. A sample space for this experiment is the set $S = \{H, T\}$ where H and T are as defined earlier. Suppose we assign a number p between 0 and 1 as the probability of the elementary event H. Since the probability for the set S has to be 1 by the definition of a probability measure, it follows that the probability to be assigned to the elementary event T should be 1-p. If the outcomes H and T are equally likely, then $p = 1/2$ and the coin is said to be a fair coin or an unbiased coin.

Example 2.8. Let us consider another experiment where a fair coin is tossed three times as discussed in the Example 2.1. A sample space for this experiment is

$$S = \{HHH, HHT, HTH, HTT, THH, THT, TTH, TTT\}.$$

Since the coin is a fair coin, it is intuitively reasonable to assume that all the elementary events in the sample space S are equally likely and a probability assignment of $\frac{1}{8}$ for each of the eight elementary events constituting the sample space S will give rise to a probability measure on S.

Example 2.9. Let us now consider an experiment where a fair coin is tossed until a head H appears and we record the number of the lost toss. A sample space for this experiment consists of countably infinite number of elements and it can be represented in the form $S = \{1, 2, \ldots, k, \ldots E\}$ where the element k denotes that the first $(k-1)$ tosses yielded tails T and that the k-th toss resulted in heads H and E denotes the event that all the tosses resulted in tails T. The question now is how to assign a probability model for this experiment. One possibility is to assign the probability $\frac{1}{2^k}$ for the elementary event k and zero probability for the event E. We leave it to the reader to check that such an assignment gives rise to a probability measure on S.

If the coin is not necessarily fair in the above example, how to assign a reasonable probability model? We will discuss this question later in this in this book.

Example 2.10. Consider the random experiment of recording the birthdays of students in a class of n students. We ignore leap year considerations and assume that all the 365 days of the year are equally likely as birthdays. It is clear that the sample space for the experiment consists of $(365)^n$ sample points and all the elementary events are equally likely to occur. This follows from the fundamental principle of counting since each of the n students can have any one of the 365 days as his or her birthday and the total number of possibilities is

$$(365)(365)\ldots(365)$$

where the product has n terms. We can assign a probability of $\frac{1}{(365)^n}$ to each of the sample points to build a probability measure on the sample space. Let us now ask the question what is the probability that at least two of the n students have the same birthday. Let E denote such an event. Then $P(E) = 1 - P(\bar{E})$ and $\bar{E}$ is the event that all of the n students have different birthdays. The number of possible ways that all the n students can have different birthdays is

$$(365)(364)\ldots(365-(n-1)).$$

Since each of these sample points have the probability $\frac{1}{(365)^n}$, it follows that the probability of the event $\bar{E}$ is

$$\frac{(365)(364)\ldots(365-(n-1))}{(365)(365)\ldots(365)}.$$

Hence the probability of the event E is

$$1-\frac{(365)(364)\ldots(365-(n-1))}{(365)(365)\ldots(365)}.$$

Check what happens if n=23! The probability that at least two or more students will have the same birthday will be more than 50% in a group of 23 students.

Example 2.11. Let us consider a box containing N objects out of which a are red and b are black. Suppose that all the objects are of the same type but differ only in their colour. Suppose we want pick a sample of n objects out of N at random, that is, in such a way that all combinations of such a selection are equally likely. Rule # 3 implies that the number of possibilities for such a selection is

$$N_n^C = \frac{N!}{(N-n)!n!}.$$

The sample space for this experiment consists of N_n^C sample points and all are equally likely. Hence an assignment of the probability $\frac{1}{N_n^C}$ to each of the N_n^C sample points gives rise to a probability measure on the sample space. Suppose now that we select n objects out of N as described above and ask the question: what is the probability of selecting k red objects in a sample of n? It is obvious that k has to be less than or equal to the sample size n and less than or equal to the number of available red objects a in the box. For such a value of k, the sample will consist of k red objects and $(n-k)$ black objects. Following the fundamental principle of counting and

Rule #3, it follows that the number of such combinations is $(a_k^C)(b_{n-k}^C)$ and each such combination has the probability $\frac{1}{N_n^C}$. Hence the probability of selecting k red objects out of n is

$$\frac{(a_k^C)(b_{n-k}^C)}{N_n^C}.$$

2.3 Conditional Probability

We will now discuss a method of a assignment of probability models for random experiments when partial information is available.

Let us consider the random experiment discussed in the Example 2.11. Suppose we are interested in assigning probabilities to different combinations of outcomes when $n = 2$ objects are drawn out of the N objects in the box. Suppose we select one object at first and then a second object from the box without replacing the object drawn at the first stage. Let A be the event that the first object selected is red and B be the event that the second object selected is red. It is clear that the combination of objects in the box at the second draw depends on what has happened in the first draw. In other words the likelihood or the probability of the event B should depend on the occurrence or the nonoccurrence of the event A at the first draw. We define the *conditional probability* of an event B given A to be

$$P(B|A) = \frac{P(B \cap A)}{P(A)}.$$

Conditional probability of an event B given an event A is defined only when $P(A) > 0$. It is easy to see that

$$P(B \cap A) = P(A)P(B|A).$$

This formula is called the *multiplicative law* of probability.

Example 2.12. Let us again consider the situation discussed in the Example 2.11 and suppose we are selecting two objects at random from the N in the box without replacement. A sample space for this experiment can be represented in the form $S = \{RR, RB, BR, BB\}$ where the first letter denotes the colour of the object on the first draw and the second letter denotes the colour of the object in the second draw. Let D be the event that the colour of the object on the first draw is R and E be the event that the color of the object on the second draw is R. It is reasonable to assign the conditional probability of E given D to be $\frac{a-1}{N-1}$ as there are only $a-1$ red objects in the box out of $N-1$ which is the total number after the first draw of a red object and all are equally probable. We can assign the probability $\frac{a(a-1)}{N(N-1)}$ to the sample point RR following the multiplicative law of probability. Similarly we can assign probabilities to the simple events RB, BR and BB and build the probability model for the experiment. Note that the event D is the subset $\{RR, RB\}$ and the event E is the subset $\{RR, BR\}$ when represented as subsets of the sample space S and the event $D \cap E$ is the subset $\{RR\}$. Hence

$$\begin{aligned}P(E|D) &= \frac{P(RR)}{P(RR)+P(RB)} \\ &= \frac{\frac{a(a-1)}{N(N-1)}}{\frac{a(a-1)}{N(N-1)}+\frac{ab}{N(N-1)}} \\ &= \frac{a(a-1)}{a(a-1)+ab} \\ &= \frac{a-1}{N-1}\end{aligned}$$

observing that $N = a+b$. We assigned the probabilities to different sample points in the sample space S by making use of the multiplicative law of probability as described above.

Example 2.13. Let us now consider the following random experiment. Suppose there are two boxes numbered '1' and '2' with the Box'1' containing a red objects and b black objects and the Box'2' containing c red objects and d black objects. The random experiment consists of selecting a box at random from the two boxes and then selecting one object at random from the box chosen earlier. We will now discuss how to build probability models for such experiments. Let E_1 denote the event that the Box '1' is selected and E_2 denote the event that the Box '2' is selected. Let us also denote the event that a red object is selected by R and the event that a black object is selected by B. The sample space for the experiment can be represented in the form $S = \{(E_1, R), (E_1, B), (E_2, R), (E_2, B)\}$. Note that

$$\begin{aligned}P(R) &= P(E_1 \cap R) + P(E_2 \cap R) \\ &= P(E_1)P(R|E_1) + P(E_2)P(R|E_2)\end{aligned}$$

by the multiplicative law and hence

$$P(R) = \frac{1}{2}\frac{a}{a+b} + \frac{1}{2}\frac{c}{c+d}.$$

We note that the sets E_1 and E_2 form a partition of the sample space S. In other words the sets E_1 and E_2 are disjoint and their union is the space S and the above example gives a general method of computing probabilities when there are two stages involved in performing the random experiment.

The basic relation, which gives the above method for building probability models, is the *law of total probability.* We now state it as a theorem.

Theorem 2.1. **(Law of total probability)** Let $\{A_i, 1 \leq i \leq n\}$ be a partition of a sample space S corresponding to a random experiment, that is, $A_i \cap A_j = \Phi$ for all $1 \leq i \neq j \leq n$ and $\cup_{i=1}^n A_i = S$. Further suppose that $P(A_i) > 0, 1 \leq i \leq n$.. Then, for any event E in S,

$$P(E) = P(A_1)P(E|A_1) + P(A_2)P(E|A_2) + \cdots + P(A_n)P(E|A_n).$$

Proof. Since

$$A_1 \cup A_2 \cup \ldots A_n = S,$$

it follows that

$$E = (E \cap A_1) \cup (E \cap A_2) \cup \cdots \cup (E \cap A_n).$$

But the sets $E \cap A_i, 1 \leq i \leq n$ are disjoint since the sets $A_i, 1 \leq i \leq n$ are disjoint. From the properties of a probability measure, it follows that

$$P(E) = P(E \cap A_1) + P(E \cap A_2) + \cdots + P(E \cap A_n).$$

Applying the definition of the conditional probability, we get that

$$P(E \cap A_i) = P(A_i)P(E|A_i), 1 \leq i \leq n.$$

Combining the above relations, we obtain that

$$P(E) = P(A_1)P(E|A_1) + P(A_2)P(E|A_2) + \cdots + P(A_n)P(E|A_n).$$

Example 2.14. Let us now consider the experiment described in the Example 2.13 again. Suppose a red object is observed at the end of the experiment. Let us ask the question what is the probability that the object observed has been selected from Box '1'? In other words we would like to find the conditional probability of the event E_1 given that the event R has occurred. Note that

$$\begin{aligned} P(E_1|R) &= \frac{P(E_1 \cap R)}{P(R)} \\ &= \frac{P(R|E_1)P(E_1)}{P(R)} \\ &= \frac{P(R|E_1)P(E_1)}{P(R|E_1)P(E_1) + P(R|E_2)P(E_2)} \end{aligned}$$

and we can compute the required probability from the above formula. Such a formula is known as the *Bayes rule.* The next theorem gives a general formulation of the Bayes rule.

Theorem 2.2. (**Bayes rule**) Let $\{E_1, E_2, \ldots, E_k\}$ be a partition of a sample space S corresponding to a random experiment, that is, a collection of subsets E_i, $1 \leq i \leq k$ of the set S such that $\cup_{i=1}^{k} E_i = S$ and $E_i \cap E_j = \Phi$ for $1 \leq i \neq j \leq k$. Then, for any event F in S with $P(F) > 0$,

$$P(E_j|F) = \frac{P(F|E_j)P(E_j)}{P(E_1)P(F|E_1) + P(E_2)P(F|E_2) + \cdots + P(E_k)P(F|E_k)}$$

for $1 \leq j \leq k$.

We now give some more examples.

Example 2.15. Suppose that the reliability of a test for the detection of a disease such as AIDS is as follows: Of those having AIDS, 90% were detected by the test but 10% go undetected. Further among those free of AIDS, 99% are detected free of the disease by the test but 1% are diagnosed as having AIDS by the test. Suppose the probability for a person to contract AIDS is .001. If a person chosen at random is diagnosed as having AIDS by the test, what is the probability that person is in fact contracted with the AIDS?

Let E denote the event that a person selected has AIDS and F be the event that the test administered on a person diagnoses the person as having AIDS. We would like to calculate $P(E|F)$. We are given that $P(E) = .001, P(F|E) = .9$, and $P(F|\bar{E}) = .01$. Applying the Bayes formula, we get that

$$\begin{aligned} P(E|F) &= \frac{P(E \cap F)}{P(F)} \\ &= \frac{P(F|E)P(E)}{P(F|E)P(E) + P(F|\bar{E})P(\bar{E})} \\ &= \frac{(.9)(.001)}{(.9)(.001) + (.01)(.999)} \\ &= .083. \end{aligned}$$

Note that even though the test used for the diagnosis of AIDS is reliable, it can only give correct diagnosis about 8% of the time for an individual who is afflicted with the AIDS.

Example 2.16. Transmission of information over communication channels is done using byte (binary units). Suppose in a communication system, a '0' is transmitted with probability p and '1' is transmitted with probability $1-p$. Due to the noise in the communication system, it is possible that a '0' transmitted can be received as '1' with probability α and a '1' transmitted can be received as '0' with probability β. Suppose '1' was received. What is the probability that '1' was transmitted from the original source?

Let E be the event that '1' was transmitted and F be the event that '1' is received. Note that

$$\begin{aligned} P(E|F) &= \frac{P(E \cap F)}{P(F)} \\ &= \frac{P(F|E)P(E)}{P(F|E)P(E) + P(F|\bar{E})P(\bar{E})} \\ &= \frac{(1-\beta)(1-p)}{(1-\beta)(1-p) + \alpha p}. \end{aligned}$$

Note that the conditional probability calculated above depends on p as well as α and β. If $\alpha = \beta = 1$ or 0, then the channel is said to be noiseless.

2.4 Independence

Let us consider again the random experiment discussed in the Example 2.11. Suppose we are interested in assigning the probabilities to different possibilities of outcomes when $n = 2$ objects are drawn out of the N in the box. Suppose we select one object at first, note down its colour, place it again in the box and then a second object is drawn from the box. Let A be the event that the first object selected is red and B be the event that the second object selected is red. It is clear that the probability assignment to the event B should not depend on the occurrence of the the event A as the composition of the red and black objects in the box has not changed. In other words the conditional probability of the event B given that the event A has occurred should be the same as the probability of the event B irrespective of the occurrence of the event A. In such a case, the events A and B are said to be independent. Note that $P(B|A) = P(B)$ implies that $P(A \cap B) = P(A)P(B)$.

Definition. Let S be a sample space corresponding to a random experiment. Two events A and B are said to be *independent events* if

$$P(A \cap B) = P(A)P(B).$$

Three events A, B, C are said to be *mutually independent events* if

(i) $P(A \cap B) = P(A)P(B)$,
(ii) $P(B \cap C) = P(B)P(C)$,
(iii) $P(C \cap A) = P(C)P(A)$, and
(iv) $P(A \cap B \cap C) = P(A)P(B)P(C)$.

Example 2.17. Suppose a fair die is thrown twice. Let E be the event that "6" appears on the first throw and F be the event that "5" appears on the second throw. We assign equal probability to all the 36 possible outcomes. Check that $P(E) = 1/6, P(F) = 1/6$ and $P(E \cap F) = 1/36$. Hence $P(E \cap F) = P(E)P(F)$ which shows that the events E and F are independent.

Example 2.18. Suppose a fair die is thrown twice and let A be the event that an odd number has appeared on the first throw and B be the event that an even number has appeared on the second throw. It is easy to check that $P(A) = 1/2, P(B) = 1/2$ and $P(A \cap B) = 1/4$ and hence A and B are independent events.

Remarks. For building probability models for random experiments in some examples earlier, we assumed that all the possible outcomes are equally likely when there are finite number of possible outcomes for the experiment. However there are several situations where this is not possible. One can build probability models for complex experiments involving several stages by applying the notions of conditional probability and independence. A complex experiment can sometimes be divided into two or many stages. If there is information about different stages of the experiment, then this can be used to build the conditional probability for models from one stage to the next stage of the experiment as in the Example 2.13.

2.5 Exercises

2.1 A group of three people is selected at random from six people named A, B, C, D, E and F. (a)Find the probability that A or B is selected.

(b) Find the probability that neither A nor B is selected.

2.2 A die is said to be fair if all of its faces are equally likely to appear. Suppose a fair die is thrown twice. What is the probability that at least one throw results in a number less than 3?

2.3 The output of a machine producing bolts is known to contain 2% defectives. From a large lot of bolts, two are drawn at random. Find the probability that at least one of them is defective.

2.4 A student applies for admission to colleges A and B. Suppose the probability that the student will be admitted to the College A is 0.7 and that he or she will be rejected by the College B is 0.5. Further suppose that the probability that at least one of the colleges rejects the student is 0.6. Find the probability that the student gets admission in at least one of the colleges A and B.

2.5 For any two events A and B in a sample space S, prove that

$$P(E \cap F) \leq P(E) \leq P(E \cup F) \leq P(E) + P(F).$$

2.6 For any three events E_1, E_2, E_3 contained in a sample space S, show that

$$P(E_1 \cup E_2 \cup E_3) = P(E_1) + P(E_2) + P(E_3) - P(E_1 \cap E_2) \\ - P(E_2 \cap E_3) - P(E_3 \cap E_1) + P(E_1 \cap E_2 \cap E_3).$$

2.7 A committee of three people is selected from a group of six people A, B, C, D, E, and F. Find the conditional probability that A and B are selected given that C and D are not selected.

2.8 An almirah contains four black, six brown and two black socks. Two socks are chosen at random from the almirah. Find the probability that both the socks chosen will be of the same colour.

2.9 In a college , there are 40% girls in the first year, 30% in the second year and 30% in the third year. It is observed that 30% of all the first year girl students, 30% of the second year girl students and 50% of the third year girl students are enrolled in a computer course. What is the probability that a girl chosen at random is enrolled in a computer course?

2.10 A student taking a true-false test marks the correct answer to a question when he or she knows it and decides true or false on the basis of tossing a fair coin when the answer is not known. Suppose the probability that the student knows the answer for a question is $\frac{3}{5}$. What is the probability that the student knew the answer to a question to a correctly marked question?

2.11 Suppose a fair coin is tossed twice. Let E be the event " not more than one head" and F be the event "at least one of each face". Show that the events E and F are not independent.

2.12 If E and F are independent events, show that E and $\bar{F}$, $\bar{E}$ and F, and $\bar{E}$ and $\bar{F}$ form pairs of independent events.

2.13 If E_1, E_2, E_3 are independent events, prove that $\bar{E}_1$ and $E_2 \cap \bar{E}_3$ are independent events.

2.14 A shot is fired from each of three guns numbered 1,2 and 3 in a shooting competition. Let E_i be the event that the target is hit by the i-th gun. Suppose that $P(E_1) = .5, P(E_2) = .6$ and $P(E_3) = .8$. Assuming that the events $E_i, i = 1, 2, 3$ are independent, find the probability that exactly one gun hits the target.

Chapter 3

DISCRETE PROBABILITY DISTRIBUTIONS

3.1 Introduction

We have seen earlier that the sample space for a random experiment might have either finite number of elements or a countably infinite number or even an uncountable number of elements in it. We called those sample spaces which have a finite or countable number of elements as *discrete sample spaces.* The subject matter for this chapter is the study of probability measures defined on discrete sample spaces or some times called *discrete probability distributions.*

3.2 Discrete Random Variables

We are not interested most of the time in the actual outcome of a random experiment we are considering but on a characteristic based on it. For instance, if we toss a coin n times, we might be interested only in the total number of heads that appeared in n tosses rather than the individual information on the outcome of each toss. A *discrete random variable* is a numerical valued function defined on the discrete sample space S corresponding to a random experiment $\mathcal{E}$.

Let us suppose that a sample space S corresponding to an experiment $\mathcal{E}$ is defined by $S = \{e_1, e_2, \dots\}$ and let $P(.)$ be a probability measure defined on S and $P(\{e_i\}) = v_i, i \geq 1$. Suppose that X is a discrete random variable defined on S taking the values $x_i, i \geq 1$. Compute the probability that X takes the value x_i. Let the $P\{X = x_i\} = p_i, i \geq 1$. This function is called the *probability function* or *probability distribution* of the discrete random variable X. For any real number x, define

$$F(x) = \sum_{i: x_i \leq x} p_i.$$

The function $F(.)$ is called the cumulative *distribution function* of the discrete random variable X. We will discuss properties of the distribution functions in detail later in this book.

Expectation

The mean or the *expectation* (also called the expected value) of a discrete random variable X, denoted by $E(X)$, is defined by

$$E(X) = \sum_{i=1}^{\infty} x_i p_i$$

whenever the infinite series $\sum_{i=1}^{\infty} x_i p_i$ is absolutely convergent. Suppose the series is absolutely convergent. Then $E(X)$ is finite and we will denote it by μ. Expectation of a random variable indicates the average value or the central tendency for X that one can expect as the name suggests. For instance, if the random variable takes the values $x_i, i = 1, \ldots, n$ and if $P(X = x_i) = \frac{1}{n}$, then

$$\begin{aligned} E(X) &= \sum_{i=1}^{n} x_i P(X = x_i) \\ &= \sum_{i=1}^{n} x_i \left(\frac{1}{n}\right) \\ &= n^{-1}(x_1 + \cdots + x_n) \end{aligned}$$

which is the usual average of the numbers $x_1, \ldots, x_n$. However the expectation of a random variable may at times be quite misleading . This can be seen from the following example.

Example 3.1. Let X_1 be a random variable taking the values 10 and -10 with probability 1/2 each and X_2 be a random variable taking the values 10000 and -10000 with probability 1/2 each. Check that $E(X_1) = 0$ and $E(X_2) = 0$ but there is wide disparity between the possible values of X_1 and those of X_2.

It is also possible that the mean of a random variable might not exist. This can be seen from the following example.

Example 3.2. Let X be a discrete random variable taking the values 2^i with probability 2^{-i} for $i = 1, 2, \ldots$. Check that $E(X)$ does not exist.

Further more if a random variable takes only nonnegative values , then its expectation is either finite and nonnegative or it is infinite. The following properties of the expectation of a random variable are easy consequences of the definition.

(i) $E(X) = c$ if $P(X = c) = 1$ for any constant c.
(ii) $E(aX + b) = aE(X) + b$ for any two constants a and b.
(iii) If X is a nonnegative random variable, then $E(X) \geq 0$.
(iv) $E(X + Y) = E(X) + E(Y)$ provided all the expectations are finite.

The last property will be called the *additivity* property of the expectation. We will prove this and other properties of the expectation later in this book.

Variance

Variability or spread or dispersion of a random variable X from its mean μ can be measured by looking at the difference of X and μ. This difference can be positive or negative or zero depending on the possible values for X. One can measure the average variability of X by taking the expectation of $X - \mu$. But this is not a meaningful way of measuring the variability as $E(X - \mu)$ is always equal to zero for any random variable X with finite expectation μ by the properties of expectation stated earlier (check?). In order to quantify the variability of a random variable X, a convenient measure is $E(X - \mu)^2$ and it is called the *variance* of the random variable X whenever it exists. It is also possible to measure the variability by using $E|X - \mu|$. It is however inconvenient to handle mathematically. We denote the variance of the random variable X by $Var(X) = \sigma^2$ whenever it is finite. It is always nonnegative if it is finite. This is obvious from the property (iii) of the expectations. Further more

$$
\begin{aligned}
Var(X) &= E(X - \mu)^2 \\
&= E(X^2 - 2\mu X + \mu^2) \\
&= E(X^2) - 2\mu E(X) + \mu^2 \\
&= E(X^2) - \mu^2 \\
&= E(X^2) - [E(X)]^2
\end{aligned}
$$

since $E(X) = \mu$. All the above equalities follow from the properties of expectations stated above. The last equality gives an alternate method for computing the variance of a random variable whenever it exists. From the fact that $Var(X) \geq 0$, it follows that

$$E(X^2) \geq [E(X)]^2$$

for any random variable X.

Moments

Just as the mean and the variance of a random variable are measures of the average and dispersion of a random variable respectively, we now introduce the concept of moment of a random variable. We will see the use of these characteristics in the study of the probability distribution of a random variable.

The k-th *raw moment* or the k-th *moment* of a random variable X is defined to be $E(X^k)$ whenever it is finite and the function $M_X(t) = E(e^{tX})$ for real t, whenever it is finite, is called the *moment generating function* of the random variable X.

We will present some examples discussing these concepts in the next section.

Independence

Let X and Y be two discrete random variables. The random variables X and Y are said to be (stochastically) *independent* if

$$P(X = x, Y = y) = P(X = x)P(Y = y)$$

for all x and y. In general, a set of discrete random variables $X_1, X_2, \ldots, X_n$ are said to be *independent* if

$$P(X_1 = x_1, X_2 = x_2, \ldots, X_n = x_n) = \Pi_{i=1}^n P(X_i = x_i)$$

for $-\infty < x_i < \infty, i = 1, \ldots, n$.

We will come across examples of such sets of random variables in the next section.

3.3 Discrete Probability Distributions

We now discuss some elementary discrete probability distributions.

Bernoulli trials

Consider a random experiment whose outcomes can be described as either a success S or a failure F. Let p be the probability for a success and $q = 1 - p$ will be the probability for a failure. Suppose the experiment is performed only once. Such a trial is called a *Bernoulli trial.* Let X be a random variable taking the value 1 if the outcome of the Bernoulli trial is a success S and 0 if the outcome is a failure F. Such a random variable is called a *Bernoulli random variable.* The sample space for the experiment consists of two elements namely S and F and

$$P(S) = P(X = 1) = p \text{ and } P(F) = P(X = 0) = 1 - p.$$

It is easy to see that $E(X) = p$ and $Var(X) = p(1 - p)$.

Binomial distribution

Suppose we consider n independent Bernoulli trials with the same probability of success p. Let $X_i = 1$ if the outcome on the i-th trial is a success S and $X_i = 0$ if the outcome on the i-th trial is a failure F. Let X be the total number of successes in n trials. It is clear that the possible values of X are $0, 1, \ldots, n$ and

$$X = X_1 + \cdots + X_n.$$

The sample space Ω for this random experiment consists of 2^n elements each of which can be represented by an ordered string with n components of 0's and 1's. The Bernoulli trials are considered to be independent in the sense that the corresponding Bernoulli random variables $X_i, i = 1, \ldots, n$ are independent. This leads to an assignment of the probability $p^r q^{n-r}$ for any elementary event in the sample space Ω consisting of r successes S and $(n - r)$ failures F irrespective of in which order they appear in the elementary event. We now count the number of ways in which such an elementary event with r successes and $n - r$ failures can occur. Think of n empty boxes which have to be filled up with either with an S alone or with an F alone. It is easy to see that the number of such ways is the same as the number of ways of picking r empty boxes and filling all of them with one S in each box and the remaining $n - r$ boxes with one F in each box. This can be done in n_r^C ways by the Rule # 3 given in the Chapter 2. Hence

$$P(X = r) = n_r^C p^r q^{n-r}, r = 0, 1, \ldots, n.$$

This probability distribution of the random variable X is called the *Binomial distribution.* The name Binomial comes from the fact that the coefficient n_r^C occurring in the probability expression given above is the Binomial coefficient present in the Binomial expansion of the expression $(p + q)^n$. The quantities p and n are called the *parameters* of the Binomial distribution.

We now compute the mean and the variance of the random variable X. Note that

$$\begin{aligned}
E(X) &= \sum_{r=0}^{n} rP(X=r) \\
&= \sum_{r=0}^{n} r n_r^C p^r q^{n-r} \\
&= \sum_{r=1}^{n} r \frac{n!}{r!n-r!} p^r q^{n-r} \\
&= np \sum_{r=1}^{n} \frac{n-1!}{r!n-r!} p^{r-1} q^{n-r} \\
&= np \sum_{j=0}^{n-1} \frac{n-1!}{j!(n-1-j)!} p^j q^{n-1-j} \\
&= np \sum_{j=0}^{n-1} (n-1)_j^C p^j q^{n-1-j} \\
&= np(p+q)^{n-1} \\
&= np.
\end{aligned}$$

There is an easier method to compute this expectation using the additive property of an expectation. Since

$$X = X_1 + X_2 + \cdots + X_n$$

where $X_1, X_2, \ldots, X_n$ are Bernoulli random variables, it follows by the additivity property of the expectation that

$$E(X) = E(X_1) + E(X_2) + \cdots + E(X_n).$$

But $E(X_1) = \cdots = E(X_n) = p$ from the definition of the Bernoulli random variables $X_i, i = 1, \ldots, n$. Hence $E(X) = np$.

Let us now compute the Var(X). As we have seen earlier that

$$Var(X) = E(X^2) - [E(X)]^2,$$

it is sufficient to compute $E(X^2)$. Observe that

$$X^2 = X(X-1) + X$$

and hence

$$E(X^2) = E[X(X-1)] + E(X).$$

We leave it to the reader to compute $E[X(X-1)]$ directly by the method used above for computing $E(X)$ and hence show that

$$Var(X) = np(1-p).$$

An alternate method of computation of Var(X) will be discussed later in this book.

Example 3.3. Suppose a fair coin is tossed six times. What is the probability of the event that at least 5 heads will appear?

Let X be the number of heads in six tosses. Then

$$\begin{aligned} P(X \geq 5) &= P(X=5) + P(X=6) \\ &= 6_5^C(\frac{1}{2})^5(\frac{1}{2})^1 + 6_6^C(\frac{1}{2})^6(\frac{1}{2})^0 \\ &= \frac{6}{64} + \frac{1}{64} = \frac{7}{64}. \end{aligned}$$

Example 3.4. Suppose the metal parts produced by a machine contains 5% defective parts. How many parts should be produced in order that the probability of at least one defective is 50% or more?

Let n be the number of parts to be produced and X be the number of defective parts. Assume that the parts are produced independently. Considering the production of a defective part as a success, we have to find n such that $P(X \geq 1) \geq \frac{1}{2}$. Note that

$$P(X \geq 1) = 1 - P(X = 0) = 1 - (.95)^n. \tag{3.1}$$

In order that $1-(.95)^n \geq \frac{1}{2}$, it is necessary that $n \geq 14$ (check?).

Example 3.5. Suppose that five missiles were fired against an aircraft carrier in an ocean and it takes at least two direct hits to sink the ship. All the five missiles are on the correct trajectory but must get through the guns of the ship. It is known that the guns of the ship can destroy a missile with the probability 0.9. What is the probability that the ship will still be afloat after the encounter?

Let E be the event that the ship is afloat after the 5 missiles were fired and F be the event of a missile getting through the defense of the guns of the ship. Then $P(F) = 0.1$ and

$$\begin{aligned} P(E) &= 1 - P(\bar{E}) \\ &= 1 - \sum_{i=2}^{5} 5_i^C (0.1)^i (0.9)^{5-i} \\ &= 0.92. \end{aligned}$$

Geometric distribution

Let us consider independent Bernoulli trials as described earlier *but* now our random experiment consists of observing the number of the trial X required to observe a success S for the first time. The sample space for this experiment can be considered to be the set $\Omega = \{1, 2, \ldots; E\}$ where the integer k denotes that k trials were needed to obtain a success S and E denotes the elementary event that a failure F was observed on every trial. The event k can be described as a sequence of $k-1$ failures F followed by a successes S on the k-th trial. Applying the independence of the Bernoulli trials, a natural assignment of probability for the elementary event k is $(1-p)^{k-1}p$ where p is the probability for a success on a single trial. In other words

$$P(X = k) = (1-p)^{k-1}p, k = 1, 2, \ldots.$$

Such an assignment of probability to the event k for $k \geq 1$ automatically leads to the assignment of probability zero to the event e since the sum of the probabilities of all the elementary events should be equal to one. The random variable X is said to have a *Geometric* distribution. Show that the mean of the random variable X is $\frac{1}{p}$ and and the variance of the random variable is $\frac{1-p}{p^2}$ under the above probability model.

Example 3.6. Suppose the probability that an engine of an automobile does not work during any one hour period is $p = 0.02$. Find the probability that a given engine will work for 2 hours.

Let X denote the number of one hour periods including the period when the first break down of the engine occurs. Then

$$\begin{aligned} P(\text{Engine works for 2 hours}) &= P(X \geq 3) \\ &= 1 - P(X = 1) - P(X = 2) \\ &= 1 - p - (1-p)p \\ &= 1 - 0.02 - (0.98)(0.02) = 0.9604. \end{aligned}$$

Example 3.7. It was known that 60% of the households in a region prefer to purchase a particular brand A of televisions. If a group of randomly selected households in the same region is interviewed, what is the probability that exactly five have to be interviewed to select the first house who prefer to purchase the television of brand A? What is the probability that at least five have to be interviewed to select the first house who prefer to purchase the television of brand A?

Let X be the number of households to be interviewed before a success occurs where a success is defined to be the encounter of a house who prefer a television of brand A. Then X follows the Geometric distribution with parameter p= Probability of a success=0.60. Then

$$P(X = 5) = (0.4)^4(0.6) = .01536$$

and

$$\begin{aligned} P(X \geq 5) &= 1 - P(X \leq 4) \\ &= 1 - (0.6) - (0.4)(0.6) - (0.4)^2(0.6) - (0.4)^3(0.6) \\ &= 0.0256. \end{aligned}$$

We note an important property of the Geometric distribution. Suppose that X is a random variable with the Geometric distribution with

$$P(X = k) = (1 - p)^{k-1}p, k = 1, 2, \ldots.$$

Then, for any two positive integers j and k,

$$\begin{aligned} P(X > j + k | X > j) &= \frac{P(X > j; X > j + k)}{P(X > j)} \\ &= \frac{P(X > j + k)}{P(X > j)} \\ &= (1 - p)^k \\ &= P(X > k). \end{aligned}$$

This property is called the *memoryless* property of the Geometric distribution.

Poisson distribution

For modelling data such as the number of telephone calls received at a particular place during a specified time period or the number of car accidents in a fixed stretch of a road during a time period or the number of printing errors found on a page of a book, Poisson distribution has been found to be the most suitable distribution. It is used for modelling rare occurrences of events.

Let X be a random variable taking the values in the set of nonnegative integers such that

$$P(X = k) = \frac{e^{-\lambda}\lambda^k}{k!}, k \geq 0 \quad \text{where } \lambda > 0.$$

The parameter λ can be shown to be the expectation or the mean of the random variable X and hence is the expected number of events. Check that the variance of the random variable X is also equal to λ. The probability distribution of the random variable X is called the *Poisson distribution* with the parameter λ.

Let us consider a random experiment where we observe the number $X(t)$ of telephone calls received at a particular place during a time period $[0,t]$. It is clear that $X(0) = 0$ and the set of possible values for the random variable $X(t)$ is the set $\{0,1,2,\ldots\}$ for $t > 0$. We would like to build a probability model on the set $\{0,1,2,\ldots\}$ for the number of calls to be received in the time period $[0,t]$. Suppose that an average of λ calls are received in a unit amount of time. Let us assume that the following conditions hold:

- C(i) Probability that there is more than one call in an interval $[t, t+\Delta t)$ is $o(\Delta t)$;
- C(ii) Probability that there is exactly one call in an interval $[t, t+\Delta t)$ is $\lambda\,\Delta t + o(\Delta t)$;
- C(iii) Probability that there are k calls in an interval $[t, t+h)$ depends only on the length h of the interval and not on t; and
- C(iv) the events that there are k calls in the interval $[t_1, t_2]$ and j calls in the interval $[t_3, t_4]$ are independent for every $k \geq 0$ and $j \geq 0$ whenever $0 \leq t_1 < t_2 \leq t_3 < t_4 < \infty$.

Recall that a function $f(t)$ is said to be $o(g(t))$ if

$$\frac{f(t)}{g(t)} \to 0 \text{ as } t \to 0.$$

The condition C(iii) listed above is known as the *stationarity* condition and the condition C(iv) is known as the condition of *independent increments*. Let $p_k(t)$ be the probability that $X(t) = k$ for $k \geq 0$. Note that

$$\begin{aligned} p_0(t+h) &= P(X(t+h) = 0) \\ &= P[X(t) = 0 \text{ and there are no calls in the interval } [t, t+h)] \\ &= p_0(t)[1 - \lambda h + o(h)] \end{aligned}$$

by the conditions C(i)- C(iv). Hence

$$p_0(t+h) = p_0(t)(1 - \lambda h + o(h))$$

which implies that

$$p_0(t+h) - p_0(t) = -\lambda p_0(t)h + o(h).$$

Dividing both sides by h and taking the limit as $h \to 0$, we obtain the differential equation

$$\frac{dp_0(t)}{dt} = -\lambda p_0(t)$$

with the initial condition $p_0(0) = 1$. It is easy to solve this differential equation and we have

$$p_0(t) = e^{-\lambda t}, \quad t \geq 0. \tag{3.2}$$

Furthermore, for $k \geq 1$,

$$\begin{aligned} p_k(t+h) &= P(X(t+h) = k) \\ &= P(X(t) = k \text{ and there are no calls} \\ &\qquad \text{in the interval } [t, t+h)]) \\ &\quad + P(X(t) = k-1 \text{ and there is exactly one call} \\ &\qquad \text{in the interval } [t, t+h)]) \\ &\quad + \sum_{j=2}^{\infty} P(X(t) = k-j \text{ and there are exactly j calls} \\ &\qquad \text{in the interval } [t, t+h)]). \end{aligned}$$

The last term is bounded by

$$P[\text{There are more than one call in the interval } [t, t+h)]$$

which is $o(h)$ by the condition C(i). Applying the conditions C(ii) to C(iv), we have the equation

$$\begin{aligned} p_k(t+h) &= p_k(t)p_0(h) + p_{k-1}(t)p_1(h) + o(h) \\ &= p_k(t)(1 - \lambda h + o(h)) + p_{k-1}(t)(\lambda h + o(h)) + o(h). \end{aligned}$$

Hence

$$p_k(t+h) - p_k(t) = -p_k(t)\lambda h + o(h) + p_{k-1}(t)(\lambda h + o(h)) + o(h).$$

Dividing by h and taking the limit as $h \to 0$, we obtain the differential equation

$$\frac{dp_k(t)}{dt} = -\lambda(p_k(t) - p_{k-1}(t)). \tag{3.3}$$

We have seen that $p_0(t) = e^{-\lambda t}$ for all $t \geq 0$.. Furthermore $p_k(0) = 0$ for all $k \geq 1$. The above differential equation can be solved recursively under these initial conditions and we obtain that

$$p_k(t) = e^{-\lambda t}\frac{(\lambda t)^k}{k!}, \quad k = 1, 2, \ldots \tag{3.4}$$

Combining the equations (3.2) and (3.4), we obtain that the random variable $X(t)$ has the Poisson distribution described at the beginning of this section. The family of random variables $\{X(t), t \geq 0\}$ is said to form a *Poisson process.*

Poisson distribution can also be considered as a limiting case of a Binomial distribution. Let us suppose that we are dealing with a random experiment where the number of trials n is large but the probability of a success p is close to zero. We have seen that the expected number of successes is np. Suppose $n \to \infty$ such that $np \to \lambda$. Note that

$$\begin{aligned} P(X = k) &= n_k^C p^k (1-p)^{n-k} \\ &= \frac{n!}{(n-k)!k!} p^k (1-p)^{n-k} \\ &= \frac{n!}{(n-k)!k!n^k} (np)^k (1 - \frac{np}{n})^n (1 - \frac{np}{n})^{-k}. \end{aligned}$$

Applying the result

$$\lim_{n \to \infty} (1 - \frac{\lambda}{n})^n \to e^{-\lambda} \text{ as } n \to \infty,$$

it can be checked that

$$P(X = k) = n_k^C p^k (1-p)^{n-k} \to e^{-\lambda} \frac{\lambda^k}{k!} \text{ as } n \to \infty.$$

This shows that the Poisson distribution can be considered as a limiting case of the Binomial distribution for large n and small p. By reversing the roles of a Success and a Failure in a Binomial distribution , it can be shown that one can obtain a suitable Poisson approximation for probabilities under a Binomial distribution if p is close to one and n is large. We will come back to this discussion again in the Chapter 7. Tables for computing probabilities under the Poisson distribution for different values of the mean λ are available.

Example 3.8. Suppose that radioactive particles strike a target according to a Poisson process at an average rate of 3 particles per minute. What is the probability that 10 or more particles will strike the target in a particular 2-minute period?

Let λ be the average number of particles that strike the target in a 1-minute period. Here $\lambda = 3$. Since the radioactive particles strike the target according to a Poisson process, the number of strikes in a 2-minute period will have a Poisson distribution with mean $2\lambda = 6$.. Let X be the number of particles that strike in a 2-minute period. It can be checked from the tables for the Poisson distribution that $P(X \geq 10) = 0.0838$.

Example 3.9. A telephone exchange receives calls at an average rate of 16 per minute. If it can handle at most 24 calls per minute, what is the probability that in any one minute the switch board of the exchange will saturate?

Let X be the number of telephone calls received in one minute. Suppose that X has a Poisson distribution with mean $\lambda = 16$. It is clear that a saturation at the telephone exchange will occur if the number of calls in a minute exceeds 24. Probability of such an event is

$$P(X \geq 25) = \sum_{k=25}^{\infty} e^{-16} \frac{(16)^k}{k!} = 0.017.$$

Since 0.017 is about $\frac{1}{60}$, a caller will experience saturation at the telephone exchange once in 60 minutes on the average.

Hypergeometric distribution

Let us consider a box containing a black balls and b red balls. Let $N = a + b$. Suppose we choose n balls out of N *with replacement* in the sense whenever one ball is drawn, its colour is noted and is placed back in the box before the next draw. Suppose we consider the outcome of the experiment as a success if a black ball is selected. It is clear that the probability for success is a/N and the number of trials is n. Suppose the draws are independent. If X denotes the number of successes , then it has the Binomial distribution with parameters n and $p = a/N$.

Let us now consider again a box containing a black balls and b red balls. Let $N = a + b$. Suppose we choose n balls out of N *without replacement* in the sense that whenever one ball is drawn, its colour is noted and the ball is removed from the box before the next draw. It is clear that the draws can no longer be considered as independent and the conditional probability of selecting a particular coloured ball, say black, depends on the composition of the black and the red balls at that stage. Let X denote the number of black balls selected. It is obvious that the possible values for X are $0, 1, \ldots, min(a, n)$. Since there are N_n^C combinations of n out of N, it is easy to see that

$$P(X = k) = \frac{a_k^C b_{n-k}^C}{N_n^C}, k = 0, 1, \ldots, min(a, n).$$

Interpreting $m_r^C = 0$ whenever $r > m$, we can rewrite the above equation in the form

$$P(X = k) = \frac{a_k^C b_{n-k}^C}{N_n^C}, k = 0, 1, \ldots, n.$$

This probability distribution of the random variable X is called the *Hypergeometric Distribution* with parameters $(n; a, N)$.

If the total number of balls N is large compared to a or b, then one can approximate the probabilities computed under the Hypergeometric distribution with parameters $(n; a, N)$ by a Binomial distribution with parameters (n, p) where $p = \frac{a}{N}$. This is intuitively clear since one can consider the draws as almost independent when N is large. Check that for any $0 \leq k \leq n$,

$$P(X = k) = \frac{a_k^C b_{n-k}^C}{N_n^C} \to n_k^C p^k (1-p)^{n-k}$$

as $N \to \infty$ such that $a/N \to p$. In other words the probabilities under sampling without replacement and sampling with replacement are almost the same when the lot size is large. Let $X_i = 1$ if the i-th draw leads to a black ball and $X_i = 0$ otherwise. Then $X = X_1 + X_2 + \cdots + X_n$. It is clear that $P(X_1 = 1) = \frac{a}{N}$ and $P(X_1 = 0) = \frac{b}{N}$ and hence $E(X_1) = \frac{a}{N}$. Let us now compute the distribution of X_2. Note that

$$\begin{aligned}
P(X_2 = 1) &= P([X_2 = 1] \cap [X_1 = 1]) + P([X_2 = 1] \cap [X_1 = 0]) \\
&= P([X_2 = 1] | [X_1 = 1]) P([X_1 = 1]) \\
&\quad + P([X_2 = 1] | [X_1 = 0]) P([X_1 = 0]) \\
&= \frac{a-1}{N-1} \frac{a}{N} + \frac{a}{N-1} \frac{b}{N} \\
&= \frac{a}{N}.
\end{aligned}$$

Hence $P(X_2 = 0) = 1 - P(X_2 = 1) = b/N$ which implies that the probability distributions of X_1 and X_2 are the same. This is an interesting fact since the sampling is done without replacement. Similarly it can be shown that the random variables $X_i, i = 3, \ldots, n$ have the same distribution as that of X_1. In particular, $E(X_i) = a/N$ for $i = 1, \ldots, n$. Therefore

$$E(X) = E(X_1) + \cdots + E(X_n) = n(a/N).$$

Note that this expectation of X matches with the expectation of a Binomial distribution with parameters n and $p = (a/N)$. However the random variables $X_i, 1 \leq i \leq n$ are not independent. We leave it to you to check that

$$Var(X) = n\frac{a}{N}(1 - \frac{a}{N})\frac{N-n}{N-1}$$

and compare it with the variance in a Binomial distribution with parameters n and $p = (a/N)$.

Example 3.10. A storage depot contains ten machines four of which are defective. A company selects five of these machines at random thinking that they are in working condition. What is the probability that all five of the machines are nondefective?

Let X be the number of non-defective machines among the five selected. Then X has the Hypergeometric distribution and

$$P(X = 5) = \frac{(6_5^C)(4_0^C)}{10_5^C} = \frac{6}{252} = \frac{1}{42}.$$

3.4 Probability Generating Function

Suppose X is a discrete random variable taking values in the set of non-negative integers $\{0, 1, 2, \ldots\}$ with $P(X = k) = p_k$. Let

$$\phi_X(s) = \sum_{k=0}^{\infty} s^k p_k$$

for all values of s for which the power series is absolutely convergent. Since $\sum_{k=0}^{\infty} p_k = 1$, the power series is absolutely convergent for all s in the interval $(-1, 1]$. The function $\phi_X(s)$ is called the *probability generating function* (p.g.f) of the random variable X. Note that the coefficient of s^k in the power series for the function $\phi(s)$ is $P(X = k)$ and hence the function $\phi_X(s)$ is called the probability generating function of the random variable X. If we are dealing with only one random variable alone, then we omit the suffix X in the notation $\phi_X(s)$ and write $\phi(s)$ for the p.g.f. Note that $\phi_X(1) = 1$ and $\phi_X(s) = E[s^X]$ following the definition of the expectation of a random variable.

Example 3.11. Suppose X is a random variable with the Binomial distribution with parameters n and p. Then

$$\begin{aligned}\phi_X(s) &= \sum_{k=0}^{n} s^k P(X = k) \\ &= \sum_{k=0}^{n} s^k n_k^C p^k (1-p)^{n-k} \\ &= \sum_{k=0}^{n} (ps)^k n_k^C (1-p)^{n-k} \\ &= (ps+q)^n\end{aligned}$$

where $q = 1 - p$. Note that the p.g.f. of X exists for every real s in this example.

Example 3.12. Suppose X is a random variable with the Poisson distribution with parameters λ. Then

$$\begin{aligned}\phi_X(s) &= \sum_{k=0}^{\infty} s^k P(X = k) \\ &= \sum_{k=0}^{\infty} s^k e^{-\lambda} \frac{\lambda^k}{k!} \\ &= \sum_{k=0}^{\infty} (\lambda s)^k e^{-\lambda} \left(\frac{1}{k!}\right) \\ &= e^{-\lambda} e^{\lambda s} \\ &= e^{\lambda(s-1)}.\end{aligned}$$

Example 3.13. Suppose X_1 and X_2 are independent random variables with Binomial distributions with parameters (n_1, p) and (n_2, p) respectively. If we interpret X_1 and X_2 as the number of successes, it is clear that $X = X_1 + X_2$ represents the number of successes in $n_1 + n_2$ independent Bernoulli trials with p as the probability of success in a single trial and hence has a Binomial distribution with parameters $n_1 + n_2$ and p. We leave it to the reader to directly check this by calculating the $P(X_1 + X_2 = k)$ for $k = 0, 1, \ldots, n_1 + n_2$. It is not true however that if X_1 and X_2 are independent random variables with binomial distributions with parameters (n_1, p_1) and (n_2, p_2), then $X_1 + X_2$ has a binomial distribution if $p_1 \neq p_2$ (check?).

Remarks. It will be shown later that $\phi_{X_1+X_2}(s) = \phi_{X_1}(s)\phi_{X_2}(s)$ whenever X_1 and X_2 are independent random variables using the properties of expectation for a product of independent random variables. This will give an alternate method for finding the distribution of sum of two independent random variables in the above example.

3.5 Exercises

3.1 Find the probability that in a family with five children girls outnumber boys assuming that births are independent trials each with probability of a boy equal to $\frac{1}{2}$.

3.2 In a 20-question true/false test, a student tosses a fair coin to determine the answer for each question. If the coin falls heads, he/she answers it as true and if it falls tails, he/she answers it as false. Find the probability that the student answers at least 12 questions correctly.

3.3 What is the probability of throwing exactly nine heads twice in five throws of ten fair coins?

3.4 Ten independent binary pulses per second arrive at a receiver. A zero received as a one and a one received as a zero is considered as an error. Suppose the probability of an error is .001.What is the probability of at least one error per second?

3.5 An oil company is allowed to drill a succession of holes in a given area to find a productive well. The probability for successful drilling in a trial is 0.2. What is the probability that the third hole drilled is the first to give a productive well? If the company can drill a maximum of ten wells, what is the probability that it will fail to discover a productive well?

3.6 Suppose that the number of car accidents during the weekend at a certain intersection in a city has a Poisson distribution with mean 0.7. What is the probability for at least three accidents during a weekend?

3.7 Suppose that a book with n pages contain on the average λ misprints per page. What is the probability that there will be at least m pages out of n containing more than k misprints?

[illegible]

[illegible]

[illegible]

[illegible]

[illegible]

[illegible]

[illegible]

[illegible]

[illegible]

Chapter 4

CONTINUOUS PROBABILITY DISTRIBUTIONS

4.1 Introduction

In the previous chapter, we have discussed probability models on discrete sample spaces and examples of some discrete probability distributions. As we have already indicated in Chapter 2, there are random experiments where the number of possible outcomes is too large or uncountably infinite. A set is said to be *uncountable* if all the elements of the set cannot be listed as a sequence. Consider a random experiment of a selecting a real number x in the closed interval $[0, 1]$. Obviously the number of possible outcomes is "very large", in fact uncountable in the sense defined above. There are other random experiments where one encounters sample spaces which have possibly uncountable number of outcomes; for instance, recording the lifetime of an electric bulb or recording the arrival time of a particular train at a specified place or measuring the height of a person etc. These measurements are all ideally continuous. However, due to the limitations of recording instruments and measuring skills, the measurements cannot be made accurately and we end up with observations measured on a discrete scale of measurement. We now discuss some methods of building probability models for such random experiments. To build such probability models, we consider an ideal situation where the measurements of the observations are assumed to be made on a continuous scale.

4.2 Distribution Function

Let us denote the sample space for a random experiment $\mathcal{E}$ by Ω and let ω be a typical element of Ω. The set Ω is a list of all possible outcomes of the random experiment $\mathcal{E}$ and ω is a typical elementary outcome. A real-valued function X defined on Ω is called a random variable. We will make this concept more precise later in this section. We would like to assign probabilities to sets of the form

$$[X \leq x] = [\omega : X(\omega) \leq x]$$

for all real numbers x. Let $\mathcal{F}$ be a collection of subsets of Ω with the following properties:

(i) $\Omega \in \mathcal{F}$;
(ii) $A \in \mathcal{F}$ implies that $\bar{A} \in \mathcal{F}$; and
(iii) if $A_i \in \mathcal{F}, i \geq 1$, then $\cup_{i=1}^{\infty} A_i \in \mathcal{F}$.

Such a collection of subsets of Ω is called a *σ-algebra* of subsets of Ω. Examples of such collections are the set of all subsets of Ω and the set consisting of the two elements Φ and Ω.

Every subset E in the σ-algebra $\mathcal{F}$ is called an *event.* We say that an *event* E has occurred if the outcome of the random experiment belongs to E.

Definition. A probability measure $P(.)$ defined on the family of sets given by the σ-algebra $\mathcal{F}$ is a real-valued function satisfying the following properties:

(i) $P(\Omega) = 1$;
(ii) $0 \leq P(E) \leq 1$ for all sets $E \in \mathcal{F}$; and
(iii) $P(\cup_{i=1}^{\infty} A_i) = \sum_{i=1}^{\infty} P(A_i)$ if $A_i \in \mathcal{F}, i \geq 1$ and $A_i \cap A_j = \Phi$ if $i \neq j$.

Remarks. A sequence of sets $A_n, n \geq 1$ is said to be *increasing* if $A_i \subset A_j$ if $i \leq j$. It can be proved that if $A_i, i \geq 1$ is an increasing sequence of sets in the family $\mathcal{F}$, then

$$\lim_{n \to \infty} P(A_n) = P(\cup_{n=1}^{\infty} A_n). \tag{4.1}$$

Definition. A real-valued function X defined on $(\Omega, \mathcal{F})$ is called a *random variable* if the σ-algebra $\mathcal{F}$ is such that the event $[X \leq x] \in \mathcal{F}$ for all x.

Let X be a random variable defined on Ω. From the properties of the σ-algebra $\mathcal{F}$, it follows that the events

$$[a < X \leq b], [a \leq X < b], [a \leq X \leq b], \text{ and } [a < X < b]$$

belong to $\mathcal{F}$ for all real numbers a and b. In particular, we can assign a probability to the event $[X \leq x]$ given a probability measure $P(.)$ defined on $\mathcal{F}$, namely, $P([X \leq x])$. For convenience, we write $P(X \leq x)$ hereafter for $P([X \leq x])$. Let

$$F(x) = P(X \leq x), -\infty < x < \infty.$$

The function $F(.)$ is called the *distribution function* of the random variable X. It is also called the *cumulative distribution function* (c.d.f.) of the random variable X.

Properties of a distribution function

Any distribution function $F(.)$ satisfies the following properties:

(i) $0 \leq F(x) \leq 1$ for all x;
(ii) if $x \leq y$, then $F(x) \leq F(y)$;
(iii) $\lim_{x\to\infty} F(x) = 1$ and $\lim_{x\to-\infty} F(x) = 0$; and
(iv) the function $F(x)$ is right continuous, that is , $F(x+h) \to F(x)$ as h decreases to zero for every x.

The property (i) is a consequence of the observation that the probability of an event $[X \leq x]$ has to be a number between zero and one from the definition of a probability measure and the property (ii) follows from the observation that for any $x \leq y$,

$$[X \leq y] = [X \leq x] \cup [x < X \leq y]$$

and the events $[X \leq x]$ and $[x < X \leq y]$ are disjoint. Hence

$$P[X \leq y] = P[X \leq x] + P[x < X \leq y]$$

which implies that

$$F(y) = F(x) + P[x < X \leq y].$$

Since $P[x < X \leq y] \geq 0$ from the definition of a probability measure, it follows that $F(x) \leq F(y)$ and, in fact, for any $x < y$,

$$P[x < X \leq y] = F(y) - F(x).$$

The property (iii) can be interpreted in the form

$$P[X < \infty] = 1 \text{ and } P[X < -\infty] = 0.$$

These relations hold since $[\omega : X(\omega) < \infty] = \Omega$ and $[\omega : X(\omega) < -\infty] = \Phi$. Furthermore $P(\Omega) = 1$ and $P(\Phi) = 0$ from the definition of a probability measure. A rigourous proof of the property (iii) can be given by using the result in (4.1) in the remarks made above. Property (iv) is again a consequence of (4.1).

Example 4.1. Suppose a random variable X takes the values 0 and 1 with probabilities p and $1-p$ respectively. Then the distribution function of X is

$$\begin{aligned} F(x) &= 0 \text{ if } x < 0 \\ &= p \text{ if } 0 \leq x < 1 \\ &= 1 \text{ if } x \geq 1. \end{aligned}$$

This distribution function can be represented graphically as shown in the Figure 4.1. Observe that the distribution function is continuous every where except at the points $x = 0$ and $x = 1$. It has a jump of size p at $x = 0$ and of size $1-p$ at $x = 1$.

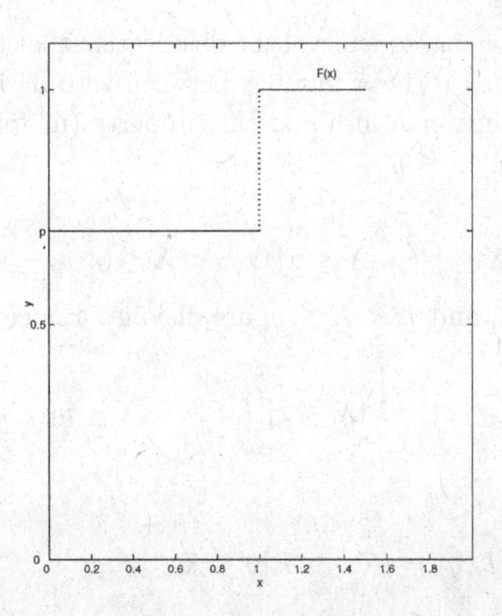

Fig. 4.1

Example 4.2. Suppose X is a random variable taking the values $1, 2, 3$ with probabilities $1/6, 2/6, 3/6$ respectively. Then the distribution function of X is given by

$$\begin{aligned} F(x) &= 0 \text{ if } x < 1 \\ &= 1/6 \text{ if } 1 \leq x < 2 \\ &= 3/6 \text{ if } 2 \leq x < 3 \\ &= 1 \text{ if } x \geq 3. \end{aligned}$$

The graph of the distribution function F in Example 4.2 is as shown in Figure 4.2.

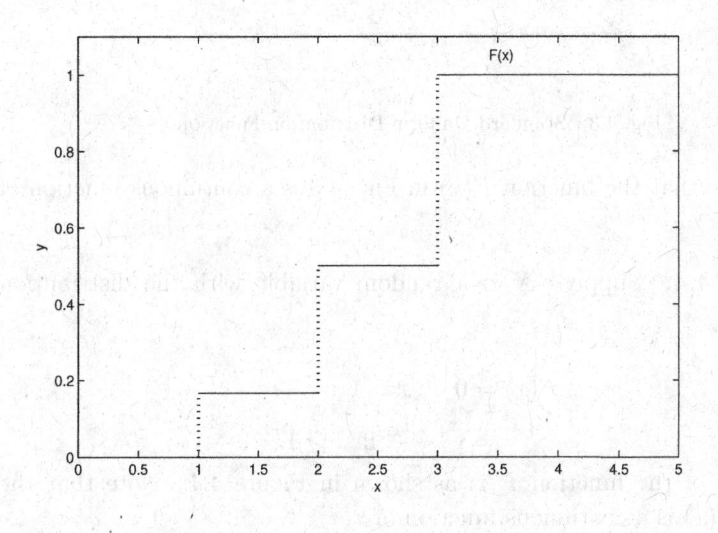

Fig. 4.2

Remarks. Observe that the graphs of the distribution functions, in both the examples given above, are step functions and the jumps of the distribution function F are at the points at which the random variables have positive probabilities.

Example 4.3. Suppose X is a random variable with the distribution function F defined by

$$\begin{aligned} F(x) &= x \text{ if } 0 \leq x \leq 1 \\ &= 0 \text{ if } x < 0 \\ &= 1 \text{ if } x > 1. \end{aligned}$$

The graph of the distribution function F is as shown in Figure 4.3.

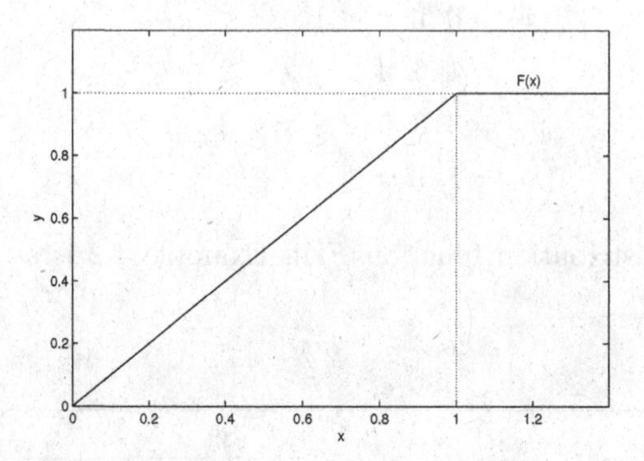

Fig. 4.3 Standard Uniform Distribution Function

We note that the function $F(x)$ in Fig. 4.3 is a continuous function of x.

Example 4.4. Suppose X is a random variable with the distribution function

$$F(x) = 0 \text{ if } x < 1$$
$$= 1 - \frac{1}{x^2} \text{ if } x \geq 1.$$

The graph of the function F is as shown in Figure 4.4. Note that the function $F(x)$ is a continuous function of x.

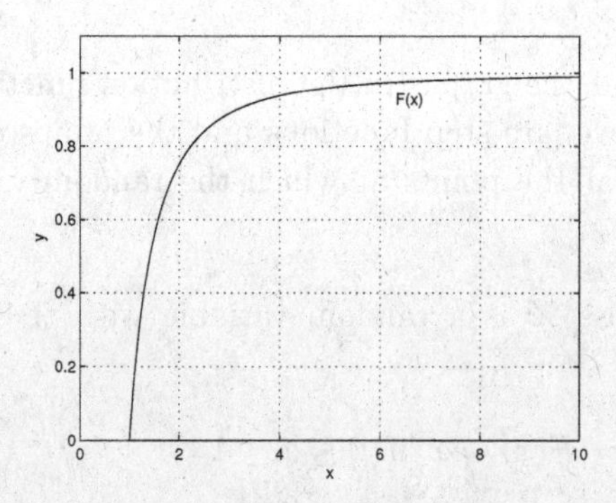

Fig. 4.4

Example 4.5. Let X be a random variable with the distribution function

$$F(x) = 0 \text{ if } x < 0$$
$$= 1 - \frac{1}{3}e^{-2x} \text{ if } x \geq 0.$$

The graph of the function F is as shown in Figure 4.5. Here is an example of a distribution function which is neither a purely step function nor a purely continuous function. This distribution function has a discontinuity at 0 with a jump of size 2/3 at that point and it is continuous everywhere else.

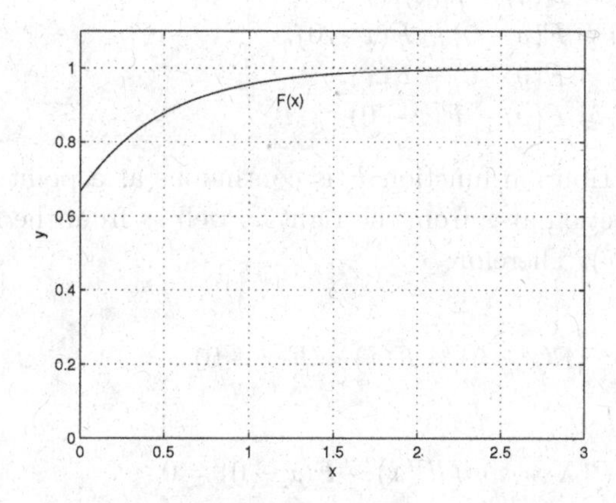

Fig. 4.5

Since a distribution function F is always nondecreasing, the function F has limits from the left and from the right at each point x. We have already seen that a distribution function F is always right continuous.

We now compute the the probability that a random variable X with the distribution function F takes the value x, that is, $P(X = x)$. We note that

$$P(X = x) = P(X \leq x) - P(X < x)$$
$$= F(x) - P(X < x).$$

Since the sets $A_n = [X \leq x - \frac{1}{n}], n \geq 1$ is an increasing sequence of subsets increasing to the event $A = [X < x]$, it follows that

$$P(X < x) = \lim_{n \to \infty} F(x - \frac{1}{n}).$$

The limit on the right side of the above equation exists from the monotonicity of the function F. We denote this limit by $F(x-0)$. Hence

$$P(X = x) = F(x) - F(x-0), -\infty < x < \infty.$$

We have now the following formulae for computing the probabilities of various events in terms of a distribution function: for any x and y,

(i) $P(X \leq x) = F(x)$,
(ii) $P(X < x) = F(x-0)$,
(iii) $P(X = x) = F(x) - F(x-0)$,
(iv) $P(x < X \leq y) = F(y) - F(x)$,
(v) $P(x \leq X < y) = F(y-0) - F(x-0)$,
(vi) $P(x < X < y) = F(y-0) - F(x)$,
(vii) $P(x \leq X \leq y) = F(y) - F(x-0)$.

Remarks. If a distribution function F is continuous at a point x, then the limits of the function at x from the right as well as from the left exist and are equal to $F(x)$. Therefore

$$F(x-0) = F(x) = F(x+0)$$

and

$$P(X = x) = F(x) - F(x-0) = 0.$$

Hence, if the distribution function F of a random variable X is continuous at a point x, then $P(X = x) = 0$. In particular, if the distribution function F of a random variable X is continuous everywhere, then $P(X = x) = 0$ for every x. In spite of this observation, note that

$$\begin{aligned} P(a < X < b) &= P(a \leq X < b) \\ &= P(a < X \leq b) \\ &= P(a \leq X \leq b) \\ &= F(b) - F(a) \end{aligned}$$

which can be positive.

Suppose a random variable X is discrete valued as described in the previous chapter. Let the $P(X = x_i) = p_i, i \geq 1$ with $\sum_{i=1}^{\infty} p_i = 1$. Then the distribution function $F(x)$ of X is given by

$$F(x) = \sum_{i:x_i \leq x} P(X = x_i) = \sum_{i:x_i \leq x} p_i.$$

This distribution function is a step function and it is called a discrete distribution. Such a random variable X is said to be of *discrete type* or a *discrete random variable* as discussed in the Chapter 3.

On the other hand, suppose F is a distribution function of a random variable X and there exists a nonnegative function $f(x)$ such that

$$F(x) = \int_{-\infty}^{x} f(y)dy$$

for all real x. Then we say that F is an *absolutely continuous distribution* or a *continuous distribution* for short and the random variable X is said to be of *continuous type* or a *continuous random variable.* The function $f(x)$ is called the *probability density function* (p.d.f.) or the *density function* in short of the random variable X.

From the properties of a distribution function, we have

$$\int_{-\infty}^{\infty} f(y)\, dy = 1.$$

The integral defined above is a Lebesgue integral of the function f over the real line. Discussion about Lebesgue integration is beyond the scope of this book. However, it is known that, if the function f is continuous, then the Lebesgue integration and the Riemann integration agree and we can interpret the integral

$$\int_{-\infty}^{\infty} f(y)\, dy = 1$$

as a Riemann integral and it is the area bounded between the curve $y = f(x)$ and the line $y = 0$.

Remarks. It is known, from the properties of integrals, that if a function F is absolutely continuous, then it is also continuous but the converse is not true. In other words, a distribution function may be a continuous function but it may not be absolutely continuous. However we will not come across such distribution functions in the course of our study in this book. The distribution functions we consider are either purely discrete or purely (absolutely) continuous. Occasionally, we might consider examples of distribution functions which are neither discrete nor absolutely continuous but a mixture of the two as in the Example 4.5 given above. With some possible abuse of terminology, here after we shall call an absolutely continuous distribution as continuous distribution. It is possible for two distinct random variables to have the same distribution function. For instance, consider the random experiment of tossing a fair coin once. Define the random variable X to be equal to 1 if a "head" appears and X to be 0 if a "tail" appears. Similarly let $Y = 1$ if a "tail" appears and $Y = 0$ if a "head" appears. Obviously X and Y are distinct random variables but both of them have the same distribution function (check?).

4.3 Probability Density Function

We have discussed the concept of a probability density function in the previous section. Note that a function $f(x)$ defined on the real line is called a *probability density function* of a random variable X if

(i) $f(x) \geq 0$ for all x,
(ii) $P(X \leq x) = \int_{-\infty}^{x} f(y)dy$ for all x, and
(iii) $\int_{-\infty}^{\infty} f(y)\ dy = 1$.

Let F denote the distribution function of a continuous random variable X. It follows that

$$F(b) - F(a) = P(a \leq X \leq b) = \int_a^b f(y)\ dy.$$

Letting $a \to -\infty$, we obtain that

$$F(b) = \int_{-\infty}^{b} f(y)\ dy \tag{4.2}$$

for all real numbers b since $F(-\infty) = 0$. Given a density function f of a continuous random variable X, we can construct its distribution function F from the above relation. Conversely, it can be shown that if F is an (absolutely) continuous distribution function, then there exists a probability density function f satisfying the relation (4.2). If the relation (4.2) holds, then the function F is differentiable and

$$\frac{dF}{dx} = f(x)$$

almost every where. Typical graphs of the functions F and f are as shown in Figure 4.6 and Figure 4.7.

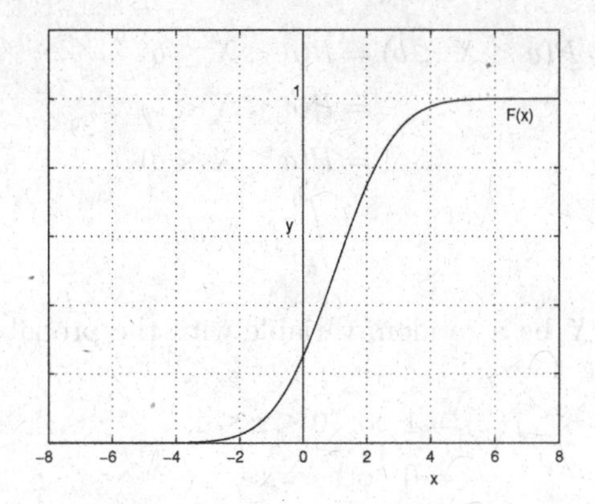

Fig. 4.6 Distribution Function $F(x)$

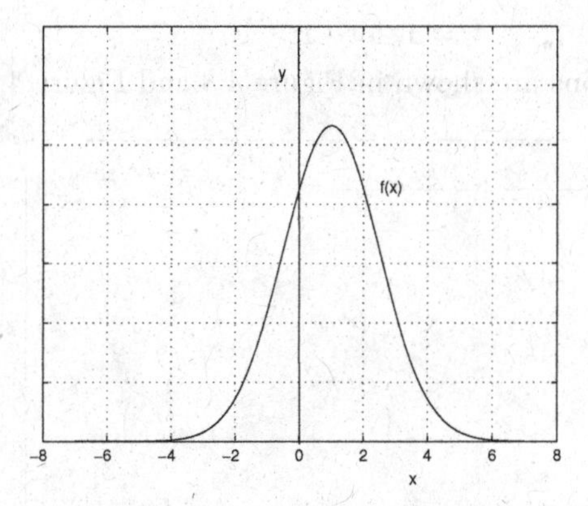

Fig. 4.7 Probability Density Function $f(x)$

Recall that

$$P(a \leq X \leq b) = \int_a^b f(y)\ dy$$

and that the probability that a random variable X takes values between a and b can be interpreted as the area between the curve $y = f(x)$ and the line $y = 0$ and the ordinates $x = a$ and $x = b$. Observe that the area enclosed between the curves $y = f(x)$ and the line $y = 0$ is unity since it is $P(-\infty < X < \infty)$. We have noticed earlier that $P(X = x) = 0$ for any random variable X with continuous distribution function F and hence for any $a < b$,

$$\begin{aligned} P(a < X < b) &= P(a < X \le b) \\ &= P(a \le X < b) \\ &= P(a \le X \le b) \\ &= \int_a^b f(y)\ dy. \end{aligned}$$

Example 4.6. Let X be a random variable with the probability density function

$$\begin{aligned} f(x) &= 1 \text{ for } 0 \le x \le 1 \\ &= 0 \quad \text{otherwise.} \end{aligned}$$

Check that

$$\begin{aligned} F(x) &= 0 \text{ for } x \le 0 \\ &= x \text{ for } 0 < x < 1 \\ &= 1 \text{ for } x \ge 1. \end{aligned}$$

Graphs of the functions are shown in Figure 4.8 and Figure 4.9.

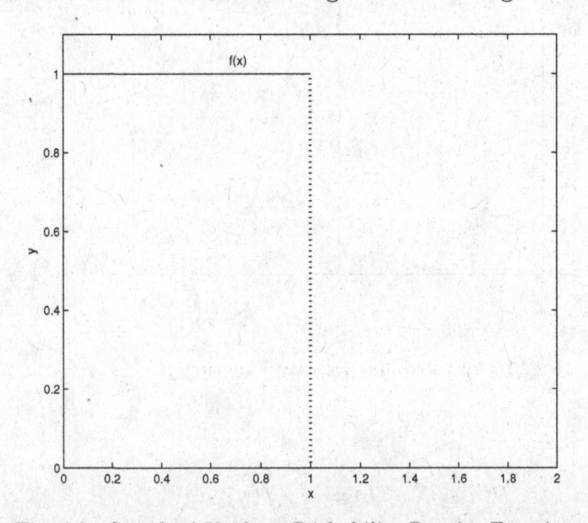

Fig. 4.8 Standard Uniform Probability Density Function

The distribution function F defined above is called the *(standard) uniform distribution* or the *rectangular distribution* from the shape of the graph of the probability density function f. Note that for any $0 \le a \le b \le 1$,

$$P(a \le X \le b) = b - a$$

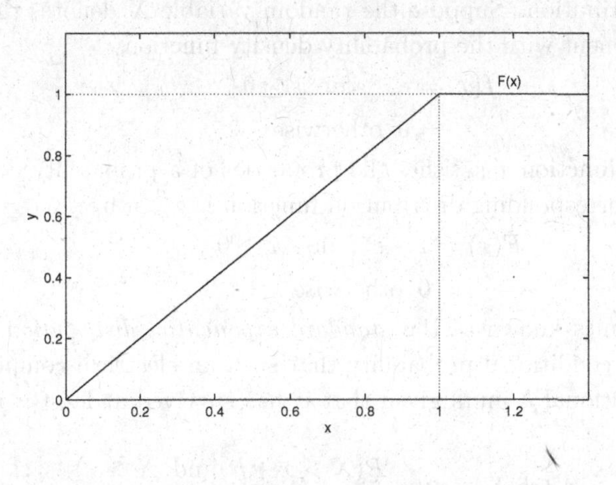

Fig. 4.9 Standard Uniform Distribution Function

and hence the probability of any interval $[a, b]$ contained in the interval $[0, 1]$ is $b - a$ which is the length of the interval.

Example 4.7. Let X be a random variable with the probability density function f equal to a constant c over an interval $[\alpha, \beta]$ and equal to zero outside the interval $[\alpha, \beta]$. In other words

$$f(x) = c \text{ for } \alpha \le x \le \beta$$
$$= 0 \quad \text{otherwise.}$$

It follows from the properties of a probability density function that $c = \frac{1}{\beta - \alpha}$. The function f is given by

$$f(x) = \frac{1}{\beta - \alpha} \quad \text{for } \alpha \le x \le \beta$$
$$= 0 \text{ otherwise}$$

and the distribution function F is

$$F(x) = 0 \text{ for } x < \alpha$$
$$= \frac{x - \alpha}{\beta - \alpha} \text{ for } \alpha \le x \le \beta$$
$$= 1 \text{ for } x > \beta.$$

This distribution is called the *uniform distribution* on the interval $[\alpha, \beta]$.

Example 4.8. A probability distribution that is frequently used in modelling the life time of an electrical component, such as a light bulb, is the *exponential* distribution. Suppose the random variable X denotes the life time of a component with the probability density function

$$\begin{aligned} f(x) &= e^{-x} \text{ for } x \geq 0 \\ &= 0 \text{ otherwise .} \end{aligned}$$

Check that the function f satisfies the properties of a probability density function. The corresponding distribution function is given by

$$\begin{aligned} F(x) &= 1 - e^{-x} \text{ for } x \geq 0 \\ &= 0 \text{ otherwise .} \end{aligned}$$

This distribution is known as the *standard exponential distribution.* Let us compute the conditional probability that such an electrical component survives an additional h units given that it has survived at least x units, that is,

$$\begin{aligned} P(X > x + h | X > x) &= \frac{P(X > x + h \text{ and } X > x)}{P(X > x)} \\ &= \frac{P(X > x + h)}{P(X > x)} \\ &= \frac{e^{-(x+h)}}{e^{-x}} \\ &= e^{-h} \\ &= P(X > h). \end{aligned}$$

The above calculations show that a component with lifetime having the exponential distribution has the property that, given that the component has worked at least for x units of time, the conditional probability that it will work for an additional h units is the same as the unconditional probability that the component will work for at least h units of time. Such a property of the exponential distribution is called the *memoryless property.* It can be shown that this is a characteristic property of the exponential distribution under some conditions. In other words, if a continuous probability distribution has the memoryless property, then it has to be exponential distribution under some technical conditions.

Example 4.9. Another probability distribution which is the most important distribution function for modelling errors is the *standard normal distribution* or some times also referred to as the *standard Gaussian distribution.* Suppose X is a random variable with the probability density function

$$\phi(x) = \frac{1}{\sqrt{2\pi}} e^{-\frac{x^2}{2}}, -\infty < x < \infty.$$

It is obvious that ϕ is a nonnegative function. It can be shown that

$$\int_{-\infty}^{\infty} \phi(y) \; dy = 1.$$

See the discussion given later in this chapter. The distribution function Φ corresponding to the probability density function ϕ is given by

$$\Phi(x) = \int_{-\infty}^{x} \frac{1}{\sqrt{2\pi}} e^{-\frac{y^2}{2}} \; dy, -\infty < x < \infty.$$

The graphs of the functions Φ and ϕ are shown in Figure 4.10 and Figure 4.11 respectively.

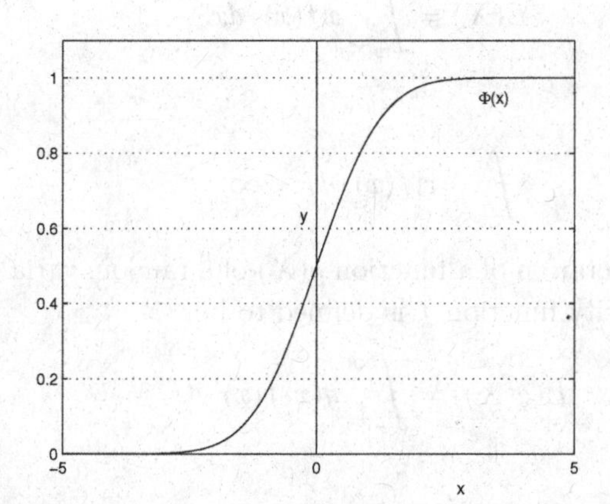

Fig. 4.10 Standard Normal Distribution $\Phi(x)$ Function

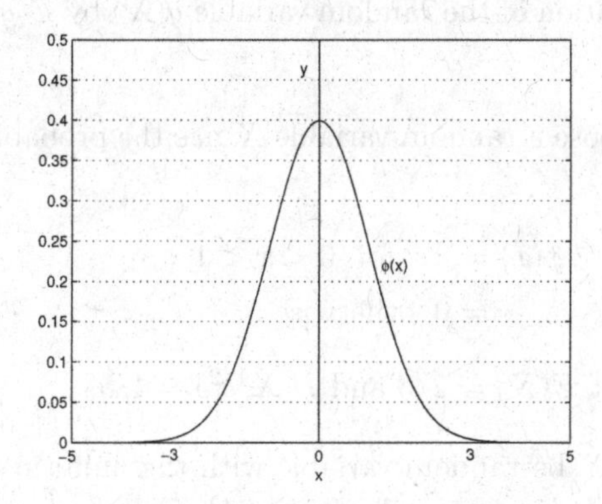

Fig. 4.11 Standard Normal Probability Density Function $\phi(x)$

Observe that the graph of the function $\phi(x)$ is symmetric about the origin with a peak at zero and it is bell-shaped.

4.4 Expectation and Variance

We have introduced the concept of the expectation or the mean of a discrete random variable in the previous chapter. We now extend it to a random variable X of continuous type with probability density function f. It is denoted by $E(X)$ and is defined by

$$E(X) = \int_{-\infty}^{\infty} x f(x)\ dx$$

provided that

$$\int_{-\infty}^{\infty} |x| f(x)\ dx < \infty.$$

In general, the expectation of a function $g(X)$ of a random variable X with the probability density function f is defined to be

$$E(g(X)) = \int_{-\infty}^{\infty} g(x) f(x)\ dx$$

provided

$$\int_{-\infty}^{\infty} |g(x)| f(x)\ dx < \infty.$$

We denote the expectation of the random variable $g(X)$ by $E(g(X))$ whenever it exists.

Example 4.10. Suppose a random variable X has the probability density function

$$\begin{aligned} f(x) &= 2x \text{ for } 0 < x < 1 \\ &= 0 \text{ otherwise .} \end{aligned}$$

It is easy to check that $E(X) = 2/3$ and $E(X^{1/2}) = 4/5$.

Example 4.11. Let X be random variable with the uniform distribution on the interval $[\alpha, \beta]$. It is easy to check that $E(X) = \frac{\alpha+\beta}{2}$ and $E(X^2) = \frac{\beta^2+\alpha^2+\alpha\beta}{3}$.

Example 4.12. Let X be a random variable with an exponential distribution, that is, X has the probability density function f given by

$$f(x) = \lambda e^{-\lambda x} \text{ for } x \geq 0$$
$$= 0 \text{ otherwise}$$

for some constant $\lambda > 0$. It is easy to check that $E(X) = \frac{1}{\lambda}$. If the random variable X is interpreted as the life time an electric bulb, then the mean or expected life time is $\frac{1}{\lambda}$.

Example 4.13. Suppose X is a random variable with the probability density function

$$f(x) = \frac{1}{\sqrt{2\pi\sigma^2}} e^{-\frac{1}{2}\frac{(x-\mu)^2}{\sigma^2}}, -\infty < x < \infty$$

where $-\infty < \mu < \infty$ and $0 < \sigma^2 < \infty$. Then

$$\begin{aligned}
E(X) &= \int_{-\infty}^{\infty} x \frac{1}{\sqrt{2\pi\sigma^2}} e^{-\frac{1}{2}\frac{(x-\mu)^2}{\sigma^2}}\, dx \\
&= \int_{-\infty}^{\infty} (\sigma y + \mu) \frac{1}{\sqrt{2\pi}} e^{-\frac{1}{2}y^2}\, dy \\
&= \frac{\sigma}{\sqrt{2\pi}} \int_{-\infty}^{\infty} y e^{-\frac{1}{2}y^2}\, dy + \mu \int_{-\infty}^{\infty} \frac{1}{\sqrt{2\pi}} e^{-\frac{1}{2}y^2}\, dy \\
&= \frac{\sigma}{\sqrt{2\pi}} [e^{-\frac{y^2}{2}}]_{-\infty}^{\infty} + \mu \\
&= \mu
\end{aligned}$$

since

$$\int_{-\infty}^{\infty} \frac{1}{\sqrt{2\pi}} e^{-\frac{1}{2}y^2}\, dy = 1.$$

The second equality given above follows from the transformation $y = \frac{x-\mu}{\sigma}$. Let $g(x) = (x-\mu)^2$. The function $g(X)$ can be considered as a measure of the deviation of X from the expectation μ of X and

$$\begin{aligned}
E[g(X)] &= \int_{-\infty}^{\infty} g(x) f(x)\, dx \\
&= \int_{-\infty}^{\infty} (x-\mu)^2 \frac{1}{\sqrt{2\pi\sigma^2}} e^{-\frac{1}{2}\frac{(x-\mu)^2}{\sigma^2}}\, dx \\
&= \sigma^2 \int_{-\infty}^{\infty} y^2 \frac{1}{\sqrt{2\pi}} e^{-\frac{1}{2}y^2}\, dy
\end{aligned}$$

by applying the transformation $y = \frac{x-\mu}{\sigma}$. Integrating by parts, check that

$$\int_{-\infty}^{\infty} y^2 \frac{1}{\sqrt{2\pi}} e^{-\frac{1}{2}y^2}\, dy = 1$$

which shows that

$$E[(X-\mu)^2] = \sigma^2.$$

If $\mu = 0$ and $\sigma = 1$, then the the probability density function f reduces to the standard normal probability density function discussed earlier in the Example 4.9.

In most of the examples discussed earlier, the random variables concerned have finite expectations. However there are random variables whose expectations do not exist.

Example 4.14. Let X be a random variable with the *(standard) Cauchy* probability density function

$$f(x) = \frac{1}{\pi(1+x^2)}, -\infty < x < \infty.$$

Then

$$\int_{-\infty}^{\infty} |x| f(x) dx = 2\int_0^{\infty} x \frac{1}{\pi(1+x^2)}\ dx$$
$$= \frac{1}{\pi}[\log(1+x^2)]_0^{\infty}$$

and $\frac{1}{\pi}[\log(1+x^2)] \to \infty$ as $x \to \infty$. Hence $E(X)$ does not exist and the Cauchy distribution does not have a finite expectation. However the Cauchy probability density function is symmetric about zero.

Properties of the expectation

(i) Suppose X is a random variable with finite expectation. Let $Y = aX + b$ where a and b are real numbers. Then $E(Y)$ is finite and

$$E(Y) = aE(X) + b.$$

(ii) Suppose X is a random variable taking nonnegative values with probability one. Then $E(X) \geq 0$. It is possible that $E(X) = \infty$.

(iii) If X and Y are random variables with finite expectations and a and b are real numbers, then $E(aX + bY)$ is finite and

$$E(aX + bY) = aE(X) + bE(Y).$$

The properties (i) and (ii) follow from the properties of integrals. We will prove property (iii) later in this book.

Let X be a random variable with expectation $E(X) = \mu$. Note that μ represents a measure of the central tendency for X or the centre of gravity of the distribution of X. The random variable X might take values different from μ. We are now interested in studying a measure of the dispersion or spread of the values X around the mean μ. A natural measure is $E(X - \mu)$. However this measure is not a good measure as $E(X - \mu) = 0$ for any random variable X with finite expectation μ. This can be seen from the properties of the expectation stated above. A natural and mathematically convenient measure is $E[(X - \mu)^2]$. Another measure of the spread is $E|X - \mu|$. The quantity $E[(X - \mu)^2]$ is called the *variance* of the random variable X and it is usually denoted by $Var(X)$ whenever it is finite. Note that variance of a random variable is always nonnegative but could be infinite. The nonnegative square root of the variance of X whenever it is finite is called the *standard deviation* of X. Check that if the variance of a random variable X is finite, then its mean is also finite.

Properties of the variance

(i) If X is a random variable such that $P(X = c) = 1$ where c is a constant, then $Var(X) = 0$. Conversely if $Var(X) = 0$ for a random variable X then, there exists a constant c such that $P(X = c) = 1$.

(ii) For any two constants a and b,

$$Var(aX + b) = a^2 Var(X)$$

whenever $Var(X)$ is finite.

The first property follows from the observation that if $P(X = c) = 1$, then $E(X) = c$ and hence $Var(X) = E(X - c)^2 = 0$ since $X - c = 0$ with probability one. Conversely, if the $Var(X) = 0$ for a random variable X with mean μ, then $E(X - \mu)^2 = 0$ and it is intuitively clear that if X differs from μ on a set of positive probability, then $E(X - \mu)^2 > 0$ which contradicts the assumption. One can give a rigourous proof of this result by using the *Chebyshev's inequality* discussed in Theorem 7.1 in Chapter 7.

We now give a proof of the Property (ii). Let $E(X) = \mu$. Note that $E(aX + b) = aE(X) + b = a\mu + b$ and

$$\begin{aligned} Var(aX + b) &= E[(aX + b) - (a\mu + b)^2] \\ &= a^2 E[(X - \mu)^2] \\ &= a^2 Var(X) \end{aligned}$$

from the properties of the expectation discussed above. In particular $Var(X) = Var(-X)$ and $Var(X + b) = Var(X)$ for any constant b whenever $Var(X)$ is finite.

An alternate convenient formula for computing the variance of a random variable X with mean μ is

$$Var(X) = E[X^2] - \mu^2.$$

This can be seen by using the properties of the expectation (check?). In particular it follows that

$$E(X^2) \geq [E(X)]^2$$

for any random variable X with finite variance since the $Var(X)$ of a random variable X is nonnegative. Note this inequality proves that if $E(X^2)$ is finite , then $E(X)$ is also finite.

All the properties of the expectation and the variance hold for the random variables be they of discrete type or of continuous type or of a mixture of the two types as long as the expectations and variances involved are finite.

Example 4.15. Let X be a random variable with the uniform distribution on the interval $[\alpha, \beta]$. We have seen earlier that $E(X) = \frac{\alpha+\beta}{2}$. Note that

$$Var(X) = E(X^2) - [E(X)]^2.$$

Check that

$$E(X^2) = \frac{1}{\beta - \alpha}\left(\frac{\beta^3 - \alpha^3}{3}\right)$$

and hence

$$Var(X) = \frac{(\beta - \alpha)^2}{12}.$$

Example 4.16. Let X be a random variable with the exponential probability density function given by

$$\begin{aligned} f(x) &= \lambda e^{-\lambda x} \quad \text{for } x \geq 0 \\ &= 0 \text{ otherwise} \end{aligned}$$

for some $\lambda > 0$. We have seen that $E(X) = \frac{1}{\lambda}$. Let us now compute $E(X^2)$. Note that

$$\begin{aligned} E(X^2) &= \int_0^\infty x^2 \lambda e^{-\lambda x} \; dx \\ &= \frac{1}{\lambda^2} \int_0^\infty y^2 e^{-y} \; dy \end{aligned}$$

by applying the transformation $y = \lambda x$. The function defined by

$$\Gamma(\alpha) = \int_0^\infty y^{\alpha-1}e^{-y}, \alpha > 0$$

is called the *Gamma function* and it is known that

$$\Gamma(n+1) = n!$$

for any positive integer n. It can now be checked that $\Gamma(3) = 2$ and

$$E(X^2) = \frac{2}{\lambda^2}.$$

Hence

$$\begin{aligned} Var(X) &= E(X^2) - [E(X)]^2 \\ &= \frac{2}{\lambda^2} - \frac{1}{\lambda^2} \\ &= \frac{1}{\lambda^2}. \end{aligned}$$

Example 4.17. Suppose X is a random variable with the probability density function

$$f(x) = \frac{1}{\sqrt{2\pi\sigma^2}}e^{-\frac{1}{2}\frac{(x-\mu)^2}{\sigma^2}}, -\infty < x < \infty$$

where $-\infty < \mu < \infty$ and $0 < \sigma^2 < \infty$. We have seen above that $E(X) = \mu$ and $E(X-\mu)^2 = \sigma^2$. Hence the mean and variance of the random variable X are μ and σ^2 respectively. The probability density function of X is known as the *normal* or the *Gaussian* density with mean μ and variance σ^2.

Definition. Let X be a random variable with finite mean μ and a positive finite variance σ^2. Define

$$Z = \frac{X-\mu}{\sigma}.$$

The random variable Z is called the *standardization* of the random variable X. Note that $E(Z) = 0$ and $Var(Z) = 1$.

4.5 Moments

Suppose X is a random variable . Let $k \geq 1$ be an integer and b be a real number. Suppose that $E(X-b)^k$ exists. It is called the k-th *moment about the point* b of the random variable X or of its probability distribution. If $b = 0$, then it is called the k-th *raw moment* or k-th *moment* and it is

$E(X^k)$ denoted by μ'_k. If we take $b = \mu = E(X)$, then $\mu_k = E(X-\mu)^k$ is called the k-th *central moment* and the number k is called the *order* of the moment. We noted earlier that $E(X-\mu)^k$ exists if and only if $E|X-\mu|^k$ exists. The quantity $E|X-\mu|^k$ is called the k-th *absolute central moment* of order k.

Suppose a random variable X has the first two moments. Observe that the first moment μ'_1 of a random variable X is its mean and the first central moment is zero and the second central moment is $\mu_2 = Var(X)$ (check?). The moments of a random variable X or of its distribution are descriptive measures computed from the distribution. The mean and variance of a random variable are two such measures. Higher order moments describe other aspects of the distribution.

Suppose X is a random variable with the normal probability density function

$$f(x) = \frac{1}{\sqrt{2\pi\sigma^2}} e^{-\frac{1}{2}\frac{(x-\mu)^2}{\sigma^2}}, -\infty < x < \infty.$$

Note that the normal density is completely determined by the mean μ and variance σ^2 in the sense that we can compute the probabilities of all events involving the random variable X if the mean μ and the variance σ^2 are known. This is not the case in general. One of the famous problems in probability is the so called *moment problem.* This problem deals with finding necessary and sufficient conditions under which the moments uniquely determine a probability distribution. Discussion on this topic is beyond the scope of this book.

Relations between the raw and central moments

We now discuss the relations between the raw and the central moments of a random variable X whenever they exist. An application of the Binomial theorem shows that

$$\begin{aligned}
\mu_k &= E[(X-\mu)^k] \\
&= E[\sum_{j=0}^{k} X^j(-\mu)^{k-j} k_j^C] \\
&= \sum_{j=0}^{k} E(X^j)(-\mu)^{k-j} k_j^C \\
&= \sum_{j=0}^{k} \mu'_j(-\mu)^{k-j} k_j^C
\end{aligned}$$

where we observe that $\mu_0' = 1$. Note that $\mu_1' = \mu$. In particular,

$$\begin{aligned}
\mu_1 &= E[X-\mu] = 0; \\
\mu_2 &= E[X-\mu]^2 = \sigma^2 = \mu_2' - \mu_1'^2 = \mu_2' - \mu^2; \\
\mu_3 &= E[X-\mu]^3 = \mu_3' - 3\mu_1'\mu_2' + 2\mu_1'^3; \\
\mu_4 &= E[X-\mu]^4 = \mu_4' - 4\mu_1'\mu_3' + 6\mu_1'^2\mu_2' - 3\mu_1'^4
\end{aligned}$$

and so on.

Example 4.18. Suppose X is a random variable with the normal probability density function with mean μ and variance σ^2. Then

$$\begin{aligned}
\mu_k &= \int_{-\infty}^{\infty} (x-\mu)^k \frac{1}{\sqrt{2\pi\sigma^2}} e^{-\frac{1}{2}\frac{(x-\mu)^2}{\sigma^2}}\ dx \\
&= \sigma^k \int_{-\infty}^{\infty} y^k \frac{1}{\sqrt{2\pi}} e^{-\frac{1}{2}y^2}\ dy.
\end{aligned}$$

If k is an odd integer, then the integrand is an odd function, that is a function with the property $h(x) = -h(-x)$ for all x. Hence the integral of such a function over the real line is zero (check?). This implies that all the central moments of odd order of a normal probability density function are equal to zero. Suppose k is an even integer. Then $k = 2m$ for some integer $m \geq 1$ and

$$\begin{aligned}
\mu_{2m} &= \int_{-\infty}^{\infty} (x-\mu)^{2m} \frac{1}{\sqrt{2\pi\sigma^2}} e^{-\frac{1}{2}\frac{(x-\mu)^2}{\sigma^2}}\ dx \\
&= \sigma^{2m} \int_{-\infty}^{\infty} y^{2m} \frac{1}{\sqrt{2\pi}} e^{-\frac{1}{2}y^2}\ dy \\
&= 2\sigma^{2m} \int_{0}^{\infty} y^{2m} \frac{1}{\sqrt{2\pi}} e^{-\frac{1}{2}y^2}\ dy \\
&= 2\sigma^{2m} \int_{0}^{\infty} (2z)^{m} \frac{1}{\sqrt{2\pi}} e^{-z} (2z)^{-\frac{1}{2}}\ dz \\
&= \frac{1}{\sqrt{2}} 2^{m+1} \sigma^{2m} \int_{0}^{\infty} (z)^{m-\frac{1}{2}} \frac{1}{\sqrt{2\pi}} e^{-z}\ dz \\
&= 2^m \sigma^{2m} \Gamma(m + \frac{1}{2}) \pi^{-\frac{1}{2}}
\end{aligned}$$

where

$$\Gamma(\alpha) = \int_{0}^{\infty} z^{\alpha-1} e^{-z}\ dz$$

as defined in the Example 4.16. It is known that

$$\Gamma(1) = 1,$$
$$\Gamma(\frac{1}{2}) = \sqrt{\pi},$$
$$\Gamma(\alpha + 1) = \alpha\Gamma(\alpha), \alpha > 0,$$
$$\Gamma(k + 1) = k! \text{ if } k \text{ is a positive integer,}$$

and

$$\frac{\Gamma(m)\Gamma(n)}{\Gamma(m+n)} = \int_0^1 u^{m-1}(1-u)^{n-1}\, du.$$

Using the above relations, check that

$$\mu_{2m} = \{(2m-1)(2m-3)\ldots 1\}\sigma^{2m}, m \geq 1.$$

In particular $\mu_2 = \sigma^2$ and $\mu_4 = 3\sigma^4$. Observe that $\mu_4 = 3\mu_2^2$.

Remarks. For any random variable X with finite fourth moment, let

$$\gamma_1 = \frac{E(X-\mu)^3}{\sigma^3} = \frac{\mu_3}{\sigma^3}$$

and

$$\gamma_2 = \frac{E(X-\mu)^4}{\sigma^4} - 3 = \frac{\mu_4}{\sigma^4} - 3.$$

The *skewness* of a probability density function is measured in terms of its third central moment. If the probability density function is symmetric, then all the odd central moments are zero if they exist and in particular the third moment is zero as in the case of the normal probability density function. The lack of symmetry is measured by the *coefficient of skewness* defined by γ_1. Observe that γ_1 is dimension free. Another measure that is used for measuring the *peakedness* of a probability density function is γ_2 defined above. It is called the *coefficient of kurtosis*. From the above calculations, we note that $\gamma_1 = 0$ and $\gamma_2 = 0$ for a normal probability density function.

4.6 Moment Generating Function

Calculation of the moments of a random variable directly is difficult at times even when they exist. An alternate method of obtaining the moments is through the moment generating function whenever it exists.

Let X be a random variable such that $M_X(t) = E[e^{tX}]$ exists for some $t \neq 0$. It is obvious that the function $M_X(t)$ is well-defined at $t = 0$ and $M_X(0) = 1$. The function $M_X(t)$ is called the *moment generating function* (m.g.f.) of the random variable X whenever it is well-defined. Suppose the function $M_x(t)$ is defined in a neighbourhood of the point $t = 0$. Let us expand the function $M_X(t)$ by using the Taylor's theorem around the value $t = 0$. Suppose that sufficient conditions hold for the validity of the following expansion:

$$M_X(t) = M_X(0) + t\frac{dM_X(t)}{dt}|_{t=0} + \cdots + \frac{t^n}{n!}\frac{d^n M_X(t)}{dt^n}|_{t=0} + \ldots.$$

On the other hand, suppose that the following computation is justified:

$$\begin{aligned} M_X(t) &= E[e^{tX}] \\ &= E[1 + tX + \frac{t^2X^2}{2!} + \ldots] \\ &= 1 + tE[X] + \frac{t^2}{2!}E[X^2] + \ldots. \end{aligned}$$

Comparing the coefficients of t^n in both the expansions given above, we have the relation

$$\frac{d^n M_X(t)}{dt^n}|_{t=0} = E[X^n], n \geq 1.$$

This relation implies that the n-th moment about zero or the n-th raw moment of the random variable X can be obtained by differentiating the m.g.f. $M_X(t)$ of X with respect to t for n times and then evaluating the n-th derivative at $t = 0$. This is the reason why the function $M_X(t)$ is called the moment generating function of X. The above arguments can be justified under some conditions on the existence of moments and the m.g.f. But such a discussion is beyond the scope of this book.

Example 4.19. Suppose X is a random variable with the uniform distribution on the interval $[\alpha, \beta]$. Then, for any $t \neq 0$,

$$\begin{aligned} M_X(t) = E[e^{tX}] &= \int_\alpha^\beta e^{tx}\frac{1}{\beta - \alpha}\, dx \\ &= \frac{1}{\beta - \alpha}[\frac{e^{tx}}{t}]_\alpha^\beta \\ &= \frac{e^{t\beta} - e^{t\alpha}}{t(\beta - \alpha)} \end{aligned}$$

and it is always true that $M_X(0) = 1$. Hence the m.g.f. $M_X(t)$ of the random variable X exists for every t.

Example 4.20. Suppose X is a random variable with the exponential probability density function

$$\begin{aligned} f(x) &= \lambda e^{-\lambda x} \text{ for } x \geq 0 \\ &= 0 \text{ otherwise} \end{aligned}$$

where $\lambda > 0$. Then

$$\begin{aligned} M_X(t) = E[e^{tX}] &= \int_0^\infty e^{tx} \lambda e^{-\lambda x} \, dx \\ &= \lambda \int_0^\infty e^{(t-\lambda)x} \, dx \end{aligned}$$

and the integral on the right side of the above equation is finite if and only if $t < \lambda$. Check that, for $t < \lambda$,

$$M_X(t) = \frac{\lambda}{\lambda - t}$$

and the m.g.f. is not defined for $t \geq \lambda$.

Example 4.21. Suppose X is a random variable with the normal probability density function with mean μ and variance σ^2. Then the m.g.f. of X is given by

$$\begin{aligned} M_X(t) = E[e^{tX}] &= \int_{-\infty}^{\infty} e^{tx} \frac{1}{\sqrt{2\pi\sigma^2}} e^{-\frac{1}{2}\frac{(x-\mu)^2}{\sigma^2}} \, dx \\ &= \int_{-\infty}^{\infty} e^{t(\sigma y + \mu)} \frac{1}{\sqrt{2\pi}} e^{-\frac{1}{2}y^2} \, dy \\ &= e^{\mu t} \int_{-\infty}^{\infty} \frac{1}{\sqrt{2\pi}} e^{-\frac{1}{2}y^2 + t\sigma y} \, dy \\ &= e^{\mu t} \int_{-\infty}^{\infty} \frac{1}{\sqrt{2\pi}} e^{-\frac{1}{2}(y - t\sigma)^2 + \frac{1}{2}t^2\sigma^2} \, dy \\ &= e^{\mu t + \frac{1}{2}t^2\sigma^2} \int_{-\infty}^{\infty} \frac{1}{\sqrt{2\pi}} e^{-\frac{1}{2}(y - t\sigma)^2} \, dy \\ &= e^{\mu t + \frac{1}{2}t^2\sigma^2} \end{aligned}$$

since the last integral on the right side of the above equation is equal to one as the integrand can be considered as the probability density function of the normal distribution with mean $t\sigma$ and variance one. Hence

$$M_X(t) = e^{\mu t + \frac{1}{2}t^2\sigma^2}, -\infty < t < \infty.$$

This is another example where the m.g.f. of the random variable exists for all t.

Remarks. Recall our discussion earlier that the moments do not determine a distribution uniquely in general. In other words, there could be two random variables with the same sequence of moments but the distributions corresponding to the random variables X and Y might be different. However the following theorem is valid. We will not discuss the proof as it is beyond the scope of this book.

Theorem 4.1. Suppose X and Y are random variables with the m.g.f.'s $M_X(t)$ and $M_Y(t)$ respectively. Suppose further that the functions $M_X(t)$ and $M_Y(t)$ exist in an open interval containing zero and $M_X(t) = M_Y(t)$ in that interval. Then the distributions of X and Y are the same.

4.7 Functions of a Random Variable

In the previous sections, we are concerned with the expectations of a random variable X or some functions of it such as $(X - \mu)^2$, $|X - \mu|, |X|^k, X^k, e^{tX}$ and so on. We will now discuss some methods of determining the distribution functions of a function $g(X)$ of a given random variable X with a known distribution function.

Let X be a random variable with a known distribution unction $F_X(x)$. Suppose $Y = aX + b$ where a and b are constants with $a > 0$. Then the distribution function F_Y of Y is given by

$$\begin{aligned}
F_Y(y) &= P(Y \leq y) \\
&= P(aX + b \leq y) \\
&= P(X \leq \frac{y-b}{a}) \\
&= F_X(\frac{y-b}{a}).
\end{aligned}$$

If $a < 0$, then the distribution function of Y is given by

$$
\begin{aligned}
F_Y(y) &= P(X \geq \frac{y-b}{a}) \\
&= 1 - P(X < \frac{y-b}{a}) \\
&= 1 - F_X(\frac{y-b}{a} - 0).
\end{aligned}
$$

If the random variable X has a distribution function $F_X(x)$ which is continuous, then we can write the above relation in the form

$$F_Y(y) = 1 - F_X(\frac{y-b}{a}).$$

Suppose the random variable X has a probability density unction f_X. We know that the function f_X can be obtained from F_X by differentiation. If the random variable has the probability density function f_X, then we get that the random variable Y has also a probability density function f_Y given by

$$f_Y(y) = \frac{1}{a} f_X(\frac{y-b}{a})$$

if $a > 0$ and

$$f_Y(y) = -\frac{1}{a} f_X(\frac{y-b}{a})$$

if $a < 0$. Combining the above relations, we get that the probability density function of f_Y of Y is given by

$$f_Y(y) = \frac{1}{|a|} f_X(\frac{y-b}{a}).$$

(What happens if $a = 0$? Check?)

Suppose that X is a random variable with the distribution function F_X. Let $Y = g(X)$ where $g(.)$ is a continuous and strictly increasing function. Since $g(x)$ is a strictly increasing function, there exists an unique inverse function $r(y)$ of $g(x)$ such that $g(r(y)) = y$ for all y. Since g is continuous, r is also continuous and $g(x) \leq y$ if and only if $x \leq r(y)$. If F_Y denotes the distribution function of Y, then

$$
\begin{aligned}
F_Y(y) &= P(Y \leq y) \\
&= P(g(X) \leq y) \\
&= P(X \leq r(y)) \\
&= F_X(r(y)).
\end{aligned}
$$

Suppose the random variable X has a probability density function f_X and $Y = g(X)$ where $g(.)$ is a continuous and strictly increasing function on the set $\{x : f_X(x) > 0\}$. Define the inverse function $r(y)$ of $g(x)$ as given above. Suppose the function $r(y)$ is differentiable in y. Then $Y = g(X)$ has a probability density function f_Y and

$$\begin{aligned} f_Y(y) = \frac{dF_Y(y)}{dy} &= \frac{dF_X(r(y))}{dy} \\ &= f_X(r(y))\frac{dr(y)}{dy}. \end{aligned}$$

If the function $g(x)$ is continuous and strictly decreasing, then again there exists an unique inverse continuous function $r(y)$ with the property that $g(x) \leq y$ if and only if $x \geq r(y)$. Furthermore

$$\begin{aligned} F_Y(y) &= P(Y \leq y) \\ &= P(g(X) \leq y) \\ &= P(X \geq r(y)) \\ &= 1 - F_X(r(y) - 0). \end{aligned}$$

Suppose the random variable X has a probability density function f_X and $Y = g(X)$ where $g(.)$ is a continuous and strictly decreasing function on the set $\{x : f_X(x) > 0\}$. Define the inverse function $r(y)$ of $g(x)$ as above. Suppose $r(y)$ is differentiable in y. Then $Y = g(X)$ has a probability density function f_Y and

$$\begin{aligned} f_Y(y) = \frac{dF_Y(y)}{dy} &= -\frac{dF_X(r(y))}{dy} \\ &= -f_X(r(y))\frac{dr(y)}{dy}. \end{aligned}$$

From the above discussion, we have the following result.

Theorem 4.2. Supposer X is a random variable with the probability density function f_X. Let $Y = g(X)$ where g is a continuous and either a strictly increasing or a strictly decreasing function. Let $r(y)$ be the inverse function of $g(x)$ and suppose that $r(y)$ is differentiable in y. Then Y has a probability density function f_Y and

$$f_Y(y) = f_X(r(y))|\frac{dr(y)}{dy}|.$$

Remarks. Note that, if the function $g(x)$ is differentiable with $\frac{dg(x)}{dx} \neq 0$, then the derivatives of g and its inverse function r are connected by the relation

$$\frac{dr(y)}{dy} = 1/\frac{dg(x)}{dx}|_{x=r(y)}.$$

The inverse function $r(y)$ of $g(x)$ is usually denoted by $g^{-1}(y)$.

Example 4.22. Suppose X is a random variable with the probability density function

$$\begin{aligned} f(x) &= 3x^2 \text{ for } 0 < x < 1 \\ &= 0 \text{ otherwise.} \end{aligned}$$

Let $Y = X^2$. Here the function $g(x) = x^2$ in the earlier notation. It is strictly increasing and continuous in the interval $(0,1)$ and $r(y) = y^{1/2}$ is the inverse function of g. The function $r(y)$ is differentiable on the interval $(0,1)$ and hence by Theorem 4.2, the random variable Y has a probability density function f_Y and it is given by

$$\begin{aligned} f_Y(y) = f(r(y))\frac{dr(y)}{dy} &= 3(y^{1/2})^2\frac{1}{2}y^{-1/2}, \quad 0 < y < 1 \\ &= 0 \text{ otherwise} \end{aligned}$$

which simplifies to

$$\begin{aligned} f_Y(y) &= \frac{3}{2}y^{1/2}, \quad 0 < y < 1 \\ &= 0 \text{ otherwise.} \end{aligned}$$

Example 4.23. Suppose X is a random variable with the standard uniform probability density function f. Define $Y = -\frac{1}{\lambda}\log X$ where $\lambda > 0$ and $\log x$ denotes the natural logarithm of x. The function

$$g(x) = -\frac{1}{\lambda}\log x$$

maps the interval $(0,1)$ to $(0,\infty)$ and it is continuous and strictly decreasing on the interval $(0,1)$ which is the support of the standard uniform probability density function. The inverse function $r(y)$ of $g(x)$ is given by

$$r(y) = e^{-\lambda y}, \quad 0 < y < \infty$$

which is differentiable. Applying Theorem 4.2, it is easy to check that the random variable Y has a probability density function and it is given by

$$\begin{aligned} f_Y(y) &= -f(r(y))\frac{dr(y)}{dy} \quad \text{for } y > 0 \\ &= \lambda e^{-\lambda y} \quad \text{for } y > 0 \end{aligned}$$

and

$$f_Y(y) = 0 \quad \text{for } y \leq 0.$$

Remarks. In both the examples discussed above, we could apply Theorem 4.2 directly . However this is not always possible and the next example gives one such case.

Example 4.24. Suppose a random variable X has the standard normal probability density function. Note that the support of the normal probability density function is the whole real line. Let $Y = X^2$. Here $g(x) = x^2$. This function is continuous but strictly increasing on $(0, \infty)$ and strictly decreasing on $(-\infty, 0)$. Further the function is not one-to-one since $g(x) = g(-x)$ for all x. Therefore we cannot apply the Theorem 4.2 given above. It is obvious that

$$F_Y(y) = P(Y \leq y) = P(X^2 \leq y) = 0 \quad \text{for } y < 0.$$

Suppose $y > 0$. Then

$$\begin{aligned} F_Y(y) = P(Y \leq y) &= P(X^2 \leq y) \\ &= P(|X| \leq \sqrt{y}) \\ &= P(-\sqrt{y} \leq X \leq \sqrt{y}) \\ &= \int_{-\sqrt{y}}^{\sqrt{y}} \phi(x)dx \end{aligned}$$

where $\phi(x)$ denotes the standard normal probability density function. From the symmetry of the probability density function $\phi(x)$, we get that

$$\begin{aligned} F_Y(y) &= 2\int_0^{\sqrt{y}} \frac{1}{\sqrt{2\pi}} e^{-\frac{1}{2}x^2} dx \\ &= \int_0^{y} \frac{1}{\sqrt{2\pi}} e^{-\frac{1}{2}z} z^{-1/2} dz. \end{aligned}$$

It is easy see that $F_Y(0) = 0$ (check?). The above derivation shows that the random variable Y has a probability density function and it is given by

$$\begin{aligned} f_Y(y) &= \frac{1}{\sqrt{2\pi}} e^{-\frac{1}{2}y} y^{-1/2} \quad \text{for } y > 0 \\ &= 0 \quad \text{for } y \leq 0. \end{aligned}$$

Recalling that $\Gamma(\frac{1}{2}) = \sqrt{\pi}$, we can rewrite the function $f_Y(y)$ in the form

$$\begin{aligned} f_Y(y) &= \frac{1}{\sqrt{2}\Gamma(\frac{1}{2})} e^{-\frac{1}{2}y} y^{(1/2)-1} \quad \text{for } y > 0 \\ &= 0 \quad \text{for } y \leq 0. \end{aligned}$$

This probability density function is known as the *Chi-square probability density function* with 1 degree of freedom. We will discuss more about this distribution later in this book.

Remarks. In order to compute the expectation of the random variable $Y = g(X)$ whenever it exists, either we can use the probability distribution of Y or that of X. For instance, suppose X has the probability density function f_X and Y has the probability density function f_Y. Then

$$E[g(X)] = \int_{-\infty}^{\infty} g(x) f_X(x) dx$$

and

$$E[Y] = \int_{-\infty}^{\infty} y f_Y(y) dy.$$

It can be shown that $E[Y]$ exists if and only if $E[g(X)]$ does and both these methods of calculation lead to the same result. The choice of the method depends on the complexity involved in finding the distribution or the probability density function of $Y = g(X)$.

We continue the discussion in the Example 4.24 to illustrate the above remark.

Example 4.25. Let X be be a random variable with standard normal probability density function as in the Example 4.24. Let $Y = X^2$. We know that $E(X^2) = 1$ from the earlier discussions. Let us now compute

$E(Y)$ directly by using the probability density of Y derived in the Example 4.24. Then

$$\begin{aligned}
E(Y) &= \int_0^\infty y \frac{1}{\sqrt{2}\Gamma(\frac{1}{2})} e^{-\frac{1}{2}y} y^{(1/2)-1} dy \\
&= \frac{1}{\sqrt{2}\Gamma(\frac{1}{2})} \int_0^\infty e^{-\frac{1}{2}y} y^{(3/2)-1} dy \\
&= \frac{1}{\sqrt{2}\Gamma(\frac{1}{2})} \sqrt{2} \int_0^\infty 2e^{-u} u^{(3/2)-1} du \\
&= \frac{2}{\Gamma(\frac{1}{2})} \Gamma(\frac{3}{2}) \\
&= 1
\end{aligned}$$

as it should be since $\Gamma(\frac{3}{2}) = \frac{1}{2}\Gamma(\frac{1}{2})$.

4.8 Standard Continuous Probability Distributions

We now discuss some more properties of the standard continuous probability distributions such as the normal, exponential, Gamma and Beta distributions in more detail. These are widely used for stochastic modelling problems.

Normal distribution

Normal distribution, also known as the Gaussian distribution, can be considered as one of the important probability distributions in view of its applicability for stochastic modelling. It was observed that a normal distribution is a good fit for a large class of data sets found in practice *after proper scaling.* For instance, it is a good approximation to the distribution of heights of people in a particular region, to the distribution of marks obtained by students from a particular university in a given examination or to the distribution of diameters of bolts produced in a certain factory. It has been empirically observed by Gauss that a normal distribution is a good fit for the distribution of errors in measurement and hence the distribution is also known as the *Gaussian distribution.*

Another reason for the importance of the normal distribution in the subject of Statistics is the *Central limit theorem.* We will discuss this result later in this book. What it essentially implies is that, even though the original data set or observations might not follow a normal distribution, the averages of these observations will be distributed approximately normal as long as the number of observations is large. More precisely if the original distribution has finite positive variance, then the sample means of random samples will have an approximately normal distribution for large samples.

A random variable X is said to have a *normal distribution* if it has a probability density function f of the form

$$f(x;\mu,\sigma) = \frac{1}{\sqrt{2\pi\sigma^2}}e^{-\frac{1}{2}\frac{(x-\mu)^2}{\sigma^2}}, -\infty < x < \infty$$

for some μ and σ with $-\infty < \mu < \infty$ and $0 < \sigma < \infty$. Let us now check whether the function f is a probability density function. Recall that a function f is a probability density function if and only if $f(x;\mu,\sigma) \geq 0$ for all x and

$$\int_{-\infty}^{\infty} f(x;\mu,\sigma)dx = 1.$$

Nonnegativity of the function f is obvious from the definition of the function f. Let us check the second condition:

$$\begin{aligned}\int_{-\infty}^{\infty} f(x;\mu,\sigma)\ dx &= \int_{-\infty}^{\infty} \frac{1}{\sqrt{2\pi\sigma^2}}e^{-\frac{1}{2}\frac{(x-\mu)^2}{\sigma^2}}\ dx \\ &= \int_{-\infty}^{\infty} \frac{1}{\sqrt{2\pi}}e^{-\frac{1}{2}y^2}\ dy\end{aligned}$$

by the transformation $y = \frac{x-\mu}{\sigma}$. Let

$$I = \int_{-\infty}^{\infty} \frac{1}{\sqrt{2\pi}}e^{-\frac{1}{2}y^2}\ dy.$$

Then

$$\begin{aligned}I^2 &= (\int_{-\infty}^{\infty} \frac{1}{\sqrt{2\pi}}e^{-\frac{1}{2}y^2}\ dy)(\int_{-\infty}^{\infty} \frac{1}{\sqrt{2\pi}}e^{-\frac{1}{2}y^2}\ dy) \\ &= \frac{1}{2\pi}\int_{-\infty}^{\infty}\int_{-\infty}^{\infty} e^{-\frac{1}{2}(y^2+z^2)}\ dydz.\end{aligned}$$

Applying the transformation $y = r\ cos\ \theta, z = r\ sin\ \theta$ and using the change of variable formula for double integrals, we have

$$\begin{aligned}I^2 &= \frac{1}{2\pi}\int_{0}^{2\pi}\int_{0}^{\infty} e^{-\frac{1}{2}r^2} r\ drd\theta \\ &= \frac{1}{2\pi}2\pi\int_{0}^{\infty} e^{-\frac{1}{2}r^2} r\ dr \\ &= 1.\end{aligned}$$

This proves that $I = 1$ since $I \geq 0$ being the value of the integral of a nonnegative function. We have seen earlier that if X is a random variable with the probability density function given by the function $f(x; \mu, \sigma)$, then $E[X] = \mu$ and $Var(X) = \sigma^2$.

Figure 4.12 shows the graph of the normal probability density function $f(x; \mu, \sigma)$ with mean μ= 1.5 and variance σ^2=4 and Figure 4.13 shows the graph of the standard normal probability density function.

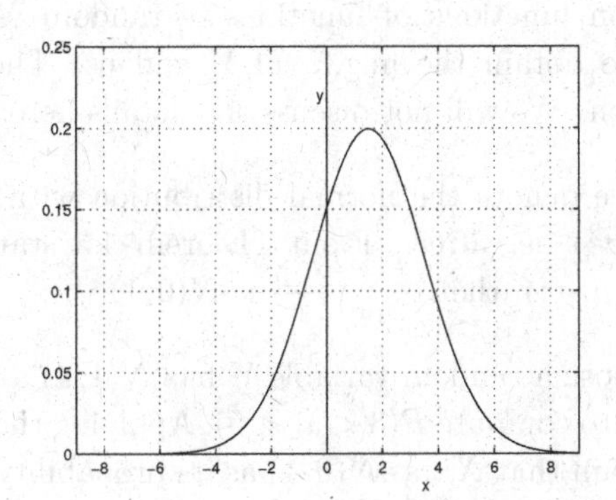

Fig. 4.12 Normal Probability Density Function $N(\mu, \sigma^2), \mu = 1.5, \sigma^2 = 4$

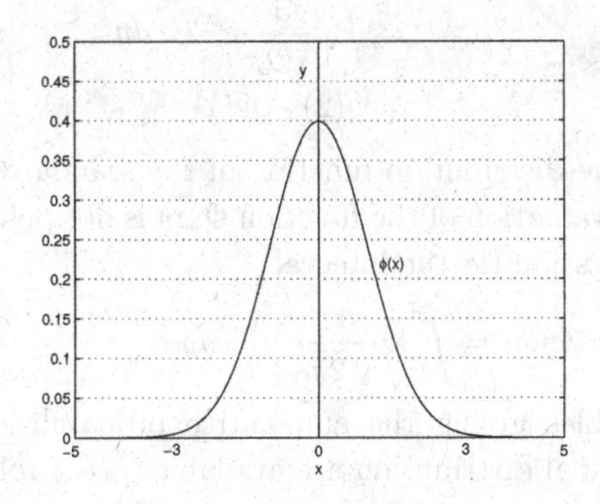

Fig. 4.13 Standard Normal Probability Density Function $\phi(x)$

It is easy to see that the probability density function $f(x; \mu, \sigma)$ is symmetric about the value μ and the graph of the function $f(x; \mu, \sigma)$ is bell-shaped. We have computed the m.g.f. $M_X(t)$ of the random variable X

earlier in this chapter. It is given by

$$M_X(t) = e^{\mu t + \frac{1}{2}t^2\sigma^2}, -\infty < t < \infty.$$

Theorem 4.3. Suppose X is a random variable with the normal probability density with mean μ and variance σ^2. Then the random variable $Y = aX + b$ has the normal distribution with mean $a\mu + b$ and variance σ^2.

This theorem can be proved by using the methods described earlier for finding the distribution functions of functions of random variables. An alternate method is to obtain the m.g.f. of Y and use Theorem 4.1 to identify the distribution. We will not discuss the details here.

For convenience, we denote the normal distribution with mean μ and variance σ^2 by $N(\mu, \sigma^2)$ hereafter. From Theorem 4.3 stated above, it follows that if Y is $N(\mu, \sigma^2)$, then $X = \frac{Y-\mu}{\sigma}$ is $N(0, 1)$.

Example 4.26. Suppose a random variable Y has $N(1, 4)$ as its distribution and we would like to compute $P(3 < Y < 5)$. Applying the transformation $X = \frac{Y-1}{2}$, we obtain that X has $N(0, 1)$ as its probability distribution from Theorem 4.3 and

$$\begin{aligned} P(3 < Y < 5) &= P(1 < X < 2) \\ &= \int_1^2 \frac{1}{\sqrt{2\pi}} e^{-\frac{1}{2}y^2}\, dy \\ &= \Phi(2) - \Phi(1) \end{aligned}$$

where $\Phi(x)$ denotes the distribution function of the standard normal random variable. Exact evaluation of the function $\Phi(x)$ is not possible as there is no closed form expression for the integral

$$\Phi(x) = \int_{-\infty}^{x} \frac{1}{\sqrt{2\pi}} e^{-\frac{1}{2}y^2}\, dy.$$

However extensive tables giving the cumulative probabilities calculated from the standard normal distribution are available (See Table C.1 in the Appendix). It is helpful to note that $\Phi(-x) = 1 - \Phi(x), x \geq 0$. This relation follows from the symmetry property of the standard normal probability density function.

Suppose a random variable Y has $N(\mu, \sigma^2)$ as its distribution. Let $X = \frac{Y-\mu}{\sigma}$. Then

$$\begin{aligned} P[\mu - 3\sigma \leq Y \leq \mu + 3\sigma] &= P[-3 \leq X \leq 3] \\ &= \Phi(3) - \Phi(-3) \\ &= 0.9974 \end{aligned}$$

from Table C.1. Similarly check that

$$P[\mu - 2\sigma \leq Y \leq \mu + 2\sigma] = 0.9546.$$

In other words, more than 99% of the total probability is carried by the interval $[\mu - 3\sigma, \mu + 3\sigma]$ and about 95% of the total probability is supported by the interval $[\mu - 2\sigma, \mu + 2\sigma]$ for any random variable Y which has $N(\mu, \sigma^2)$ as its distribution.

Example 4.27. It has been observed that that the marks obtained by the students in an examination for a particular subject generally follow a normal distribution. A teacher uses the percentages of marks obtained by different students and obtains the mean μ and the variance σ^2. Suppose the letter Grade A is assigned to a student if the percentage of marks is greater than $\mu + \sigma$, B to those whose percentage is between μ and $\mu + \sigma$, C to those between $\mu - \sigma$ and μ, D to those between $\mu - 2\sigma$ and $\mu - \sigma$ and F to those getting a percentage below $\mu - 2\sigma$. (This method of grading is some times called "grading on a curve".) Let Y denote the random variable corresponding to the percentage of marks obtained in the examination and suppose that Y has a $N(\mu, \sigma^2)$ as its probability distribution. Check that

$$\begin{aligned} P(Y > \mu + \sigma) &= 1 - \Phi(1) = 0.1587, \\ P(\mu < Y \leq \mu + \sigma) &= \Phi(1) - \Phi(0) = 0.3413 \\ P(\mu - \sigma < Y \leq \mu) &= \Phi(0) - \Phi(-1) = 0.3413 \\ P(\mu - 2\sigma < Y \leq \mu - \sigma) &= \Phi(-1) - \Phi(-2) = 0.1359 \\ P(Y \leq \mu - 2\sigma) &= \Phi(-2) = 0.0228 \end{aligned}$$

from Table C.1. In other words, out of all students who took the examination approximately 16% of the students will receive grade A, 34% grade B, 34% grade C, 14% grade D and 2% will receive grade F.

Exponential distribution

We have discussed some properties of the exponential distribution earlier. Recall that a random variable X is said to have an *exponential* distribution if it has a probability density function of the form

$$f(x) = \lambda e^{-\lambda x} \text{ for } x \geq 0$$
$$= 0 \text{ otherwise}$$

where $\lambda > 0$. The exponential distribution can be considered for building a stochastic model whenever the data consists of lifetimes of some components such as light bulbs or the waiting times for a specific event to occur. An important property of the exponential distribution is the *memoryless property* discussed in the Example 4.8. This property is a characteristic property of the exponential distribution. In other words, if a probability distribution has this property, then it has to be an exponential distribution under reasonable conditions. This can be seen from the following observation.

Suppose that X is a nonnegative random variable such that

$$P(X > s + t | X > t) = P(X > s)$$

for all $s \geq 0$ and for all $t \geq 0$. Let $\bar{F}(x) = P(X > x)$. Then the above equation can be written in the form

$$\bar{F}(s + t) = \bar{F}(s)\bar{F}(t)$$

for all $s, t \geq 0$.. Note that $\bar{F}(0) = 1$. It can be shown that the only continuous solution of this equation subject to the condition $\bar{F}(0) = 1$ and $F(\infty) = 1$ is

$$\bar{F}(s) = e^{-\lambda s}, s \geq 0$$

for some $\lambda > 0$.

For any random variable X with the distribution function F and the probability density function f, the function

$$\lambda(x) = \frac{f(x)}{1 - F(x)}$$

is defined for all x such that $F(x) < 1$ and it is called the *hazard rate* or *failure rate* of the random variable X or of the corresponding distribution function F. It is easy to see that the failure rate is a constant for the exponential distribution (check?). If we interpret the random variable X as the lifetime of a component, then the quantity $\lambda(x)\Delta x$ is an approximation to the conditional probability that the component will work for an additional Δx units of time given that it has worked for at least x units of time. If the hazard rate $\lambda(x)$ of a random variable X is a constant, then the distribution of the random variable has to be an exponential distribution.

Relation between the Poisson and the exponential distributions

We have seen in the previous chapter that the Poisson distribution can be considered as a stochastic model for the distribution of the number of occurrences of a specified event during a fixed time interval. Let $N(0) = 0$ and $N(t)$ denote the number of occurrences of the event (such as receiving a telephone call or involved in an automobile accident) in the interval $[0, t]$. We have seen that $N(t)$ has a Poisson distribution with mean λt where $\lambda > 0$ and the process $\{N(t), t \geq 0\}$ is a process with stationary independent increments under some conditions.

The parameter λ can be interpreted as the average number of occurrences in a unit time interval. Let X denote the waiting time for the first occurrence. It is clear that

$$\begin{aligned} P(X > t) &= P[N(t) = 0] \\ &= e^{-\lambda t}, \quad t > 0 \end{aligned}$$

and

$$\begin{aligned} P(X \leq t) &= 1 - e^{-\lambda t}, \quad t > 0 \\ &= 0, \quad t \leq 0. \end{aligned}$$

Hence the random variable X has the exponential distribution with parameter λ. Suppose that we are interested in studying the waiting time for r occurrences. Let Y be the waiting time that elapses before the r-th occurrence of the event. Then

$$\begin{aligned} P(Y > t) &= P[N(t) < r] \\ &= \sum_{j=0}^{r-1} e^{-\lambda t} \frac{(\lambda t)^j}{j!}, \quad t > 0 \end{aligned}$$

and the distribution function $F_Y(t)$ of Y for $t > 0$ is given by

$$F_Y(t) = 1 - \sum_{j=0}^{r-1} e^{-\lambda t} \frac{(\lambda t)^j}{j!}$$

and it is obvious that

$$F_Y(t) = 0, \quad t \leq 0.$$

We can compute the probability density function of Y by differentiating the function F_Y with respect to t. Check that

$$\begin{aligned} f_Y(t) &= \frac{\lambda^r}{(r-1)!} t^{r-1} e^{-\lambda t}, \quad t > 0 \\ &= 0, \quad t \leq 0. \end{aligned}$$

Note that r is a positive integer in the above discussion. If $r = 1$, then the above probability density function reduces to the exponential distribution studied earlier. The probability density function given above is a special case of the Gamma probability density function which we will discuss next in this section.

Gamma distribution

Let X be random variable with the probability density function

$$\begin{aligned} f(x) &= \frac{\lambda^\alpha}{\Gamma(\alpha)} x^{\alpha-1} e^{-\lambda x}, \quad x > 0 \\ &= 0, \quad x \leq 0 \end{aligned}$$

where $\alpha > 0, \lambda > 0$ and $\Gamma(.)$ is the Gamma function discussed earlier in this chapter. Then the random variable X is said to have the *Gamma distribution* with parameters α and λ. Check that the function f satisfies the properties of a density function.

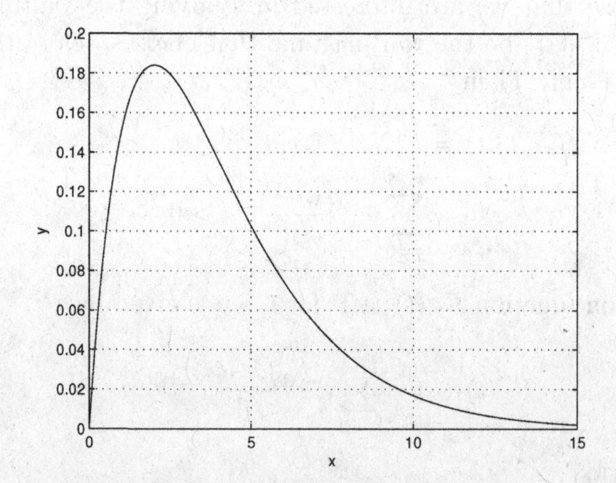

Fig. 4.14 Gamma Probability Density Function for $\alpha = 2, \lambda = 0.5$

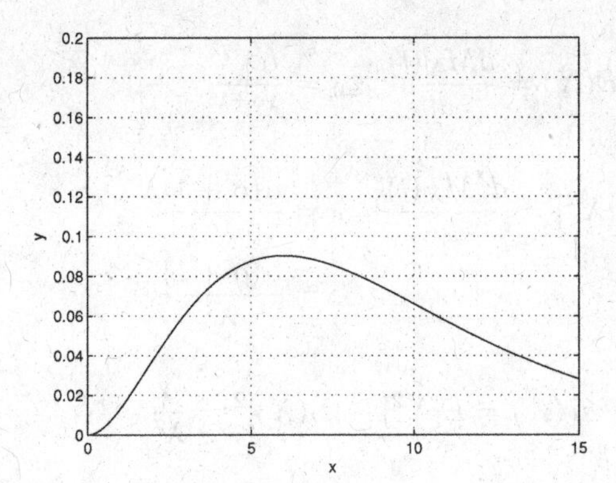

Fig. 4.15 Gamma Probability Density Function for $\alpha = 3, \lambda = \frac{1}{3}$

Let us now compute the m.g.f. of a random variable X with the Gamma probability density function with parameters α and λ. Note that

$$\begin{aligned} M_X(t) = E[e^{tX}] &= \int_0^\infty e^{tx} \frac{\lambda^\alpha}{\Gamma(\alpha)} x^{\alpha-1} e^{-\lambda x} dx \\ &= \frac{\lambda^\alpha}{\Gamma(\alpha)} \int_0^\infty x^{\alpha-1} e^{-(\lambda-t)x} dx. \end{aligned}$$

From the properties of the Gamma function, the integral given above is finite if and only if $\alpha > 0$ and $\lambda - t > 0$. Hence the m.g.f. $M_X(t)$ exists if and only if $t < \lambda$ and in such a case

$$\begin{aligned} M_X(t) &= \frac{\lambda^\alpha}{\Gamma(\alpha)} \int_0^\infty (\frac{y}{\lambda - t})^{\alpha-1} \frac{e^{-y}}{(\lambda - t)} dy \\ &= \frac{\lambda^\alpha}{(\lambda - t)^\alpha \Gamma(\alpha)} \int_0^\infty y^{\alpha-1} e^{-y} dy \\ &= \frac{\lambda^\alpha}{(\lambda - t)^\alpha}. \end{aligned}$$

It can be checked that

$$\begin{aligned} \frac{dM_X(t)}{dt} &= \lambda^\alpha(-\alpha)(\lambda - t)^{-\alpha-1}(-1) \\ &= \alpha\lambda^\alpha(\lambda - t)^{-\alpha-1} \end{aligned}$$

and

$$\frac{d^2 M_X(t)}{dt^2} = \alpha(\alpha+1)\lambda^\alpha(\lambda - t)^{-\alpha-2}.$$

In particular,

$$E(X) = \frac{dM_X(t)}{dt}|_{t=0} = \frac{\alpha\lambda^\alpha}{\lambda^{\alpha+1}} = \frac{\alpha}{\lambda}$$

and

$$\begin{aligned} E[X^2] = \frac{d^2M_X(t)}{dt^2}|_{t=0} &= \frac{\alpha(\alpha+1)\lambda^\alpha}{\lambda^{\alpha+2}} \\ &= \frac{\alpha(\alpha+1)}{\lambda^2}. \end{aligned}$$

Hence

$$Var(X) = E[X^2] - [E(X)]^2 = \frac{\alpha}{\lambda^2}.$$

Remarks. We have seen earlier that the Gamma distribution is a good probability model for the study of waiting times. It has also been used as a model for the income data as the parameters α and λ provide a flexibility and can be suitably chosen for fitting the model to the data. If X is a random variable with the Gamma probability density function with the parameters $\alpha = \frac{n}{2}$ and $\lambda = \frac{1}{2}$ where n is a positive integer, then the random variable X is said to have the *Chi-square* distribution with n degrees of freedom. Check that $E(X) = n$ and $Var(X) = 2n$ for such a random variable X.

Beta distribution

We now discuss another class of distributions which are useful for stochastic modelling. A random variable X is said to have the *Beta distribution* with parameters α and β if it has the probability density function of the form

$$\begin{aligned} f(x) &= \frac{1}{B(\alpha,\beta)}x^{\alpha-1}(1-x)^{\beta-1}, \;\; 0 < x < 1 \\ &= 0 \text{ otherwise} \end{aligned}$$

where $\alpha > 0, \beta > 0$ and

$$B(\alpha,\beta) = \int_0^1 x^{\alpha-1}(1-x)^{\beta-1}\; dx.$$

The function $B(\alpha,\beta)$ defined for $\alpha > 0$ and for $\beta > 0$ is called the *Beta function.*The Beta function and the Gamma function are related by the equation

$$B(\alpha,\beta) = \frac{\Gamma(\alpha)\Gamma(\beta)}{\Gamma(\alpha+\beta)}$$

for all $\alpha > 0$ and for all $\beta > 0$. We now prove this relation.

From the definition of the Gamma function,

$$\begin{aligned}\Gamma(\alpha)\Gamma(\beta) &= \left(\int_0^\infty x^{\alpha-1}e^{-x}\ dx\right)\left(\int_0^\infty y^{\beta-1}e^{-y}\ dy\right)\\ &= \int_0^\infty\int_0^\infty x^{\alpha-1}y^{\beta-1}e^{-(x+y)}\ dydx.\end{aligned}$$

Apply the transformation $u = \frac{x}{x+y}$ and $v = x + y$. Then $x = uv$ and $y = (1-u)v$ and the Jacobian of the transformation is the determinant of the matrix

$$\frac{\partial(x,y)}{\partial(u,v)} = \begin{pmatrix} \frac{\partial x}{\partial u} & \frac{\partial x}{\partial v} \\ \frac{\partial y}{\partial u} & \frac{\partial y}{\partial v} \end{pmatrix}.$$

This matrix in turn reduces to

$$\begin{pmatrix} v & u \\ -v & 1-u \end{pmatrix}$$

in the present case and the determinant of this matrix is $v(1-u) + uv = v$. Note that the determinant is nonnegative as the range of u is $(0,1)$ and that of v is $(0,\infty)$ as x and y take values in $(0,\infty)$. Hence, applying the change of variable formula for double integrals, it follows that

$$\begin{aligned}\Gamma(\alpha)\Gamma(\beta) &= \int_0^1\int_0^\infty u^{\alpha-1}(1-u)^{\beta-1}v^{\alpha+\beta-1}e^{-v}\ dvdu\\ &= \Gamma(\alpha+\beta)\int_0^1 u^{\alpha-1}(1-u)^{\beta-1}\ du\\ &= \Gamma(\alpha+\beta)\ B(\alpha,\beta).\end{aligned}$$

which proves the required relation. Following this observation, an alternate representation for the Beta probability density function is

$$\begin{aligned}f(x) &= \frac{\Gamma(\alpha+\beta)}{\Gamma(\alpha)\Gamma(\beta)}x^{\alpha-1}(1-x)^{\beta-1},\ \ 0 < x < 1\\ &= 0 \text{ otherwise.}\end{aligned}$$

We now compute the moments of a random variable X with the Beta distribution with parameters α and β. For any integer $k \geq 1$,

$$\begin{aligned}E(X^k) &= \frac{\Gamma(\alpha+\beta)}{\Gamma(\alpha)\Gamma(\beta)} \int_0^1 x^k x^{\alpha-1}(1-x)^{\beta-1}\ dx\\ &= \frac{\Gamma(\alpha+\beta)}{\Gamma(\alpha)\Gamma(\beta)} B(k+\alpha,\beta)\\ &= \frac{\Gamma(\alpha+\beta)}{\Gamma(\alpha)\Gamma(\beta)}\frac{\Gamma(k+\alpha)\Gamma(\beta)}{\Gamma(k+\alpha+\beta)\Gamma(\beta)}\\ &= \frac{\alpha(\alpha+1)\dots(\alpha+k-1)}{(\alpha+\beta)(\alpha+\beta+1)\dots(\alpha+\beta+k-1)}\end{aligned}$$

since $\Gamma(x+1) = x\,\Gamma(x)$ for any $x > 0$. In particular, by choosing $k = 1$ and $k = 2$, we have

$$E(X) = \frac{\alpha}{\beta}$$

and

$$E[X^2] = \frac{\alpha(\alpha+1)}{(\alpha+\beta)(\alpha+\beta+1)}.$$

Hence

$$Var(X) = \frac{\alpha\beta}{(\alpha+\beta)^2(\alpha+\beta+1)}$$

Remarks. If $\alpha = 1$ and $\beta = 1$, then the corresponding Beta distribution reduces to the uniform distribution on the interval $(0, 1)$. Typical graphs of the Beta probability density function are shown in Figure 4.16, Figure 4.17 and Figure 4.18.

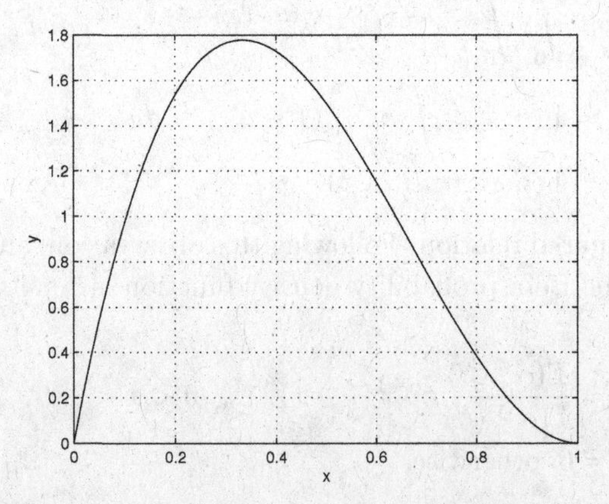

Fig. 4.16 Beta Probability Density Function for $\alpha < \beta$

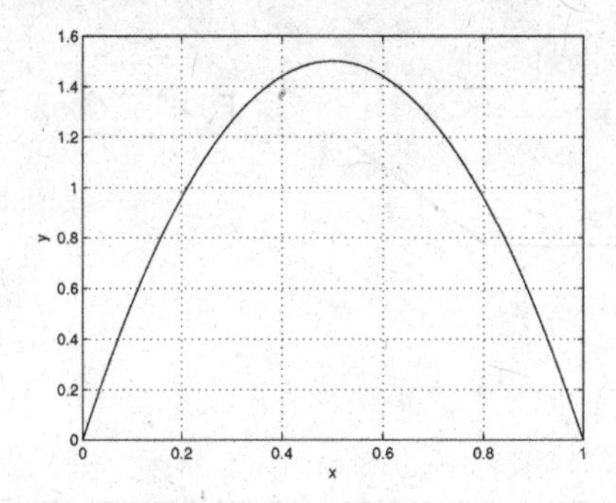

Fig. 4.17 Beta Probability Density Function for $\alpha = \beta$

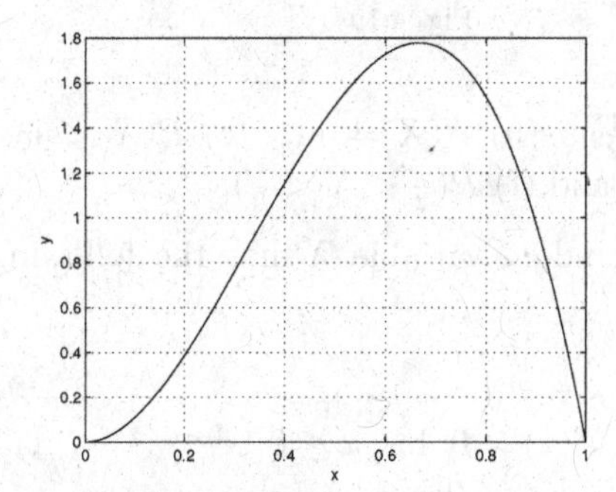

Fig. 4.18 Beta Probability Density Function for $\alpha > \beta$

4.9 Exercises

4.1 Show that the function

$$\begin{aligned} F(x) &= 0 \text{ for } x < -1 \\ &= \frac{x+2}{4} \text{ for } \quad -1 \leq x < 1 \\ &= 1 \text{ for } \; x \geq 1 \end{aligned}$$

is a distribution function of a random variable X. Sketch the graph of F and compute the following: (a)$P(-\frac{1}{2} < X \leq \frac{1}{2})$ $(b)P(X = 0)$ (c)$P(X = 1)$ $(d)P(2 < X \leq 3)$.

4.2 A random variable X has the distribution function F as shown in Figure 4.19.

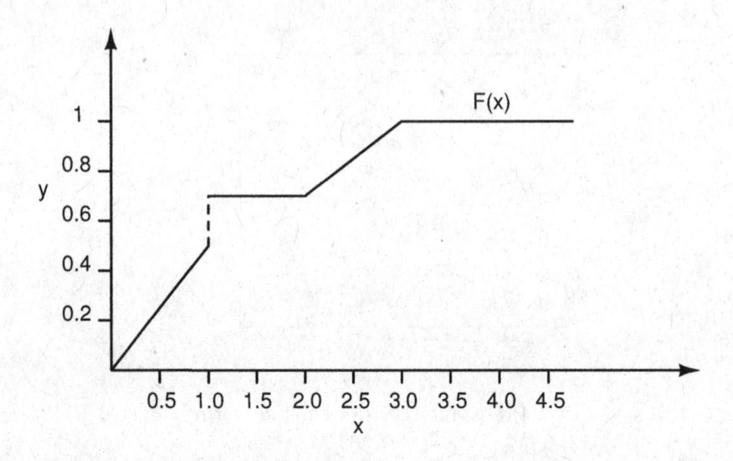

Fig. 4.19

Find (a) $P(X = \frac{1}{2})$ (b) $P(X = 1)$ (c) $P(X < 1)$ (d) $P(X \leq 1)$ (e) $P(X > 2)$ and (f) $P(\frac{1}{2} < X < \frac{5}{2})$.

4.3 Suppose that a random variable X has the following distribution function:

$$\begin{aligned} F(x) &= 0 \text{ for } x \leq 0 \\ &= \frac{2}{5}x \text{ for } 0 < x \leq 1 \\ &= \frac{3}{5}x - \frac{1}{5} \text{ for } 1 < x \leq 2 \\ &= 1 \text{ for } x > 2. \end{aligned}$$

Show that the random variable X is of continuous type and find its probability density function.

4.4 Suppose that a random variable X has the probability density function

$$f(x) = \frac{1}{2}e^{-|x|}, -\infty < x < \infty.$$

Find the value x_0 such that $F(x_0) = 0.5$.

4.5 Suppose that a random variable X has the following distribution function:

$$F(x) = 0 \text{ for } x < 0$$
$$= x^2 \text{ for } 0 \le x \le 1$$
$$= 1 \text{ for } x \ge 1.$$

Show that the random variable X is of continuous type and find its probability density function.

4.6 A random variable X has the distribution function

$$F(x) = \frac{1}{\pi}(\frac{\pi}{2} + \tan^{-1} x), -\infty < x < \infty$$

where $tan^{-1}x$ is taken in the interval $[-\pi/2, \pi/2]$. Find the probability density function of X and compute the probability $P(|X| < 1)$.

4.7 A random variable X has the probability density function

$$f(x) = 1 - |x| \text{ for } |x| < 1$$
$$= 0 \text{ for } |x| \ge 1.$$

Find the distribution function of X and sketch it. Determine $P(|X| > \frac{1}{2})$.

4.8 Show that the function

$$f(x) = \frac{1}{\pi(1 + x^2)}, -\infty < x < \infty$$

is a probability density function. This probability density function is called the standard *Cauchy probability density function.*

4.9 Buses arrive at a particular bus stop at 15-minute intervals starting at 8.00 a.m. If a passenger arrives at the stop at a time that is uniformly distributed between 8.00 a.m and 8.30 a.m., find the probability that he or she waits for less than 5 minutes for the arrival of a bus.

4.10 Suppose the speed of a molecule in a uniform gas at equilibrium is a random variable with the probability density function

$$f(x) = cx^2 e^{-gx^2} \text{ for } x \ge 0$$
$$= 0 \text{ otherwise}$$

where $g = \frac{m}{2kT}$ and k, T and m denote the Boltzmann constant, absolute temperature and the mass of the molecule respectively. Find the constant c in terms of g.

4.11 Consider the function

$$f(x) = c(2x - x^3) \text{ for } 0 < x < \frac{5}{2}$$
$$= 0 \text{ otherwise.}$$

Check whether the function f is a probability density function? If yes, calculate the constant c.

4.12 Consider the distribution function

$$F(x) = \frac{1}{1 + e^{-x}}, -\infty < x < \infty.$$

This distribution function is known as the *Logistic distribution*. Find its probability density function and show that the graph of the probability density function is symmetric about zero.

4.13 Find the expected value of a random variable Y with the probability density function

$$f(y) = \frac{1}{2} e^{-|y|}, -\infty < y < \infty.$$

4.14 Compute the expectation and the variance of a random variable Y whose probability density function is given by

$$f(x) = 1 - |y| \text{ for } |y| < 1$$
$$= 0 \text{ otherwise.}$$

4.15 Determine the mean of a random variable U whose distribution function is given by

$$F(u) = 1 - \cos u \text{ for } 0 \leq u \leq \frac{\pi}{2}$$
$$= 0 \text{ for } u < 0$$
$$= 1 \text{ for } u > \frac{\pi}{2}.$$

4.16 Find the variance of a random variable X with the probability density function

$$
\begin{aligned}
f(u) &= \frac{3}{4} \text{ for } 0 \le x \le 1 \\
&= \frac{1}{4} \text{ for } 2 \le x \le 3 \\
&= 0 \text{ otherwise .}
\end{aligned}
$$

4.17 Let X be a random variable with the probability density function

$$
\begin{aligned}
f(x) &= \frac{2}{x^3} \text{ for } x \ge 1 \\
&= 0 \text{ otherwise.}
\end{aligned}
$$

Show that the $E(X)$ is finite but the $Var(X)$ is not finite.

4.18 Let X be a random variable such that $h(a) = E[(X - a)^2]$ is finite for all real numbers a. Show that $h(a)$ is minimum when $a = E(X)$.

4.19 A *mode* of the distribution of a random variable with the probability density function $f(x)$ is a value x_0 of x at which f is maximum. If there is only one such value x_0, then x_0 is called the mode of the density f and the density is said to be *unimodal.* Find the mode of the probability density function

$$
\begin{aligned}
f(x) &= \frac{1}{2} x^2 e^{-x} \text{ for } 0 < x < \infty \\
&= 0 \text{ otherwise.}
\end{aligned}
$$

4.20 A *median* of the distribution of a random variable X is a value m such that $P(X \le m) \ge \frac{1}{2}$ and $P(X \ge m) \ge \frac{1}{2}$. If there is only one such value m, then it is the unique median of the distribution. Find the median corresponding to the probability density function

$$
f(x) = \frac{1}{\pi(1 + x^2)}, -\infty < x < \infty.
$$

4.21 For a standard normal probability density function, show that the mean, mode and median are all equal to zero.

4.22 Let $0 < p < 1$. A $100p$-th percentile (quantile of order p) of the distribution of a random variable X is a value ψ_p such that $P(X \le$

$\psi_p) \geq p$ and $P(X \geq \psi_p) \geq 1 - p$. Find the 25-th percentile of the distribution with the probability density function

$$f(x) = 4x^3 \text{ for } 0 < x < 1$$
$$= 0 \text{ otherwise.}$$

4.23 Let X be a random variable with the probability density function f. If m is the unique median of X, show that

$$E(|X - a|) = E(|X - m|) + 2\int_m^a (a - x)f(x)\ dx$$

for any real number a provided the expectations exist. Hence prove that $E(|X - a|)$ is minimum when $a = m$.

4.24 Suppose X has the uniform distribution on the interval $[\alpha, \beta]$. Let μ be its mean. Find $E(X^r)$ and $E(X - \mu)^r$ for positive integers $r \geq 1$.

4.25 Suppose X has the standard exponential probability density function. Find the coefficient of skewness.

4.26 If $Y = aX + b$ where a and b are constants, show that the random variables X and Y have the same coefficients of skewness and kurtosis whenever they are finite.

4.27 Let X be a random variable with the probability density function

$$f(x) = x \text{ for } 0 \leq x \leq 1$$
$$= 2 - x \text{ for } 1 < x \leq 2$$
$$= 0 \text{ otherwise.}$$

Determine the m.g.f. of X whenever it exists.

4.28 Calculate the moments of the standard normal probability density function from its m.g.f.

4.29 Let X be a random variable with the probability density function

$$f(x) = xe^{-x} \text{ for } x > 0$$
$$= 0 \text{ otherwise.}$$

Determine the m.g.f. of X wherever it exists.

4.30 Let $Y = aX + b$ where a and b are constants. Show that the m.g.f.'s of X and Y satisfy the relation

$$M_Y(t) = e^{bt} M_X(at).$$

4.31 Suppose X is a random variable with the m.g.f.

$$M_X(t) = e^{t^2+3t}, -\infty < t < \infty.$$

Find the mean and variance of X.

4.32 Find the probability density function of the random variable $Y = X^2$ when the random variable X has the uniform density on the interval $[-1, 1]$.

4.33 Suppose a random variable X has the probability density function

$$\begin{aligned} f(x) &= \frac{1}{2}x \text{ for } 0 < x < 2 \\ &= 0 \text{ otherwise.} \end{aligned}$$

Let $Y = X(2 - X)$. Find the probability density function of Y.

4.34 Suppose a random variable θ is uniformly distributed on the interval $(-\frac{\pi}{2}, \frac{\pi}{2})$. Find the probability density function of $R = a \sin \theta$ where a is a constant.

4.35 Let X be a random variable with uniform density on the interval $(-1, -2)$. Find the probability density function of $Y = X^2$.

4.36 If X is a random variable with a distribution function $F(x)$ which is continuous and strictly increasing, show that $Y = F(X)$ has a uniform distribution on the interval $[0, 1]$.

4.37 Suppose a random variable X has the standard uniform distribution. Let $F(x)$ be a distribution function which is continuous and strictly increasing. Let F^{-1} denote the inverse function of F. Show that $Y = F^{-1}(X)$ has the distribution function $F(x)$.

4.38 Suppose a random variable X has the probability density function

$$f_X(x) = \frac{e^{-x}}{(1 + e^{-x})^2}, -\infty < x < \infty.$$

This probability density function is known as the *Logistic probability density function.* Show that

$$Y = \frac{1}{1 + e^{-X}}$$

has the uniform probability density function on the interval $[0, 1]$.

4.39 If a random variable X has the standard normal distribution, find (a)$P(0 \leq X \leq 0.87)$, (b)$P(-2.64 \leq X \leq 0)$, (c)$P(-2.13 \leq X \leq -0.56)$, and (d)$P(|X| > 1.39)$.

4.40 If the m.g.f. of a random variable X is $M_X(t) = e^{-6t+32t^2}$, find $P(-4 \leq X < 16)$.

4.41 Show that if a random variable X has a normal distribution with mean zero and variance σ^2, so does $-X$.

4.42 If X has $N(\mu, \sigma^2)$ as its distribution, determine the probability density function of $Y = |X - \mu|$. Further prove that $E[Y] = \sigma\sqrt{\frac{2}{\pi}}$.

4.43 If X has $N(\mu, \sigma^2)$ as its distribution, determine the probability density function of $Y = \frac{(X-\mu)^2}{\sigma^2}$.

4.44 If X has $N(5, 10)$ as its distribution, find $P(0.04 < (X-5)^2 < 38.4)$.

4.45 A random variable X is said to have the *log-normal* distribution if $\log X$ has the normal distribution. Suppose $\log X$ has $N(\mu, \sigma^2)$ as its distribution. Show that

$$E[X] = \exp(\mu + \frac{\sigma^2}{2})$$

and

$$Var(X) = \{\exp(\sigma^2) - 1\} \exp(2\mu + \sigma^2).$$

4.46 Show that if a random variable X has the log-normal distribution , so does X^r for $r \neq 0$.

4.47 Let F be the distribution function of a nonnegative random variable which is continuous and satisfying the equation

$$\bar{F}(s + t) = \bar{F}(s)\bar{F}(t)$$

for all $s, t \geq 0$ where $\bar{F}(x) = 1 - F(x)$. Show that

$$\begin{aligned} F(x) &= 1 - e^{-\lambda x} \text{ for } x \geq 0 \\ &= 0 \text{ otherwise} \end{aligned}$$

for some $\lambda > 0$.

4.48 Let the hazard rate $\lambda(x)$, of a nonnegative random variable X with a probability density function f, be a constant. Show that the random variable X has an exponential distribution.

4.49 Suppose that the telephone calls arrive at a telephone exchange following an exponential distribution with parameter $\lambda = 5$ per hour. What are the probabilities that the waiting time for a call at the exchange is (a)at least 15 minutes (b)not more than 10 minutes and (c)exactly 5 minutes?

4.50 Let X be a random variable with the Gamma probability density function with parameters α and λ. Compute $E[X^m]$ for all $m \geq 1$ and hence find $E[X]$ and $Var(X)$.

4.51 Suppose X is a random variable with the Gamma probability density function with parameters α and λ and $E[X] = 2$ and $Var(X) = 7$. Find the parameters α and λ.

4.52 Suppose X is a random variable with the standard normal distribution. Show that $Y = X^2$ has the Gamma distribution with the parameters $\alpha = \frac{1}{2}$ and $\lambda = \frac{1}{2}$.

4.53 Find the mode of the Beta probability density function with the parameters $\alpha > 1$ and $\beta > 1$.

4.54 If X is a random variable with the Beta probability density function with the parameters α and β, show that $1 - X$ is a random variable with the Beta probability density function with the parameters α and β replaced by β and α respectively.

4.55 Determine the constant c such that the function

$$\begin{aligned} f(x) &= cx^3(1-x)^6 \text{ for } 0 < x < 1 \\ &= 0 \text{ otherwise} \end{aligned}$$

is a probability density function.

4.56 If k and m are integers greater than or equal to 1 and $0 < p < 1$, show that

$$\int_0^p \frac{\Gamma(k+m)}{\Gamma(k)\Gamma(m)} x^{k-1}(1-x)^{m-1}\, dx = \sum_{j=k}^{n} n_j^C p^j (1-p)^{n-j}$$

where $n = k + m - 1$.

[illegible]

[illegible]

[illegible]

[illegible]

[illegible]

[illegible]

[illegible]

[illegible]

[illegible]

[illegible]

Chapter 5

MULTIVARIATE PROBABILITY DISTRIBUTIONS

5.1 Introduction

We have studied univariate probability distributions in the previous two chapters. These are probability models for one-dimensional random variables. They deal with the one characteristic at a time of the experiment or phenomenon under consideration. There are situations where it is important to study more than one characteristic at the same time. For instance in medical studies dealing with cardiac problems, the health of a patient might depend on the blood pressure, cholesterol level, nutrition intake and several other factors. In meteorological studies, the amount of rain fall in a particular region at a particular time depends on the temperature, atmospheric pressure besides several other known and unknown variables. In marketing a consumer product, the sales might depend on the price of the product, prices of the competing brands for a similar product, demand for the product, and the purchasing power of the people to whom the product is directed. In other words, simultaneous study of the characteristics involved is needed at times instead of study of them on an individual basis so that the interrelationships between the different characteristics can be observed. This leads to the notion of a multivariate distribution. We consider the case of bivariate distributions in detail as the generalization of the concepts from the bivariate case to the multivariate case is not difficult but cumbersome in notation.

5.2 Bivariate Probability Distributions

Let Ω be a sample space corresponding to a random experiment $\mathcal{E}$ and $\mathcal{F}$ be a σ-algebra of subsets of Ω associated with it. Suppose that P is a probability measure defined on the class $\mathcal{F}$. Let X and Y be real valued functions defined on Ω such that for every x and y,

$$[\omega : X(\omega) \leq x, Y(\omega) \leq y] \in \mathcal{F}.$$

Such a function (X, Y) from Ω into the real line is called a *bivariate random vector* or a *bivariate random variable.*

In analogy with the univariate probability distributions, we define the *joint distribution* or the *bivariate distribution* of a *bivariate random vector* (X, Y) by

$$F_{X,Y}(x, y) = P[X \leq x, Y \leq y], \quad -\infty < x, y < \infty.$$

For any x and y, the event $[X \leq x, Y \leq y]$ can also be written in the form $[X \leq x] \cap [Y \leq y]$ and it is the set of all sample points $\omega \in \Omega$ such that $X(\omega) \leq x$ *and* $Y(\omega) \leq y$.

We now discuss some properties of the joint distribution functions. Let F_X and F_Y denote the distribution functions of X and Y respectively. From the definition of the function F_X, we have

$$\begin{aligned} F_X(x) &= P[X \leq x] \\ &= P[X \leq x, Y < \infty] \\ &= \lim_{y \to \infty} P[X \leq x, Y \leq y] = \lim_{y \to \infty} F_{X,Y}(x, y). \end{aligned}$$

The last statement is intuitively clear that as the set $[X \leq x, Y \leq y]$ increases to the set $[X \leq x, Y < \infty]$ as y increases to ∞ and hence the $P(X \leq x, Y \leq y)$ increases to $P(X \leq x)$ as y increases to ∞. However a rigourous argument can be given by using the result that if the sequence $A_n \to A$ as $n \to \infty$, then $P(A_n) \to P(A)$ as $n \to \infty$. We will not go into this discussion here. Let us denote the limit of the function $F_{X,Y}(x, y)$ as $y \to \infty$ by $F_{X,Y}(x, +\infty)$. Hence

$$F_X(x) = F_{X,Y}(x, +\infty).$$

Similarly

$$F_Y(y) = F_{X,Y}(+\infty, y).$$

In view of the above relations, the distribution functions $F_X(x)$ and $F_Y(y)$ are called the *marginal distribution functions* of X and Y respectively. It is clear that

$$F_{X,Y}(-\infty, -\infty) = \lim_{x \to -\infty, y \to -\infty} F_{X,Y}(x, y) = 0$$

and

$$F_{X,Y}(\infty, \infty) = \lim_{x\to\infty, y\to\infty} F_{X,Y}(x, y) = 1.$$

Let us now compute the probability that $X > x$ and $Y > y$. Let $A = [X > x]$ and $B = [Y > y]$. Then

$$\begin{aligned}
P([X > x, Y > y]) &= P(A \cap B) \\
&= 1 - P(\overline{(A \cap B)}) \\
&= 1 - P(\bar{A} \cup \bar{B}) \\
&= 1 - \{P(\bar{A}) + P(\bar{B}) - P(\bar{A} \cap \bar{B})\} \\
&= 1 - \{P[X \leq x] + P[Y \leq y] - P[X \leq x, Y \leq y]\} \\
&= 1 - F_X(x) - F_Y(y) + F_{X,Y}(x, y).
\end{aligned}$$

Using similar calculations, check that, for any $-\infty < a_1 \leq a_2 < \infty$ and $-\infty < b_1 \leq b_2 < \infty$,

$$\begin{aligned}
P(a_1 < X \leq a_2, b_1 < Y \leq b_2) = F_{X,Y}(a_2, b_2) - F_{X,Y}(a_1, b_2) \\
-F_{X,Y}(a_2, b_1) + F_{X,Y}(a_1, b_1).
\end{aligned}$$

Since the probability of any event is nonnegative, it follows that for any $-\infty < a_1 \leq a_2 < \infty$ and $-\infty < b_1 \leq b_2 < \infty$,

$$F_{X,Y}(a_2, b_2) - F_{X,Y}(a_1, b_2) - F_{X,Y}(a_2, b_1) + F_{X,Y}(a_1, b_1) \geq 0.$$

Let us first consider the case of bivariate discrete probability distributions.

Bivariate discrete distributions

Let X and Y be discrete random variables taking the values $x_i, i \geq 1$ and $y_j, j \geq 1$ respectively with

$$P(X = x_i, Y = y_j) = p_{ij} \geq 0$$

for $i \geq 1$ and $j \geq 1$ with $p_{..} = \sum_{i=1}^{\infty} \sum_{j=i}^{\infty} p_{ij} = 1$. The function $f_{X,Y}(x_i, y_j) = P(X = x_i, Y = y_j) = p_{ij}, i \geq 1, j \geq 1$ is called the *bivariate probability function* of (X, Y) and the random vector (X, Y) is said to have a *bivariate discrete distribution.* Let us display the values of the probability function $f_{X,Y}(x_i, y_j) = P(X = x_i, Y = y_j), i \geq 1, j \geq 1$ in a tabular form.

Table 5.1

			X			
Y	x_1	x_2	$\dots$	x_i	$\dots$	
y_1	p_{11}	p_{21}	$\dots$	p_{i1}	$\dots$	$p_{.1}$
y_2	p_{12}	p_{22}	$\dots$	p_{i2}	$\dots$	$p_{.2}$
$\dots$	$\dots$	$\dots$	$\dots$	$\dots$	$\dots$	$\dots$
y_j	p_{1j}	p_{2j}	$\dots$	p_{ij}	$\dots$	$p_{.j}$
$\dots$	$\dots$	$\dots$	$\dots$	$\dots$	$\dots$	$\dots$
	$p_{1.}$	$p_{2.}$	$\dots$	$p_{i.}$	$\dots$	$p_{..} = 1$

It is obvious that

$$\begin{aligned} P(X = x_i) &= \sum_{j=1}^{\infty} P(X = x_i, Y = y_j) \\ &= \sum_{j=1}^{\infty} p_{ij} \\ &= p_{i.}, \quad \text{(say)} \end{aligned}$$

for $i \geq 1$, and

$$\begin{aligned} P(Y = y_j) = \sum_{i=1}^{\infty} P(X = x_i, Y = y_j) &= \sum_{i=1}^{\infty} p_{ij} \\ &= p_{.j}, \quad \text{(say)} \end{aligned}$$

for $j \geq 1$. Note that the probability that $X = x_i$ can be obtained by summing the entries in the i-th column of the Table 5.1 and the probability that $Y = y_j$ can be obtained by summing the entries in the j-th row of the Table 5.1. The function $f_X(x_i) = P(X = x_i) = p_{i.}, i \geq 1$ is called the *marginal probability function* of X and the function $f_Y(y_j) = P(Y = y_j), j \geq 1$ is called the *marginal probability function* of Y. The *conditional probability function* of X given $Y = y_j$ is defined by

$$P(X = x_i | Y = y_j) = \frac{P(X = x_i, Y = y_j)}{P(Y = y_j)}, i \geq 1$$

for any y_j such that $P(Y = y_j) > 0$ and the *conditional probability function* of Y given $X = x_i$ is defined by

$$P(Y = y_j | X = x_i) = \frac{P(X = x_i, Y = y_j)}{P(X = x_i)}, j \geq 1$$

for any x_i such that $P(X = x_i) > 0$. The random components X and Y of the bivariate random vector (X, Y) are said to be *independent* if the events $[X = x_i]$ and $[Y = y_j]$ are independent for every $i \geq 1$ and $j \geq 1$. In other words, the random variables X and Y are independent if

$$P(X = x_i, Y = y_j) = P(X = x_i)P(Y = y_j), i \geq 1, j \geq 1.$$

Let us consider an example to illustrate these concepts.

Example 5.1. Suppose the bivariate probability distribution of a random vector (X, Y) is given according to the entries in the following table.

Table 5.2

	X		
Y	0	1	2
0	1/9	2/9	1/9
1	2/9	2/9	0
2	1/9	0	0

By adding the entries in the Table 5.2 row wise, we get that

$$P(Y = 0) = 4/9, P(Y = 1) = 4/9, P(Y = 2) = 1/9$$

which gives the marginal probability function of Y and by adding the entries in the Table 5.2 column wise, we get that

$$P(X = 0) = 4/9, P(X = 1) = 4/9, P(X = 2) = 1/9$$

which gives the marginal probability function of X. The random variables X and Y are not independent since

$$P(X = 2, Y = 2) \neq P(X = 2)P(Y = 2)$$

for instance. Let us compute the conditional probability function of X given $Y = 2$. Observe that

$$P(X = 0|Y = 2) = \frac{P(X = 0, Y = 2)}{P(Y = 2)} = \frac{(1/9)}{(1/9)} = 1,$$

$$P(X = 1|Y = 2) = \frac{P(X = 1, Y = 2)}{P(Y = 2)} = \frac{0}{(1/9)} = 0, \text{ and}$$

$$P(X = 2|Y = 2) = \frac{P(X = 2, Y = 2)}{P(Y = 2)} = \frac{0}{(1/9)} = 0.$$

Bivariate continuous distributions

Recall that, if a random variable X has a distribution function F_X and a probability density function f_X, then

$$F_X(x) = \int_{-\infty}^{x} f_Y(y)dy, -\infty < x < \infty$$

and

$$f_X(x) = \frac{dF_X(x)}{dx}$$

almost everywhere.

A bivariate random vector (X, Y) is said to be of *absolutely continuous type* if there exists a nonnegative function $f_{X,Y}(x, y)$ such that

$$F_{X,Y}(x, y) = \int_{-\infty}^{x} \int_{-\infty}^{y} f_{X,Y}(u, v)\ dudv, -\infty, x, y < \infty.$$

The function $f_X(x, y)$ is called the *joint probability density function* of (X, Y).

Suppose (X, Y) is a bivariate random vector with the joint distribution function $F_{X,Y}(x, y)$ and the joint probability density function $f_{X,Y}(x, y)$. From the general properties of double integrals, it follows that the joint distribution function $F_{X,Y}(x, y)$ has second order partial derivatives with respect to x and y for almost all x and y and

$$\frac{\partial^2 F_{X,Y}(x, y)}{\partial x \partial y} = f_{X,Y}(x, y).$$

Furthermore

$$\int_{-\infty}^{\infty} \int_{\infty}^{\infty} f_{X,Y}(u, v)\ dudv = 1.$$

Note that

$$F(x, y) = P((X, Y) \in (-\infty, x] \times (-\infty, y]) = \int\int_{(x,y)\in C} f_{X,Y}(u, v)dudv,$$

where $C = (-\infty, x] \times (-\infty, y]$. However this relation holds for more general class of Borel Sets C in R^2. In other words, for any Borel set C contained in R^2,

$$P((X, Y) \in C) = \int\int_{(x,y)\in C} f_{X,Y}(u, v)dudv.$$

A set C is said to be a *Borel set* in R^2 if it belongs to the smallest σ-algebra containing all rectangles of the form $[a_1, a_2] \times [b_1, b_2]$ where $-\infty < a_1 \leq$

$a_2 < \infty$ and $-\infty < b_1 \leq b_2 < \infty$. We will not discuss more in detail about this in this book.

We have seen earlier that the marginal distribution function of X can be obtained by using the relation

$$\begin{aligned} F_X(x) &= F_{X,Y}(x, \infty) \\ &= \int_{-\infty}^{x} \int_{-\infty}^{\infty} f_{X,Y}(u, v) du dv \\ &= \int_{-\infty}^{x} [\int_{-\infty}^{\infty} f_{X,Y}(u, v) dv] du \\ &= \int_{-\infty}^{x} f_X(u) du \end{aligned}$$

where

$$f_X(u) = \int_{-\infty}^{\infty} f_{X,Y}(u, v) dv.$$

(The above relations can be justified by using a theorem known as Fubini's theorem.) The representation given above proves that if (X, Y) has a joint probability density function $f_{X,Y}(x, y)$, then the random variable X has a probability density function and it is given by

$$f_X(x) = \int_{-\infty}^{\infty} f_{X,Y}(x, y) dy, -\infty < x < \infty.$$

Similarly the random variable Y has the probability density function

$$f_Y(y) = \int_{-\infty}^{\infty} f_{X,Y}(x, y) dx, -\infty < y < \infty.$$

The functions $f_X(x)$ and $f_Y(y)$ are called the *marginal density functions* of X and Y respectively.

Example 5.2. Suppose that a bivariate random vector (X, Y) has the joint probability density function

$$\begin{aligned} f(x, y) &= 2e^{-x}e^{-2y}, \quad 0 < x, y < \infty \\ &= 0 \text{ otherwise.} \end{aligned}$$

Let us compute the marginal probability densities $f_X(x)$ and $f_Y(y)$ of X

and Y respectively. From the definition of the function $f_X(x)$, we have

$$\begin{aligned} f_X(x) &= \int_{-\infty}^{\infty} f(x,y)dy \\ &= 2e^{-x} \int_0^{\infty} e^{-2y} dy \\ &= 2e^{-x}[-\frac{1}{2}e^{-2y}]_0^{\infty} \\ &= e^{-x} \end{aligned}$$

for $x > 0$, and

$$f_X(x) = 0 \text{ for } x \leq 0.$$

Similarly check that

$$\begin{aligned} f_Y(y) &= 2e^{-2y} \text{ for } y > 0 \\ &= 0 \text{ for } y \leq 0. \end{aligned}$$

Suppose that we would like to calculate $P(X < Y)$ in this example. Let $C = [X < Y]$. Then

$$\begin{aligned} P(C) = P(X < Y) &= \int\int_{[(u,v):u<v]} f(u,v)dudv \\ &= \int_0^{\infty} \{\int_0^{v} 2e^{-u}e^{-2v}du\}dv \\ &= \int_0^{\infty} 2e^{-2v}[-e^{-u}]_0^{v}dv \\ &= \int_0^{\infty} 2e^{-2v}[1-e^{-v}]dv \\ &= \int_0^{\infty} 2e^{-2v}dv - \int_0^{\infty} 2e^{-3v}dv \\ &= [-e^{-2v}]_0^{\infty} - [-\frac{2}{3}e^{-3v}]_0^{\infty} \\ &= 1 - \frac{2}{3} = \frac{1}{3}. \end{aligned}$$

Example 5.3. Suppose that a random vector (X, Y) has the uniform probability density function over the unit circle of radius r with centre at the point $(0, 0)$. In other words, the random vector (X, Y) has the joint probability density function

$$\begin{aligned} f(x,y) &= c \text{ if } 0 \leq x^2 + y^2 \leq r^2 \\ &= 0 \text{ otherwise} \end{aligned}$$

where c is a constant. Since

$$\int_{-\infty}^{\infty}\int_{-\infty}^{\infty} f(x,y)dxdy = 1,$$

it follows that

$$c\int\int_D dxdy = 1$$

where $D = \{(x,y) : x^2 + y^2 \leq r^2\}$. Note that the double integral given above represents the area of the circle with centre at the point $(0,0)$ and the radius r which is equal to πr^2. Hence

$$c = \frac{1}{\pi r^2}.$$

Let us compute the marginal density function of X. It is given by

$$\begin{aligned} f_X(x) &= \int_{-\infty}^{\infty} f(x,y)dy \\ &= \frac{1}{\pi r^2}\int_{\{(x,y):x^2+y^2\leq r^2\}} dy \\ &= \frac{1}{\pi r^2}\int_{-\sqrt{r^2-x^2}}^{\sqrt{r^2-x^2}} dy \\ &= \frac{2}{\pi r^2}\sqrt{r^2-x^2} \end{aligned}$$

for $0 \leq x^2 \leq r^2$ and

$$f_X(x) = 0 \text{ otherwise.}$$

Let $Z = \sqrt{X^2+Y^2}$. Note that Z represents the distance of the point (X,Y) from the centre $(0,0)$. Then, for any $0 \leq d \leq r$,

$$\begin{aligned} P(Z \leq d) &= P(X^2+Y^2 \leq d^2) \\ &= \int\int_{\{(x,y):x^2+y^2\leq d^2\}} \frac{1}{\pi r^2} dxdy \\ &= \frac{\pi d^2}{\pi r^2} = \frac{d^2}{r^2}. \end{aligned}$$

It is easy to see that

$$\begin{aligned} P(Z \leq d) &= 0 \text{ for } d < 0 \\ &= 1 \text{ for } d \geq r. \end{aligned}$$

Bivariate normal distribution

One of the important bivariate distributions is the bivariate normal distribution. It is an important probability model which is widely used for fitting a bivariate data. For instance, the joint distribution of heights of two individuals measured at different times in a population can be considered to be approximately bivariate normal. The joint distribution of the marks obtained by students in a class in two different but related subjects can also be modelled by a bivariate normal distribution as an approximation.

Let (X, Y) be a bivariate random vector with the joint probability density function given by

$$f(x,y) = \frac{1}{2\pi\sigma_X\sigma_Y\sqrt{1-\rho^2}}\exp\{-Q/2\}, -\infty < x, y < \infty$$

where $-\infty < \mu_X, \mu_Y < \infty$, $0 < \sigma_X, \sigma_Y < \infty$, $-1 < \rho < 1$ and

$$Q = \frac{1}{1-\rho^2}[(\frac{x-\mu_X}{\sigma_X})^2 - 2\rho(\frac{x-\mu_X}{\sigma_X})(\frac{y-\mu_Y}{\sigma_Y}) + (\frac{y-\mu_Y}{\sigma_Y})^2].$$

Note that a bivariate normal distribution is completely determined once the parameters $\mu_X, \mu_Y, \sigma_X, \sigma_Y$ and ρ are specified. Then the marginal density function of X is

$$\begin{aligned} f_X(x) &= \int_{-\infty}^{\infty} f(x,y)dy \\ &= \int_{-\infty}^{\infty} \frac{1}{2\pi\sigma_X\sigma_Y\sqrt{1-\rho^2}}\exp\{-Q/2\}dy \end{aligned}$$

where Q is as defined above. Check that Q can also be written in the form

$$Q = \frac{1}{\sigma_Y^2(1-\rho^2)}[y - (\mu_Y + \frac{\rho\sigma_Y}{\sigma_X}(x-\mu_X))]^2 + (\frac{x-\mu_X}{\sigma_X})^2.$$

Hence

$$\begin{aligned} f_X(x) &= \frac{1}{2\pi\sigma_X\sigma_Y\sqrt{1-\rho^2}}e^{-\frac{1}{2}(\frac{x-\mu_X}{\sigma_X})^2} \\ &\quad \times \int_{-\infty}^{\infty} \exp\{-\frac{1}{2\sigma_Y^2(1-\rho^2)}[y-(\mu_Y + \frac{\rho\sigma_Y}{\sigma_X}(x-\mu_X))]^2\}\ dy \\ &= \frac{1}{\sqrt{2\pi\sigma_X^2}}e^{-\frac{1}{2}(\frac{x-\mu_X}{\sigma_X})^2}\int_{-\infty}^{\infty} g(x,y)\ dy \end{aligned}$$

where

$$g(x,y) \equiv \frac{1}{\sqrt{2\pi\sigma_Y^2(1-\rho^2)}}\exp\{-\frac{1}{2\sigma_Y^2(1-\rho^2)}[y-(\mu_Y + \frac{\rho\sigma_Y}{\sigma_X}(x-\mu_X))]^2\}.$$

The integral of the function $g(x, y)$ with respect to y over the real line is equal to one since it is the integral of the normal density function with mean

$$\mu_Y + \frac{\rho \sigma_Y}{\sigma_X}(x - \mu_X)$$

and variance

$$\sigma_Y^2(1 - \rho^2).$$

Hence

$$f_X(x) = \frac{1}{\sqrt{2\pi\sigma_X^2}} e^{-\frac{1}{2}\left(\frac{x-\mu_X}{\sigma_X}\right)^2}.$$

The above derivation shows that the marginal density function of X is normal with mean μ_X and variance σ_X^2. Similarly it can be shown that the marginal probability density function of Y is normal with mean μ_Y and variance σ_Y^2.

Remarks. We will show later in this chapter that if (X, Y) has a bivariate normal distribution, then *every* linear combination $cX + dY$ is normally distributed where c and d are real numbers (as a convention, we consider a probability distribution degenerate at zero as a normal distribution with mean zero and variance zero).

5.3 Conditional Distributions

Let (X, Y) be a bivariate random vector. There are situations when we would like to know whether a change in the value of Y has influence on the value of X. For instance, if Y increases, does X also increase or does it decrease or there is no connection or relation between X and Y? For example, suppose that X denotes the height of a person and Y is his or her weight. A natural question is to examine whether there is any relation between the height and the weight. Suppose X denotes the level of education of a person and Y his annual income. Are X and Y related in any way? Suppose X denotes the total sale proceeds of a specified consumer product in a period of one month and Y is the price of the product per unit. Clearly X depends not only on Y but is also affected by the sales of other competing brands of a similar product. In all these examples, it is important to know the behaviour of X for a given value of Y and vice versa. In order to study problems of this nature, we now introduce the concept of the conditional distributions.

Suppose Y is a discrete valued random variable taking values $y_i, i \geq 1$ with $P(Y = y_i) = p_i > 0$ and $\sum_{i=1}^{\infty} p_i = 1$. Then it follows from the definition of the conditional probability discussed in the Chapter 2 that

$$P(X \leq x | Y = y_i) = \frac{P(X \leq x, Y = y_i)}{P(Y = y_i)}$$

and the function

$$F_{X|Y}(x|y) = P(X \leq x | Y = y)$$

is called the *conditional distribution function* of X given $Y = y$. It is obvious that this function can be defined only for those values of y for which $P(Y = y) > 0$. Suppose a bivariate random vector (X, Y) has the probability density function $f_{X,Y}(x, y)$. Then Y has the marginal density function

$$f_Y(y) = \int_{-\infty}^{\infty} f_{X,Y}(x, y) dx$$

and it is known that $P(Y = y) = 0$ for any y. Hence the conditional distribution function

$$F_{X|Y}(x|y) = P(X \leq x | Y = y)$$

cannot be defined directly by using the above definition. In order to circumvent this problem, we at first define the *conditional probability density function* $f_{X|Y}(x|y)$ of X given $Y = y$ as

$$f_{X|Y}(x|y) = \frac{f_{X,Y}(x, y)}{f_Y(y)}, -\infty < x < \infty$$

for any y such that $f_Y(y) > 0$.

Let us see the reasoning involved behind the above definition of the conditional probability density function of X given $Y = y$. Consider the conditional probability

$$P(x \leq X \leq x + \Delta x | y \leq Y \leq y + \Delta y)$$

for sufficiently small Δx and Δy. Applying the definition of the conditional probability, we get that

$$\begin{aligned} P(x \leq X \leq x + \Delta x | y \leq Y \leq y + \Delta y) &= \frac{P(x \leq X \leq x + \Delta x, y \leq Y \leq y + \Delta y)}{P(y \leq Y \leq y + \Delta y)} \\ &= \frac{\int_x^{x+\Delta x} \int_y^{y+\Delta y} f_{X,Y}(u, v) du dv}{\int_y^{y+\Delta y} f_Y(v) dv}. \end{aligned}$$

The numerator in the above expression can be approximated by

$$f_{X,Y}(x,y)\Delta x\Delta y$$

and the denominator can be approximated by

$$f_Y(y)\Delta y$$

and hence

$$P(x \leq X \leq x+\Delta x | y \leq Y \leq y+\Delta y)$$

is approximately equal to

$$\frac{f_{X,Y}(x,y)\Delta x\Delta y}{f_Y(y)\Delta y} = \frac{f_{X,Y}(x,y)}{f_Y(y)}\Delta x$$

which in turn is an approximation for the integral

$$\int_x^{x+\Delta x} \frac{f_{X,Y}(u,y)}{f_Y(y)} du = \int_x^{x+\Delta x} f_{X|Y}(u|y) du.$$

In other words, the quantity $f_{X|Y}(x|y)\Delta x$ represents an approximation to the conditional probability that $x \leq X \leq x+\Delta x$ given that $y \leq Y \leq y+\Delta y$ for sufficiently small Δx and Δy. This reasoning gives the intuitive justification for defining the conditional probability density function of X given $Y = y$ as given above even though $P(Y = y) = 0$ for all y when the random variable Y has a probability density function. Similarly we define the conditional density of Y given $X = x$ to be

$$f_{Y|X}(y|x) = \frac{f_{X,Y}(x,y)}{f_X(x)}, -\infty < y < \infty$$

for all x for which $f_X(x) > 0$. The conditional probability density functions $f_{X|Y}(x|y)$ and $f_{Y|X}(y|x)$ are nonnegative functions and it is easy to check that

$$\int_{-\infty}^{\infty} f_{X|Y}(x|y) dx = 1$$

for all y for which $f_Y(y) > 0$ and

$$\int_{-\infty}^{\infty} f_{Y|X}(y|x) dy = 1$$

for all x for which $f_X(x) > 0$. Since we have defined the conditional density functions, we can define the corresponding conditional distribution functions. The *conditional distribution function* of X given $Y = y$ is defined by

$$F_{X|Y}(x|y) = \int_{-\infty}^{x} f_{X|Y}(u|y) du, -\infty < x < \infty$$

for y for which $f_Y(y) > 0$ and the *conditional distribution function* of Y given $X = x$ is defined by

$$F_{Y|X}(y|x) = \int_{-\infty}^{y} f_{Y|X}(v|x)dv, -\infty < y < \infty$$

for x for which $f_X(x) > 0$. The mean of the conditional distribution of Y given $X = x$, when it exists, is called the *conditional expectation* or the *conditional mean* of Y given $X = x$ and is denoted by $E(Y|X = x)$. Similarly the variance of the conditional distribution of Y given $X = x$, when it exists, is called the *conditional variance* of Y given $X = x$ and it is denoted by $Var(Y|X = x)$. Similarly we define the conditional expectation and conditional variance of X given $Y = y$ whenever they exist. We will discuss more about these concepts later in this chapter.

Example 5.4. Suppose a bivariate random vector (X, Y) has the joint probability density function

$$\begin{aligned} f_{X,Y}(x,y) &= 2 \text{ if } 0 < x < y < 1 \\ &= 0 \text{ otherwise.} \end{aligned}$$

Then the marginal density function $f_X(x)$ of X is given by

$$\begin{aligned} f_X(x) &= \int_{-\infty}^{\infty} f_{X,Y}(x,y)dy \\ &= \int_{x}^{1} 2\,dy = 2(1-x) \text{ if } 0 < x < 1 \end{aligned}$$

and

$$f_X(x) = 0 \text{ otherwise.}$$

Similarly the marginal density function $f_Y(y)$ of Y is given by

$$\begin{aligned} f_Y(y) &= \int_{-\infty}^{\infty} f_{X,Y}(x,y)dx \\ &= \int_{0}^{y} 2\,dx = 2y \text{ if } 0 < y < 1 \end{aligned}$$

and

$$f_Y(y) = 0 \text{ otherwise.}$$

Hence the conditional probability density function of X given $Y = y$ is

$$\begin{aligned} f_{X|Y}(x|y) &= \frac{f_{X,Y}(x,y)}{f_Y(y)} \\ &= \frac{2}{2y} = \frac{1}{y} \text{ if } 0 < x < y \\ &= 0 \text{ otherwise} \end{aligned}$$

whenever $0 < y < 1$. Similarly the conditional probability density function of Y given $X = x$ is

$$\begin{aligned} f_{Y|X}(y|x) &= \frac{f_{X,Y}(x,y)}{f_X(x)} \\ &= \frac{2}{2(1-x)} = \frac{1}{(1-x)} \quad \text{if } x < y < 1 \\ &= 0 \text{ otherwise} \end{aligned}$$

whenever $0 < x < 1$. Let us now compute the conditional probability

$$\begin{aligned} P(0 < X < \frac{1}{2}|Y = \frac{3}{4}) &= \int_0^{\frac{1}{2}} f_{X|Y}(x|y = \frac{3}{4})\ dx \\ &= \int_0^{1/2} \frac{4}{3}\ dx = \frac{2}{3}. \end{aligned}$$

Note that this value is different from the unconditional probability

$$\begin{aligned} P(0 < X < \frac{1}{2}) &= \int_0^{\frac{1}{2}} f_X(x) dx \\ &= \int_0^{1/2} 2(1-x) dx = \frac{3}{4}. \end{aligned}$$

Example 5.5. Suppose that a bivariate random vector (X, Y) has the bivariate normal distribution

$$f(x,y) = \frac{1}{2\pi\sigma_X\sigma_Y\sqrt{1-\rho^2}} \exp\{-Q/2\}, -\infty < x, y < \infty$$

where $-\infty < \mu_X, \mu_Y < \infty$ and $-1 < \rho < 1$ and

$$Q = \frac{1}{1-\rho^2}[(\frac{x-\mu_X}{\sigma_X})^2 - 2\rho(\frac{x-\mu_X}{\sigma_X})(\frac{y-\mu_Y}{\sigma_Y}) + (\frac{y-\mu_Y}{\sigma_Y})^2].$$

Then the conditional probability density function of Y given $X = x$ is

$$\begin{aligned} f_{Y|X}(y|x) &= \frac{f_{X,Y}(x,y)}{f_X(x)} \\ &= \frac{1}{\sqrt{2\pi\sigma_Y^2(1-\rho^2)}} \exp\{-\frac{1}{2\sigma_Y^2(1-\rho^2)}[y - (\mu_Y + \frac{\rho\sigma_Y}{\sigma_X}(x-\mu_X))]^2\} \end{aligned}$$

which is the normal probability density function with mean $\mu_Y + \frac{\rho\sigma_Y}{\sigma_X}(x - \mu_X)$ and variance $\sigma_Y^2(1-\rho^2)$. Similarly check that the conditional probability density function of X given $Y = y$ is the normal probability density function with mean $\mu_X + \frac{\rho\sigma_X}{\sigma_Y}(y - \mu_Y)$ and variance $\sigma_X^2(1-\rho^2)$. In particular, it follows that

$$
\begin{aligned}
E(Y|X = x) &= \mu_Y + \frac{\rho\sigma_Y}{\sigma_X}(x - \mu_X), \\
E(X|Y = y) &= \mu_X + \frac{\rho\sigma_X}{\sigma_Y}(y - \mu_Y), \\
Var(Y|X = x) &= \sigma_Y^2(1 - \rho^2),
\end{aligned}
$$

and

$$Var(X|Y = y) = \sigma_X^2(1 - \rho^2).$$

These relations show that the conditional expectation of Y given $X = x$ as well as the conditional expectation of X given $Y = y$ are *linear* and the conditional variance of Y given $X = x$ and the conditional variance of X given $Y = y$ are constants whenever the random vector (X, Y) has a bivariate normal distribution. We will come back to this discussion again later in this chapter.

5.4 Independence

Suppose (X, Y) is a bivariate random vector with the joint probability density function $f_{X,Y}(x, y)$ and the marginal density functions $f_X(x)$ and $f_Y(y)$ for the random variables X and Y respectively. Suppose the conditional probability density function $f_{X|Y}(x|y)$ does not depend on y and $f_Y(y) > 0$ for all y. Then, for any x,

$$
\begin{aligned}
f_X(x) &= \int_{-\infty}^{\infty} f_{X,Y}(x, y)dy \\
&= \int_{-\infty}^{\infty} \frac{f_{X,Y}(x, y)}{f_Y(y)} f_Y(y)dy \\
&= \int_{-\infty}^{\infty} f_{X|Y(x|y)} f_Y(y)dy \\
&= f_{X|Y(x|y)} \int_{-\infty}^{\infty} f_Y(y)dy \\
&= f_{X|Y(x|y)}
\end{aligned}
$$

since the function $f_{X|Y(x|y)}$ does not depend on y. Hence

$$f_X(x) = f_{X|Y(x|y)} = \frac{f_{X,Y}(x, y)}{f_Y(y)}. \tag{5.1}$$

In other words, the marginal density function of the random variable X and the conditional probability density function of X given $Y = y$ are the same for every y with $f_Y(y) > 0$. In particular the conditional distribution function of X given $Y = y$ does not depend on y since

$$\begin{aligned} F_{X|Y}(x|y) &= \int_{-\infty}^{x} f_{X|Y}(u|y)du \\ &= \int_{-\infty}^{x} f_X(u)du \\ &= F_X(x). \end{aligned}$$

These observations indicate that the probabilistic behaviour of the random variable X does not depend on the value of Y that is observed. The equation (5.1) can also be written in the form

$$f_{X,Y}(x, y) = f_X(x)f_Y(y), -\infty < x, y < \infty \tag{5.2}$$

which is symmetric in x and y. Such a pair of random variables X and Y are said to be *independent.*

Suppose now that X and Y are independent random variables as defined above with the joint probability distribution function $F_{X,Y}(x, y)$ and the marginal distribution functions $F_X(x)$ and $F_Y(y)$. Then

$$\begin{aligned} F_{X,Y}(x, y) &= \int_{-\infty}^{x}\int_{-\infty}^{y} f_{X,Y}(u, v)dudv \\ &= \int_{-\infty}^{x}\int_{-\infty}^{y} f_X(u)f_Y(v)dudv \\ &= \{\int_{-\infty}^{x} f_X(u)du\}\{\int_{-\infty}^{y} f_Y(v)dv\} \\ &= F_X(x)F_Y(y), \quad -\infty < x, y < \infty. \end{aligned}$$

Hence, if X and Y are independent random variables, then

$$F_{X,Y}(x, y) = F_X(x)F_Y(y), -\infty < x, y < \infty. \tag{5.3}$$

In other words

$$P(X \leq x, Y \leq y) = P(X \leq x)P(Y \leq y), -\infty < x, y < \infty.$$

This relation shows that the events $[X \leq x]$ and $[Y \leq y]$ are independent for all x and y in the sense defined in Chapter 2. This discussion leads us to the following definition for independence of random variables X and Y irrespective of whether their joint distribution function is discrete or they have a joint probability density function as we assumed above.

Definition. Suppose X and Y are random variables with joint probability distribution function $F_{X,Y}(x, y)$ and the marginal distribution functions $F_X(x)$ and $F_Y(y)$ respectively. The random variables X and Y are said to be *independent* if

$$F_{X,Y}(x, y) = F_X(x)F_Y(y), -\infty < x, y < \infty. \tag{5.4}$$

Remarks. It is easy to check that if (X, Y) is a bivariate random vector with the joint distribution function $F_{X,Y}(x, y)$ and the joint probability density function $f_{X,Y}(x, y)$, then X and Y satisfy the relation (5.4) if and only if the relation (5.2) is satisfied.

Example 5.6. Suppose that a bivariate random vector (X, Y) has the joint probability density function

$$\begin{aligned} f(x, y) &= e^{-x-y}, \quad x > 0, y > 0 \\ &= 0 \text{ otherwise.} \end{aligned}$$

Check that the marginal density function of X is

$$\begin{aligned} f_X(x) &= e^{-x}, \quad x > 0 \\ &= 0 \text{ otherwise.} \end{aligned}$$

Similarly the marginal density function of Y is

$$\begin{aligned} f_Y(y) &= e^{-y}, \quad y > 0 \\ &= 0 \text{ otherwise.} \end{aligned}$$

It is easily seen that

$$f_{X,Y}(x, y) = f_X(x)f_Y(y), -\infty < x, y < \infty$$

and hence X and Y are independent random variables.

Example 5.7. Suppose that a bivariate random vector (X, Y) has the joint probability density function

$$f(x, y) = \frac{1}{2\pi}e^{-(\frac{x^2+y^2}{2})}, \quad -\infty < x, y < \infty.$$

Check that the marginal density function of X is

$$f_X(x) = \frac{1}{\sqrt{2\pi}}e^{-\frac{x^2}{2}}, \quad -\infty < x < \infty.$$

Similarly the marginal density function of Y is

$$f_Y(y) = \frac{1}{\sqrt{2\pi}} e^{-\frac{y^2}{2}}, \quad -\infty < y < \infty.$$

It is easily seen that

$$f_{X,Y}(x,y) = f_X(x) f_Y(y), -\infty < x, y < \infty$$

and hence X and Y are independent random variables.

Example 5.8. Suppose that a bivariate random vector (X,Y) has the joint probability density function

$$\begin{aligned} f(x,y) &= x+y, \quad 0 < x, y < 1 \\ &= 0 \text{ otherwise.} \end{aligned}$$

Check that the marginal density function of X is

$$\begin{aligned} f_X(x) &= x + \frac{1}{2}, \quad 0 < x < 1 \\ &= 0 \text{ otherwise.} \end{aligned}$$

Similarly the marginal density function of Y is

$$\begin{aligned} f_Y(y) &= y + \frac{1}{2}, \quad 0 < y < 1 \\ &= 0 \text{ otherwise.} \end{aligned}$$

It is obvious that $f_{X,Y}(x,y) \neq f_X(x) f_Y(y)$, for some $0 < x, y < 1$ and hence X and Y are not independent random variables.

Example 5.9. Suppose that a bivariate random vector (X,Y) has the joint probability density function

$$\begin{aligned} f(x,y) &= 8xy, \quad 0 < x < y < 1 \\ &= 0 \text{ otherwise.} \end{aligned}$$

Check that the marginal density function of X is

$$\begin{aligned} f_X(x) &= 8x(1 - \frac{x^2}{2}), \quad 0 < x < 1 \\ &= 0 \text{ otherwise} \end{aligned}$$

and the marginal density function of Y is

$$\begin{aligned} f_Y(y) &= 4y^3, \quad 0 < y < 1 \\ &= 0 \text{ otherwise.} \end{aligned}$$

It is clear that $f_{X,Y}(x,y) \neq f_X(x)f_Y(y)$ for some $0 < x, y < 1$ and hence X and Y are not independent random variables. This example indicates that the random variables X and Y need not be independent even though the joint probability density function $f_{X,Y}(x,y)$ is the product of a function of x and a function of y. Note that the the support, that is the set $[(x,y) : 0 < x < y < 1]$ where the joint probability density function is positive, is not a product set of the form $A_1 \times A_2$ where A_1 is a set defined by x alone and A_2 is a set defined by y alone.

Remarks. If X and Y are independent random variables, then it can be shown that the functions $g(X)$ and $h(Y)$ are independent random variables for any two (Borel-measurable) functions $g(.)$ and $h(.)$. We omit the proof. We will not go into the discussion of Borel-measurable functions in this book but most of the functions which we will come across in this book are Borel-measurable.

5.5 Expectation of a Function of a Random Vector

Let (X, Y) be a bivariate random vector. It would be useful to define the expectation of a functions $Z = g(X, Y)$ of (X, Y) such as $X + Y, XY$ and $|X - Y|$. Either one can find the probability distribution function of Z directly and then compute $E[Z]$ whenever it is finite or use the approach discussed below. Both the methods of course lead to the same result. We will not discuss the reasoning for the same in this book. If (X, Y) has a discrete probability distribution with

$$P(X = x_i, Y = y_j) = p_{ij}, i = 1, 2, \ldots; j = 1, 2, \ldots; \sum_{i,j} p_{ij} = 1,$$

then we define

$$E[g(X, Y)] = \sum_{i,j} g(x_i, y_j) p_{ij}$$

provided

$$E[|g(X, Y)|] = \sum_{i,j} |g(x_i, y_j)| p_{ij} < \infty.$$

Similarly if the random vector (X, Y) has the joint probability density function $f_{X,Y}(x, y)$, we define

$$E[g(X, Y)] = \int_{-\infty}^{\infty} \int_{-\infty}^{\infty} g(x, y) f_{X,Y}(x, y) \ dxdy$$

provided

$$E[|g(X,Y)|] = \int_{-\infty}^{\infty}\int_{-\infty}^{\infty} |g(x,y)| f_{X,Y}(x,y) \; dxdy < \infty.$$

Suppose (X, Y) is a bivariate random vector with joint probability density function $f_{X,Y}(x, y)$. Further suppose that $E(X)$ and $E(Y)$ are finite. Let $g(x, y) = x + y$. Then

$$\begin{aligned}
E[g(X,Y)] &= \int_{-\infty}^{\infty}\int_{-\infty}^{\infty} (x+y) f_{X,Y}(x,y) \; dxdy \\
&= \int_{-\infty}^{\infty}\int_{-\infty}^{\infty} x f_{X,Y}(x,y) dxdy + \int_{-\infty}^{\infty}\int_{-\infty}^{\infty} y f_{X,Y}(x,y) dxdy \\
&= \int_{-\infty}^{\infty} x [\int_{-\infty}^{\infty} f_{X,Y}(x,y) dy] dx + \int_{-\infty}^{\infty} y [\int_{-\infty}^{\infty} f_{X,Y}(x,y) dx] dy \\
&= \int_{-\infty}^{\infty} x f_X(x) \; dx + \int_{-\infty}^{\infty} y f_Y(y) \; dy \\
&= E[X] + E[Y].
\end{aligned}$$

All the above calculations can be justified, under the assumption that $E(X)$ and $E(Y)$ are finite, by using the Fubini's theorem. We will not discuss this theorem here. We have proved the above result when the random vector (X, Y) has a joint probability density function. Similar analysis will prove the result if (X, Y) has a bivariate discrete distribution (check?) and we have the following important property of the expectations.

Theorem 5.1. If X and Y are random variables with finite expectations $E[X]$ and $E[Y]$, then $E[X + Y]$ is finite and

$$E[X+Y] = E[X] + E[Y].$$

Remarks. The above property can be extended to any finite number of random variables $X_1, X_2, \ldots, X_n$ by mathematical induction.

Covariance

In order to study the relationship between two random variables, we now introduce the concept of covariance.

Definition. Let (X, Y) be a bivariate random vector with $E(X) = \mu_X$ and $E(Y) = \mu_Y$ finite. The *covariance* between the random variables X and Y is defined to be

$$E[(X - \mu_X)(Y - \mu_Y)]$$

whenever it is finite and is denoted by $Cov(X, Y)$.

We have the following alternate expression for the covariance between X and Y whenever it exists. In fact

$$\begin{aligned}
Cov(X,Y) &= E[(X-\mu_X)(Y-\mu_Y)] \\
&= E[XY - \mu_X Y - \mu_Y X + \mu_X \mu_Y] \\
&= E[XY] - \mu_X E[Y] - \mu_Y E[X] + \mu_X \mu_Y \\
&= E[XY] - \mu_X \mu_Y \\
&= E[XY] - E[X]E[Y].
\end{aligned}$$

It is easy to see from the definition that the $Cov(X, Y) = Cov(Y, X)$ and hence the covariance is symmetric in X and Y. It is convenient to note for computational purposes that $Cov(X, X) = Var(X)$. In general the covariance between the random variables X and Y is a measure of association between them. If X tends to be large when Y is large and small when Y is small, then the covariance is likely to be positive. On the other hand if large values of X correspond to small values of Y or small values of X correspond to large values of Y, then the covariance is likely to be negative. However the covariance between the variables X and Y depends on the unit of measurement and it is not dimension free. We will define another measure, related to the covariance which is dimension free but which also reflects the relationship between the variables X and Y, later in this chapter.

Example 5.10. Suppose that a bivariate random vector (X, Y) has the joint probability density function

$$\begin{aligned}
f(x,y) &= 8xy, \;\; 0 < x < y < 1 \\
&= 0 \;\; \text{otherwise.}
\end{aligned}$$

We have computed the marginal probability density functions of X and Y in the Example 5.9. Check that $E[X] = \frac{8}{15}, E[Y] = \frac{4}{15}, Var(X) = \frac{11}{25}$ and $Var(Y) = \frac{2}{75}$. Let us compute the $Cov(X, Y)$. Now

$$\begin{aligned}E[XY] &= \int_0^1 \int_0^y xy(8xy)dxdy \\ &= \int_0^1 8y^2\{\int_0^y x^2dx\}dy \\ &= \int_0^1 8y^2\frac{y^3}{3}dy = \frac{8}{3}[\frac{y^6}{6}]_0^1 \\ &= \frac{4}{9}.\end{aligned}$$

Hence

$$\begin{aligned}Cov(X,Y) &= E[XY] - E[X]E[Y] \\ &= \frac{4}{9} - (\frac{8}{15})(\frac{4}{15}).\end{aligned}$$

Note that the covariance between X and Y is not zero. Further more check that X and Y are not independent random variables.

Suppose X and Y are independent random variables with finite variances. What can we say about their covariance?

Theorem 5.2. Let X and Y be independent random variables with $E(X)$ and $E(Y)$ finite. Then $E(XY)$ exists and

$$E[XY] = E[X]E[Y]$$

and hence $Cov(X,Y) = 0$.

Proof. We prove the result when the random variables X and Y have probability density functions $f_X(x)$ and $f_Y(y)$ respectively. However the result is true for any two independent random variables with finite expectations. From the definition of independence, it follows that the joint probability density function of (X,Y) is $f_{X,Y}(x,y) = f_X(x)f_Y(y)$. Hence

$$\begin{aligned}E[XY] &= \int_{-\infty}^{\infty}\int_{-\infty}^{\infty} xyf_{X,Y}(x,y)dxdy \\ &= \int_{-\infty}^{\infty}\int_{-\infty}^{\infty} xyf_X(x)f_Y(y)dxdy \\ &= \int_{-\infty}^{\infty} xf_X(x)[\int_{-\infty}^{\infty} yf_Y(y)dy]dx \\ &= E[Y]\int_{-\infty}^{\infty} xf_X(x)dx \\ &= E[Y]E[X].\end{aligned}$$

Remarks. All the equalities in the above system of equations can be justified under the assumption that $E[Y]$ and $E[X]$ are finite by using the Fubini's theorem.

As a consequence of the above theorem, we have the following result concerning the variance of sums of independent random variables.

Theorem 5.3. If X and Y are independent random variables with finite variances, then

$$Var(X+Y) = Var(x) + Var(Y).$$

Proof. Let $E(X) = \mu_X$ and $E(Y) = \mu_Y$. Then $E(X+Y) = \mu_X + \mu_Y$ and

$$\begin{aligned} Var(X+Y) &= E[X+Y-(\mu_X+\mu_Y)]^2 \\ &= E[(X-\mu_X)+(Y-\mu_Y)]^2 \\ &= E[(X-\mu_X)]^2 + E[(Y-\mu_Y)]^2 + 2E[(X-\mu_X)(Y-\mu_Y)] \\ &= Var(X) + Var(Y) + 2\ Cov(X,Y). \end{aligned}$$

Since the random variables X and Y are independent, it follows that $Cov(X,Y) = 0$ and

$$Var(X+Y) = Var(X) + Var(Y).$$

Remarks. Note that it is not necessary that the random variables X and Y be independent for Theorem 5.3 to hold. It is sufficient if $Cov(X,Y)$ is zero. If $Cov(X,Y) = 0$ for two random variables X and Y, then they are said to be *uncorrelated*. Theorem 5.3 can be extended to more than two random variables. In fact, if X_i, X_j are uncorrelated random variables for every $1 \leq i \neq j \leq n$, then

$$Var(X_1 + \cdots + X_n) = Var(X_1) + \cdots + Var(X_n).$$

We have seen above that if X and Y are independent random variables with finite expectations, then $Cov(X,Y) = 0$. However it is not true that if $Cov(X,Y) = 0$ between two random variables, then the random variables are independent as the following example illustrates.

Example 5.11. Suppose that a bivariate random vector (X,Y) has the joint probability density function

$$\begin{aligned} f(x,y) &= 1 \text{ if } \ -y < x < y,\ 0 < y < 1 \\ &= 0 \text{ otherwise.} \end{aligned}$$

Check that the marginal probability density function of X is

$$\begin{aligned} f_X(x) &= 1+x \text{ if } -1 < x < 0 \\ &= 1-x \text{ if } 0 \leq x < 1 \\ &= 0 \text{ otherwise} \end{aligned}$$

and the marginal probability density function of Y is

$$\begin{aligned} f_Y(y) &= 2y \text{ if } 0 < y < 1 \\ &= 0 \text{ otherwise.} \end{aligned}$$

It is clear that $f(x,y) \neq f_X(x)f_Y(y)$ for some x and y such that $-y < x < y, 0 < y < 1$ and hence the random variables X and Y are not independent. It is easy to check that $E[X] = 0, E[Y] = 2/3$ and

$$\begin{aligned} E[XY] &= \int_0^1 \int_{-y}^{y} xyf(x,y)dxdy \\ &= \int_0^1 y[\int_{-y}^{y} xdx]dy \\ &= \int_0^1 y[\frac{x^2}{2}]_{-y}^{y} dy \\ &= \int_0^1 y[\frac{y^2}{2} - \frac{y^2}{2}]dy \\ &= 0. \end{aligned}$$

Hence $Cov(X,Y) = 0$ and the random variables X, Y are uncorrelated but they are not independent.

5.6 Correlation and Regression

Correlation

We have seen above that the covariance between two random variables depends on their units of measurement.

Definition. The *correlation coefficient* between two random variables X and Y is defined to be

$$\frac{Cov(X,Y)}{\sqrt{Var(X)Var(Y)}}$$

whenever it exists.

The correlation coefficient is dimension free and it is usually denoted by $Corr(X, Y)$ or by $\rho_{X,Y}$. Let us denote the $Var(X)$ by σ_X^2 , $Var(Y)$ by σ_Y^2 and $Cov(X, Y)$ by σ_{XY}. Then the correlation coefficient between X and Y is

$$\rho_{X,Y} = \frac{\sigma_{XY}}{\sigma_X \sigma_Y}$$

and it is symmetric in X and Y.

Remarks. We remark that the correlation between two independent random variables with finite variances is zero but if the correlation between two random variables is zero then they need not be independent. This can be seen from the Example 5.11. If the covariance and hence the correlation coefficient between any two random variables is zero, then they are said to be *uncorrelated.*

Before we discuss an important property of the correlation coefficient, we prove an inequality known as *Cauchy-Schwartz inequality* which is not only useful in obtaining the bounds on the correlation coefficient but also useful as a technical tool in several other areas of Statistics and Probability.

Cauchy-Schwartz inequality

Let (W, Z) be a bivariate random vector with $E(W^2) < \infty$ and $0 < E(Z^2) < \infty$. Consider the random variable

$$g(W, Z) = (W - kZ)^2$$

where k is any real number. From the general properties of expectations, we get that

$$\begin{aligned} 0 &\leq E[(W - kZ)^2] \\ &= E[W^2 - 2kWZ + k^2 Z^2] \\ &= E[W^2] - 2kE[WZ] + k^2 E[Z^2]. \end{aligned}$$

The last expression on the right side of the above equation is a quadratic function in k. Since the quadratic function is nonnegative for all k, the discriminant of the function has to be negative or zero. Therefore

$$4[E(WZ)]^2 - 4E(W^2)E(Z^2) \leq 0$$

or equivalently

$$[E(WZ)]^2 \leq E(W^2)E(Z^2).$$

This inequality is known as the *Cauchy-Schwartz inequality.* If the equality occurs in the above inequality, then it follows that the discriminant is zero and hence the quadratic equation

$$E[W^2] - 2kE[WZ] + k^2E[Z^2] = 0$$

has equal roots. In particular, there exists a real value k_0 such that

$$E[W^2] - 2k_0E[WZ] + k_0^2E[Z^2] = 0,$$

that is

$$E[(W - k_0Z)^2] = 0$$

which implies that $P(W = k_0Z) = 1$. Check that the Cauchy-Schwartz inequality holds trivially if $E(Z^2) = 0$.

As a consequence of the Cauchy-Schwartz inequality, we note that if $E(W^2)$ and $E(Z^2)$ are finite, then $E[WZ]$ is also finite.

Let us now get back to our discussion about the properties of the correlation coefficient between two random variables. Let (X, Y) be a bivariate random vector with $E(X) = \mu_X, E(Y) = \mu_Y, Var(X) = \sigma_X^2 > 0, Var(Y) = \sigma_Y^2 > 0$ and $Cov(X, Y) = \sigma_{XY}$. Let us choose

$$W = X - \mu_X, Z = Y - \mu_Y$$

in the above discussion. Applying the Cauchy-Schwartz inequality, we get that

$$[Cov(X, Y)]^2 \leq Var(X)Var(Y)$$

and hence

$$\sigma_{XY}^2 \leq \sigma_X^2\sigma_Y^2$$

and equality occurs in the above inequality if and only if there exists a value k_0 such that

$$P[X - \mu_X = k_0(Y - \mu_Y)] = 1,$$

that is $X = \mu_X + k_0(Y - \mu_Y)$ with probability one. This proves that X and Y are linearly related and $|\rho_{X,Y}| = 1$. We have now the following result.

Theorem 5.4. Let (X, Y) be a bivariate random vector with $E(X) = \mu_X, E(Y) = \mu_Y, Var(X) = \sigma_X^2, Var(Y) = \sigma_Y^2$ and $Cov(X, Y) = \sigma_{XY}$. Then the correlation coefficient $\rho_{X,Y}$ between X and Y satisfies the relation

$$-1 \leq \rho_{X,Y} \leq 1$$

and $\rho_{X,Y}^2 = 1$ if and only if X and Y are linearly related with probability one.

Regression

Let (X, Y) be a bivariate random vector. There are situations when we would be interested in the expected value of Y given $X = x$ or in the expected value of X given that $Y = y$. For instance if X is the height and Y is the weight of an individual chosen at random, we would be interested to know the average weight of a person whose height is 150 centimeters. If the joint probability distribution of (X, Y) is known , then we should be able to obtain the conditional distribution of Y given $X = x$ and hence find the expected value of the conditional distribution of Y given $X = x$. Suppose (X, Y) is a bivariate random vector with the joint probability density function $f_{X,Y}(x, y)$. The *conditional expectation* of a function $g(Y)$ given $X = x$ is defined to be

$$\int_{-\infty}^{\infty} g(y) f_{Y|X}(y|x) dy$$

whenever it is finite. It is denoted by $E[g(Y)|X = x]$. It is the expectation of the function $g(Y)$ with respect to the conditional distribution of Y given $X = x$. Similarly we define the conditional expectation of $h(X)$ given $Y = y$ by the relation

$$E[h(X)|Y = y] = \int_{-\infty}^{\infty} h(x) f_{X|Y}(x|y) dx.$$

All these expectations make sense only when the corresponding integrals are finite. If we choose $g(y) = y$ and $h(x) = x$, then we have

$$E[Y|X = x] = \int_{-\infty}^{\infty} y f_{Y|X}(y|x) dy$$

and

$$E[X|Y = y] = \int_{-\infty}^{\infty} x f_{X|Y}(x|y) dx.$$

Similar notions can be defined if the random vector (X, Y) has a discrete bivariate distribution by replacing the integrals by appropriate sums.

Let us now consider the conditional expectation of Y given $X = x$. It is a function of x. Let us denote it by $\eta(x)$. The function $\eta(x)$ is called the *regression function* of Y on X. If the function $\eta(x)$ is a linear function of x, say $a + bx$, then the random variable Y is said to have *linear regression* on X. The constants a and b are called the regression coefficients. Similarly the function $\psi(y) = E[X|Y = y]$ is called the regression function of X on

Y and if it is a linear function, then the random variable X is said to have *linear regression* on Y.

Example 5.12. Suppose that (X, Y) has the bivariate probability density function

$$f(x,y) = 2, \text{ if } 0 < x < y < 1$$
$$= 0 \text{ otherwise.}$$

Check that

$$f_{X|Y}(x|y) = \frac{1}{y} \text{ if } 0 < x < y$$
$$= 0 \text{ otherwise}$$

for $0 < y < 1$ and

$$f_{Y|X}(y|x) = \frac{1}{1-x} \text{ if } x < y < 1$$
$$= 0 \text{ otherwise}$$

for $0 < x < 1$. Therefore

$$E[Y|X = x] = \int_x^1 y(\frac{1}{1-x})\ dy = \frac{1+x}{2}$$

for $0 < x < 1$ and

$$E[X|Y = y] = \int_0^y x(\frac{1}{y})\ dx = \frac{y}{2}$$

for $0 < y < 1$. Hence the regression function of Y on X is linear as well as the regression function of X on Y is linear.

Example 5.13. Suppose that a bivariate random vector (X, Y) has the joint probability density function

$$f(x,y) = 1 \text{ if } -y < x < y,\ 0 < y < 1$$
$$= 0 \text{ otherwise}$$

(see Example 5.11). Let us compute the regression function of X on Y and that of Y on X. Following the calculations made in the Example 5.11, it can be checked that

$$E[X|Y = y] = \int_{-y}^y x\left(\frac{1}{2y}\right) dx$$
$$= 0$$

for $0 < y < 1$. On the other hand

$$\begin{aligned} E[Y|X=x] &= \int_{-x}^{1} y\left(\frac{1}{1+x}\right)\, dy \text{ if } -1 < x < 0 \\ &= \int_{x}^{1} y\left(\frac{1}{1-x}\right)\, dy \text{ if } 0 < x < 1. \end{aligned}$$

Hence

$$\begin{aligned} E[Y|X=x] &= \frac{1-x}{2} \text{ if } -1 < x < 0 \\ &= \frac{1+x}{2} \text{ if } 0 < x < 1. \end{aligned}$$

Therefore the random variable X has linear (in fact constant) regression on Y but the random variable Y does not have linear regression on X.

Let (X, Y) be a bivariate random vector with the joint probability density function given by

$$f(x,y) = \frac{1}{2\pi\sigma_X\sigma_Y\sqrt{1-\rho^2}} \exp\{-Q/2\}, -\infty < x, y < \infty$$

where $-\infty < \mu_X, \mu_Y < \infty$, $0 < \sigma_X, \sigma_Y < \infty$ and $-1 < \rho < 1$ and

$$Q = \frac{1}{1-\rho^2}\left[\left(\frac{x-\mu_X}{\sigma_X}\right)^2 - 2\rho\left(\frac{x-\mu_X}{\sigma_X}\right)\left(\frac{y-\mu_Y}{\sigma_Y}\right) + \left(\frac{y-\mu_Y}{\sigma_Y}\right)^2\right].$$

We have seen earlier that

$$\begin{aligned} E(Y|X=x) &= \mu_Y + \frac{\rho\sigma_Y}{\sigma_X}(x-\mu_X), \\ E(X|Y=y) &= \mu_X + \frac{\rho\sigma_X}{\sigma_Y}(y-\mu_Y), \\ Var(Y|X=x) &= \sigma_Y^2(1-\rho^2), and \\ Var(X|Y=y) &= \sigma_X^2(1-\rho^2) \end{aligned}$$

which shows that the regression of Y on X is linear as well as the regression of X on Y is linear. Further more the conditional variance of Y given $X = x$ does not depend on x and the the conditional variance of X given $Y = y$ does not depend on y. This property is some times referred to as *homoscedasticity*.

Let us now compute the correlation coefficient of the bivariate normal distribution. From the properties of the conditional expectation (See Exercise 5. 27), we have

$$\begin{aligned} E[(X-\mu_X)Y|X] &= (X-\mu_X)E[Y|X] \\ &= (X-\mu_X)(\mu_Y + \frac{\rho\sigma_Y}{\sigma_X}(X-\mu_X), \\ &= (X-\mu_X)\mu_Y + \frac{\rho\sigma_Y}{\sigma_X}(X-\mu_X)^2. \end{aligned}$$

Taking the expectations on both sides of the above equation and applying the result in Exercise 5.23, we get that

$$\begin{aligned} E[(X-\mu_X)Y] &= E[(X-\mu_X)\mu_Y] + \frac{\rho\sigma_Y}{\sigma_X}E[(X-\mu_X)^2] \\ &= \frac{\rho\sigma_Y}{\sigma_X}\sigma_X^2 \\ &= \rho\sigma_X\sigma_Y. \end{aligned}$$

Hence

$$\begin{aligned} Cov(X,Y) &= E[(X-\mu_X)(Y-\mu_Y)] \\ &= E[(X-\mu_X)Y] - \mu_Y E[X-\mu_X] \\ &= E[(X-\mu_X)Y] \\ &= \rho\sigma_X\sigma_Y \end{aligned}$$

and

$$\begin{aligned} Corr(X,Y) &= \frac{Cov(X,Y)}{\sqrt{Var(X)Var(Y)}} \\ &= \frac{\rho\sigma_X\sigma_Y}{\sigma_X\sigma_Y} = \rho. \end{aligned}$$

This shows that the parameter ρ in the bivariate normal distribution defined above is in fact the correlation coefficient between X and Y.

Suppose $\rho = 0$. Then

$$f(x,y) = \frac{1}{2\pi\sigma_X\sigma_Y}\exp\{-J/2\}, -\infty < x, y < \infty$$

where $-\infty < \mu_X, \mu_Y < \infty$ and $0 < \sigma_X, \sigma_Y < \infty$ and

$$J = [(\frac{x-\mu_X}{\sigma_X})^2 + (\frac{y-\mu_Y}{\sigma_Y})^2]$$

for all x and y. Hence X and Y are independent random variables. We have proved the following important fact: if (X,Y) has a bivariate normal distribution and if the random variables X and Y are uncorrelated, then they are independent.

Remarks. We will come back to this discussion on correlation and regression in more detail in the Chapter 10.

5.7 Moment Generating Function

Suppose (X, Y) is a bivariate random vector. Define

$$M_{X,Y}(t_1, t_2) = E[e^{t_1 X + t_2 Y}]$$

whenever the expectation exists. Here t_1 and t_2 are real numbers. The function $M_{X,Y}(t_1, t_2)$ is called the *moment generating function* of the random vector (X, Y). We will write $M(t_1, t_2)$ for $M_{X,Y}(t_1, t_2)$ for simplicity. As in the univariate case, it is not necessary that the moment generating function $M(t_1, t_2)$ exists at all points $(t_1, t_2) \in R^2$. It is easy to see that $M(0,0) = 1$, $M(t_1, 0) = E[e^{t_1 X}]$, and $M(0, t_2) = E[e^{t_2 Y}]$. In other words, the moment generating functions of the marginal distributions of X and Y can be recovered from the moment generating function of the joint probability distribution of (X, Y) whenever it exists.

Suppose that (X, Y) is a bivariate random vector with the joint probability density function $f(x, y)$. Then

$$M(t_1, t_2) = \int_{-\infty}^{\infty} \int_{-\infty}^{\infty} e^{t_1 x + t_2 y} f(x, y) dx dy.$$

We will now study how to generate *product moments* of the form $E[X^r Y^s]$ from the moment generating function when r and s are nonnegative integers. Suppose that we can partial differentiate the function $M(t_1, t_2)$ for r times with respect to t_1 and for s times with respect to t_2 and that differentiation under the integral sign is valid in the above integral. Then it follows that

$$\begin{aligned}
\frac{\partial^{r+s} M(t_1, t_2)}{\partial t_1^r \partial t_2^s} &= \frac{\partial^{r+s}}{\partial t_1^r \partial t_2^s} \int_{-\infty}^{\infty} \int_{-\infty}^{\infty} e^{t_1 x + t_2 y} f(x, y) \ dx dy \\
&= \int_{-\infty}^{\infty} \int_{-\infty}^{\infty} [\frac{\partial^{r+s} e^{t_1 x + t_2 y}}{\partial t_1^r \partial t_2^s}] f(x, y) \ dx dy \\
&= \int_{-\infty}^{\infty} \int_{-\infty}^{\infty} x^r y^s e^{t_1 x + t_2 y} f(x, y) \ dx dy \\
&= E[e^{t_1 X + t_2 Y} X^r Y^s].
\end{aligned}$$

In particular, if we choose $t_1 = 0$ and $t_2 = 0$, then we have

$$[\frac{\partial^{r+s} M(t_1, t_2)}{\partial t_1^r \partial t_2^s}]_{t_1=0, t_2=0} = E[X^r Y^s]$$

which shows that the product moment $E[X^r Y^s]$ can be generated from the moment generating function $M(t_1, t_2)$ by partially differentiating appropriate number of times whenever it is permissible and evaluating the partial derivative so obtained at the point (0,0) .

Suppose X and Y are independent random variables. Then

$$\begin{aligned}M(t_1,t_2) &= E[e^{t_1X+t_2Y}]\\ &= E[e^{t_1X}]E[e^{t_2Y}]\\ &= M_X(t_1)M_Y(t_2).\end{aligned}$$

Hence the moment generating function of (X,Y) evaluated at (t_1,t_2) is the product of the moment generating function of X evaluated at t_1 and the moment generating function of Y evaluated at t_2 whenever X and Y are independent random variables whenever they exist.

Example 5.14. Suppose that a random vector (X,Y) has the joint probability density function

$$\begin{aligned}f(x,y) &= e^{-y},\ \ 0<x<y<\infty\\ &= 0 \text{ otherwise.}\end{aligned}$$

Then

$$\begin{aligned}M(t_1,t_2) &= \int_{-\infty}^{\infty}\int_{-\infty}^{\infty} e^{t_1x+t_2y} f(x,y)\ dxdy\\ &= \int_0^{\infty}\{\int_x^{\infty} e^{t_1x+t_2y}e^{-y}\ dy\}\ dx\\ &= \int_0^{\infty}\{\int_x^{\infty} e^{-(1-t_2)y}dy\}e^{t_1x}\ dx\\ &= \int_0^{\infty}[\frac{e^{-(1-t_2)y}}{-(1-t_2)}]_x^{\infty}e^{t_1x}\\ &= \frac{1}{1-t_2}\int_0^{\infty} e^{-(1-t_2-t_1)x}\ dx\\ &= \frac{1}{(1-t_1-t_2)(1-t_2)}\end{aligned}$$

for $t_2<1$ and $t_1+t_2<1$. In particular

$$M_Y(t_2)=M(0,t_2)=\frac{1}{(1-t_2)^2}, t_2<1$$

is the m.g.f. of Y and

$$M_X(t_1)=M(t_1,0)=\frac{1}{1-t_1}, t_1<1$$

is the m.g.f. of X.

Example 5.15. Let (X, Y) be a bivariate random vector with the joint probability density function given by

$$f(x,y) = \frac{1}{2\pi\sigma_X\sigma_Y\sqrt{1-\rho^2}}\exp\{-Q/2\}, -\infty < x, y < \infty$$

where $-\infty < \mu_X, \mu_Y < \infty$, $0 < \sigma_X, \sigma_Y < \infty$,$-1 < \rho < 1$ and

$$Q = \frac{1}{1-\rho^2}[(\frac{x-\mu_X}{\sigma_X})^2 - 2\rho(\frac{x-\mu_X}{\sigma_X})(\frac{y-\mu_Y}{\sigma_Y}) + (\frac{y-\mu_Y}{\sigma_Y})^2].$$

It needs a little bit of algebra to show (check?) that the moment generating function of (X, Y) is

$$M_{X,Y}(t_1, t_2) = \exp\{t_1\mu_X + t_2\mu_Y + \frac{1}{2}(\sigma_X^2 t_1^2 + \sigma_Y^2 t_2^2 + 2\rho\sigma_X\sigma_Y t_1 t_2)\}$$

for all $-\infty < t_1, t_2 < \infty$.

In particular the m.g.f. of X is

$$M_X(t) = \exp\{t\mu_X + \frac{1}{2}\sigma_X^2 t^2\}$$

for $-\infty < t < \infty$. Note that the above function is also the m.g.f. of a normal distribution with mean μ_X and variance σ_X^2. From the earlier observation (see Theorem 4.1) that the m.g.f. uniquely determines the corresponding probability distribution if two moment generating functions exist and are equal in a neighbourhood of zero, it follows that the marginal probability distribution of X is $N(\mu_X, \sigma_X^2)$ and similarly the marginal probability distribution of Y is $N(\mu_Y, \sigma_Y^2)$. We have noted this fact earlier by direct computation of the marginal density functions of X and Y from the joint probability density function of (X, Y).

Even though it is true that the marginal density functions of a bivariate normal distribution are themselves (univariate) normal distributions, the converse is not true. There are joint probability density functions $f(x, y)$ for which the marginal density functions are normal but the joint probability density function is not bivariate normal (see Exercise 5.30).

We will derive another important property of a bivariate normal distribution using the moment generating functions.

Theorem 5.5. Let (X, Y) be a random vector with the bivariate normal probability density function given by

$$f(x,y) = \frac{1}{2\pi\sigma_X\sigma_Y\sqrt{1-\rho^2}}\exp\{-Q/2\}, -\infty < x, y < \infty$$

where $-\infty < \mu_X, \mu_Y < \infty$, $0 < \sigma_X, \sigma_Y < \infty$, $-1 < \rho < 1$ and

$$Q = \frac{1}{1-\rho^2}[(\frac{x-\mu_X}{\sigma_X})^2 - 2\rho(\frac{x-\mu_X}{\sigma_X})(\frac{y-\mu_Y}{\sigma_Y}) + (\frac{y-\mu_Y}{\sigma_Y})^2].$$

Then every linear combination $cX + dY$, where c and d are arbitrary constants, has a one-dimensional or univariate normal distribution.

Proof. We know that the m.g.f. of (X, Y) is

$$M_{X,Y}(t_1, t_2) = \exp\{t_1\mu_X + t_2\mu_Y + \frac{1}{2}(\sigma_X^2 t_1^2 + \sigma_Y^2 t_2^2 + 2\rho\sigma_X\sigma_Y t_1 t_2)\}.$$

Let us compute the m.g.f. of $Z = cX + dY$. By definition,

$$\begin{aligned} M_Z(t) &= E[e^{tZ}] \\ &= E[e^{tcX+tdY}] \\ &= M_{X,Y}(tc, td) \\ &= \exp\{tc\mu_X + td\mu_Y + \frac{1}{2}(\sigma_X^2 t^2 c^2 + \sigma_Y^2 t^2 d^2 + 2\rho\sigma_X\sigma_Y t^2 cd)\} \\ &= \exp\{t(c\mu_X + d\mu_Y) + \frac{1}{2}t^2(\sigma_X^2 c^2 + \sigma_Y^2 d^2 + 2\rho\sigma_X\sigma_Y cd)\}. \end{aligned}$$

We note that the function $M_Z(t)$ is finite for all t and it is also the m.g.f. of a normal distribution with mean $(c\mu_X + d\mu_Y)$ and variance $(\sigma_X^2 c^2 + \sigma_Y^2 d^2 + 2\rho\sigma_X\sigma_Y cd)$. Recall that the the m.g.f. of a random variable uniquely determines the corresponding distribution if two m.g.f.'s exist and are equal in a neighbourhood of zero. Hence we obtain that the random variable $Z = cX + dY$ has the normal distribution with mean $(c\mu_X + d\mu_Y)$ and variance $(\sigma_X^2 c^2 + \sigma_Y^2 d^2 + 2\rho\sigma_X\sigma_Y cd)$.

Remarks. The converse of the above theorem is also true, that is, if *every* linear combination $cX + dY$, for arbitrary constants c, d, of a random vector (X, Y) is normally distributed, then the random vector (X, Y) must have a bivariate normal distribution. We will not give the proof here.

In view of the above remarks, it is some times convenient to define that a random vector (X, Y) has a bivariate normal distribution if and only if *every* linear combination $cX + dY$ has a normal distribution instead of specifying the joint probability density function of the random vector (X, Y) directly. Both approaches are of course equivalent.

Example 5.16. Let X denote the height of the wife and Y denote the height of the husband, measured in centimetres, of a couple selected from the population of married couples. Suppose that the random vector (X, Y) has the bivariate normal probability distribution with mean $\mu_X = 150, \mu_Y = 160, \sigma_X = \sigma_Y = 10$, and $\rho = 0.68$. Let us compute the probability that wife is taller than her husband, that is, $P(X > Y)$ or equivalently $P(X - Y > 0)$. Since the random variable $X - Y$ is a linear combination of the random vector (X, Y) which has a bivariate normal distribution, it follows that the random variable $X - Y$ has the normal distribution with mean

$$E(X - Y) = \mu_X - \mu_Y = -10$$

and

$$\begin{aligned} Var(X - Y) &= Var(X) + Var(Y) - 2\ Cov(X, Y) \\ &= \sigma_X^2 + \sigma_Y^2 - 2\rho\sigma_X\sigma_Y \\ &= 64. \end{aligned}$$

Hence the random variable $X - Y$ has the normal distribution with mean 10 and variance 64. Therefore

$$\begin{aligned} P(X > Y) &= P(X - Y > 0) \\ &= P(\frac{X - Y + 10}{8} > \frac{10}{8}) \\ &= P(W > 1.25) = 0.106 \end{aligned}$$

since the random variable W has the standard normal distribution. Hence there is more than 10% chance that the wife is taller than her husband.

5.8 Multivariate Probability Distributions

In the previous section, we considered bivariate random vectors (X, Y) and their joint probability distributions. There are situations when it is necessary to consider more than two variables simultaneously. Suppose we are interested in the study of prevalence of a particular disease like diabetes or in the study of conditions for a "heart attack". Several factors might be involved in such studies. For instance the cardiac arrest might be due to the high bad cholesterol in the body, high blood pressure of the patient and possible smoking habits or alcohol intake of the patient besides several other factors such as pollution in the atmosphere. All of these factors might influence the length or time of survival of the patient after the attack. In meteorological studies, the amount of rain fall in a particular region is affected by the temperature, atmospheric pressure, wind velocity and other factors. In order to analyze a data of this nature, we need to consider possibly more than two characteristics simultaneously and develop methods for the analysis of such data. The subject area of study of such methods is known as the *Multivariate Statistical Analysis*. We do not propose to discuss this topic in detail here. However we will introduce the basic concepts and definitions.

Let Ω be a sample space corresponding to a random experiment $\mathcal{E}$ and $\mathcal{F}$ be a σ-algebra of subsets of Ω associated with it. Suppose that P is a probability measure defined on the class $\mathcal{F}$. Let $X_i, 1 \leq i \leq n$ be real valued functions defined on Ω such that for all $x_i, 1 \leq i \leq n$ in R ,

$$[\omega : X_i(\omega) \leq x_i, 1 \leq i \leq n] \in \mathcal{F}.$$

Such a function $\boldsymbol{X} = (X_1, \ldots, X_n)$ from Ω into R^n is called a *multivariate random vector* or a *n-variate random vector* in the present case.

In analogy with the *bivariate distribution* of a *bivariate random vector* (X, Y), we define,

$$F_{\boldsymbol{X}}(\boldsymbol{x}) = F_{X_1,\ldots,X_n}(x_1, \ldots, x_n) = P[X_i \leq x_i, 1 \leq i \leq n],$$

for $\boldsymbol{x} = (x_1, \ldots, x_n) \in R^n$, as the *joint probability distribution function* or the *multivariate probability distribution function* of the random vector $\boldsymbol{X} = (X_1, \ldots, X_n)$.

The random vector $\boldsymbol{X} = (X_1, \ldots, X_n)$ is said to have a *discrete probability distribution* if the component X_i takes discrete values $x_{ij}, j = 1, 2, \ldots$ for $i = 1, \ldots, n$ such that

$$P(X_i = x_{ij_i}, i = 1, \ldots, n) = p_{1j_1,\ldots,nj_n} > 0$$

and

$$\sum_{1 \leq j_i < \infty, 1 \leq i \leq n} p_{1j_1,\ldots,nj_n} = 1.$$

The random vector $\boldsymbol{X} = (X_1, \ldots, X_n)$ is said to have an *(absolutely) continuous probability distribution* if there exists a function $f_{\boldsymbol{X}}(\boldsymbol{x}) = f(x_1, \ldots, x_n)$ such that

(i) $f(x_1, \ldots, x_n) \geq 0$ for $-\infty < x_i < \infty, 1 \leq i \leq n$,
(ii) $\int_{-\infty}^{\infty} \cdots \int_{-\infty}^{\infty} f(x_1, \ldots, x_n) \; dx_1 \ldots dx_n = 1$, and
(iii) $F_{X_1,\ldots,X_n}(x_1, \ldots, x_n) = \int_{-\infty}^{x_1} \cdots \int_{-\infty}^{x_n} f(u_1, \ldots, u_n) \; du_1 \ldots du_n$, for $-\infty < x_i < \infty, 1 \leq i \leq n$.

The function $f(x_1, \ldots, x_n)$ is called the *joint probability density function* of the random vector $\boldsymbol{X} = (X_1, \ldots, X_n)$.

The integral in the above definition is an n-fold Lebesgue integral. Whenever a random vector $(X_1, \ldots, X_n)$ has an absolutely continuous probability distribution, it can be shown that its joint distribution function and density function are connected by the relation

$$f(x_1, \ldots, x_n) = \frac{\partial^n F_{X_1,\ldots,X_n}(x_1, \ldots, x_n)}{\partial x_1 \ldots \partial x_n}$$

for almost every $x_i, 1 \leq i \leq n$.

We can extend the notions of marginal distributions, conditional distributions, moment generating functions and other concepts from a bivariate case to a multivariate case without any problem. The vector of means of the components of $\mathbf{X}$ is called the *mean vector* of $\mathbf{X}$ and the matrix of order $n \times n$ with $Cov(X_i, X_j)$ as the (i, j)-th element is called the *covariance matrix* or *dispersion matrix* of $\mathbf{X}$ whenever they exist. We will briefly discuss two examples of multivariate distributions. The following example deals with a discrete multivariate probability distribution.

Multinomial distribution

Consider a random experiment which consists of n identical independent trials. Suppose that the outcomes of the experiment on a single trial can be classified into k types and the probability that the outcome on a single trial is of j-th type is p_j. It is clear that $p_1 + \cdots + p_k = 1$. Let X_j denote the number of trials for which the outcome was of the j-th type. It is again obvious that $X_1 + \cdots + X_k = n$. The probability distribution of the random vector $\boldsymbol{X} = (X_1, \ldots, X_k)$ is called the *Multinomial distribution.* We call such a random experiment as *a Multinomial experiment.* Using the results on combinations discussed in the Chapter 2, check that

$$P(X_1 = x_1, X_2 = x_2, \ldots, X_k = x_k) = \frac{n!}{x_1!x_2!\ldots x_k!} p_1^{x_1} p_2^{x_2} \ldots p_k^{x_k}$$

for $x_i = 0, 1, \ldots, n, i = 1, \ldots, k$ such that $x_1 + \cdots + x_k = n$.

If $k = 2$, then the corresponding Multinomial distribution reduces to the case of the Binomial distribution discussed in Chapter 3.

Since every trial leads either to an outcome of the j-th type or an outcome which is not of the j-th type, the random variable X_j has the Binomial distribution with parameters n and p_j. Hence

$$E(X_j) = np_j \text{ and } Var(X_j) = np_j(1 - p_j).$$

Fix r and s for $1 \leq r,\ s \leq k$. Let

$$\begin{aligned} Z_i &= 1 \text{ if the outcome of the } i\text{-th trial is of } r\text{-th type} \\ &= 0 \text{ otherwise.} \end{aligned}$$

Similarly define

$$\begin{aligned} W_j &= 1 \text{ if the outcome of the } j\text{-th trial is of } s\text{-th type} \\ &= 0 \text{ otherwise.} \end{aligned}$$

Then $X_r = \sum_{i=1}^{n} Z_i$ and $X_s = \sum_{j=1}^{n} W_j$. Check that $E(Z_i) = p_r, E(W_j) = p_s$ and $Cov(Z_i, W_j) = 0$ for $i \neq j$ since Z_i and W_j are independent random variables for $i \neq j$. If $i = j$, then

$$Cov(Z_i, W_i) = E(Z_i W_i) - E(Z_i)E(W_i) = -p_r p_s$$

since $Z_i W_i = 0$ for all i. Hence

$$\begin{aligned} Cov(X_r, X_s) &= Cov(\sum_{i=1}^{n} Z_i, \sum_{j=1}^{n} W_j) \\ &= \sum_{i=1}^{n} Cov(Z_i, W_i) + \sum \sum_{1 \leq i \neq j \leq n} Cov(Z_i, W_j) \\ &= -np_r p_s. \end{aligned}$$

We note that the covariance between X_r and X_s is negative. This is the case since large values of X_r are likely to lead to small values of X_s for $r \neq s$.

A large number of random experiments involving classification of data lead to a Multinomial distribution. For example, if we classify the people into say three classes according to the level of their annual income, we have a Multinomial experiment with three classes. If we classify items manufactured by a company according to their quality as excellent, good, acceptable and defective, then we have a Multinomial experiment with four classes.

Multivariate normal distribution

A real symmetric matrix $\mathbf{A}$ of order $n \times n$ is said to be *positive definite* if $\mathbf{x}'\mathbf{A}\mathbf{x} \geq \mathbf{0}$ for all $\mathbf{x} \in R^n$ equality occurring if and only if $\mathbf{x} = \mathbf{0}$. Here $\mathbf{x}'$ denotes the transpose of a column vector $\mathbf{x}$. Let $\mathbf{\Sigma}$ be a real symmetric positive definite matrix of order $n \times n$. A random vector $\mathbf{X}$ is said to have a *multivariate normal distribution* if it has a joint probability density function of the form

$$f(\boldsymbol{x}) = \frac{1}{(2\pi)^{n/2}|\mathbf{\Sigma}|^{1/2}} \exp[-\frac{1}{2}(\mathbf{x}-\mu)'\mathbf{\Sigma}^{-1}(\mathbf{x}-\mu)]$$

for all $\boldsymbol{x} = (x_1, \ldots, x_n)$. This distribution is denoted by $N_n(\mu, \mathbf{\Sigma})$.

It can be shown that μ is the mean vector, $\mathbf{\Sigma}$ is the covariance matrix of $\mathbf{X}$ and the moment generating function of $\mathbf{X}$ is given by

$$M_{\mathbf{X}}(\mathbf{t}) = \mathbf{E}\left(\mathbf{e}^{\mathbf{t}'\boldsymbol{x}}\right) = \exp\{\mathbf{t}'\mu + \frac{\mathbf{1}}{\mathbf{2}}\mathbf{t}'\mathbf{\Sigma}\mathbf{t}\}.$$

The following properties hold for a multivariate normal distribution:

(i) the marginal distributions of the components $X_i, 1 \leq i \leq n$ of $\boldsymbol{X}$ have normal distributions and the distribution of any finite-dimensional sub-vector of the random vector $\mathbf{X}$ has a multivariate normal distribution;

(ii) the conditional distributions are normal distributions;

(iii) regression of any component X_i on the rest of the components of $\boldsymbol{X}$ is linear;

(iv) covariance between X_i and X_j is the (i, j)-th element of the matrix $\mathbf{\Sigma}$; and

(v) every linear function of the components of $\mathbf{X}$ with at least one nonzero coefficient has a univariate normal distribution.

Conversely, if every linear combination of the components of a multivariate random vector $\mathbf{X}$ is normally distributed, then the random vector $\mathbf{X}$ itself has a multivariate normal distribution.

We do not go into the proofs of above results.

Independence

Let $\boldsymbol{X} = (X_1, \ldots, X_n)$ be a random vector with the joint distribution function $F_{\mathbf{X}}(\mathbf{x})$ and let $F_{X_i}(x_i)$ be the marginal distribution function of X_i for $1 \leq i \leq n$. The random variables $X_i, 1 \leq i \leq n$ are said to be *independent* if for all $x_i, 1 \leq i \leq n$,

$$P(X_1 \leq x_1, \ldots, X_n \leq x_n) = P(X_1 \leq x_1) \ldots P(X_n \leq x_n),$$

that is,

$$F_{\boldsymbol{X}}(\boldsymbol{x}) = F_{X_1}(x_1) \ldots F_{X_n}(x_n)$$

for all $x_i, 1 \leq i \leq n$. If the random vector $\mathbf{X}$ has a joint probability density function $f_{\mathbf{X}}(\mathbf{x})$ and if $f_{X_i}(x_i)$ denotes the marginal probability density function of X_i for $1 \leq i \leq n$, then the random variables $X_i, 1 \leq i \leq n$ are independent if, for all $x_i, 1 \leq i \leq n$,

$$f_{\boldsymbol{X}}(\boldsymbol{x}) = f_{X_1}(x_1) \ldots f_{X_n}(x_n).$$

Suppose a random vector $\mathbf{X}$ has a multivariate normal distribution with mean vector μ and covariance matrix $\mathbf{\Sigma}$. If the covariance matrix is a diagonal matrix, then the components $X_i, 1 \leq i \leq n$ are uncorrelated and it can be checked that the joint probability density function of $\mathbf{X}$ factors into the product of the marginal probability density functions of $X_i, 1 \leq i \leq n$ and hence the components $X_i, 1 \leq i \leq n$ of $\mathbf{X}$ are independent. It is obvious that if the components $X_i, 1 \leq i \leq n$ of a random vector $\mathbf{X}$ with a finite covariance matrix are independent, then the corresponding covariance matrix of $\mathbf{X}$ is diagonal.

A set of random variables $X_i, 1 \leq i \leq n$ is said to form a *random sample* of size n if $X_i, 1 \leq i \leq n$ are independent random variables with a common probability distribution function. This set of random variables $X_i, 1 \leq i \leq n$ is also referred to as independent and identically distributed (i.i.d.) random variables. If the random variables $X_i, 1 \leq i \leq n$ are i.i.d. with a probability density function $f(x)$, then the joint probability density function of the random vector $\boldsymbol{X} = (X_1, \ldots, X_n)$ is $f_{\boldsymbol{X}}(\boldsymbol{x}) = f(x_1) \ldots f(x_n)$.

5.9 Exercises

5.1 Two jobs for the execution of some projects are randomly allotted to three companies A, B and C. Let X denotes the number of jobs allotted to A and Y denote the number of jobs allotted to B. Find the joint probability function of the bivariate random vector (X, Y).

5.2 Let X denote the amount of time a child watches television and Y denote the amount spent by the child on studies in a day. Suppose (X, Y) has the joint probability density function

$$\begin{aligned} f(x, y) &= xye^{-\lambda(x+y)}, \quad x > 0, y > 0 \\ &= 0 \ \text{ otherwise.} \end{aligned}$$

Find the probability that a child chosen at random spends at least twice as much time watching television as he or she does on studies.

5.3 Consider an electronic system with two components. Suppose the system is such that one component is on the reserve and it is activated only if the other component fails. The system fails if and only if both the components fail. Let X and Y denote the life times of these components. Suppose (X, Y) has the joint probability density function

$$\begin{aligned} f(x,y) &= \lambda^2 e^{-\lambda(x+y)}, \quad x \geq 0, y \geq 0 \\ &= 0 \text{ otherwise.} \end{aligned}$$

What is the probability that the system will last for more than 500 hours?

5.4 Suppose that a random vector (X, Y) has the joint distribution function

$$\begin{aligned} F(x,y) &= (1-e^{-\lambda x})(1-e^{-\lambda y}), \quad x > 0, y > 0 \\ &= 0 \text{ otherwise.} \end{aligned}$$

Find the joint probability density function of (X, Y).

5.5 Suppose that a random vector (X, Y) has the joint probability density function

$$\begin{aligned} f(x,y) &= y^2 e^{-y(x+1)}, \quad x \geq 0, y \geq 0 \\ &= 0 \text{ otherwise.} \end{aligned}$$

Determine the marginal density functions of X and Y.

5.6 Suppose that a random vector (X, Y) has the joint probability density function

$$\begin{aligned} f(x,y) &= 6x, \quad 0 < x < 1, 0 < y < 1-x \\ &= 0 \text{ otherwise.} \end{aligned}$$

Determine the marginal density functions of X and Y.

5.7 Suppose that the bivariate random vector has the joint probability density function

$$\begin{aligned} f(x,y) &= \frac{12}{5} x(2-x-y), \quad 0 < x, y < 1 \\ &= 0 \text{ otherwise.} \end{aligned}$$

Find the conditional probability density function of X given $Y = y$ for $0 < y < 1$.

5.8 Suppose that the bivariate random vector has the joint probability density function

$$f(x, y) = c(x + y^2), \ 0 < x, y < 1$$
$$= 0 \text{ otherwise.}$$

(a) Find the conditional probability density function of X given $Y = y$ for $0 < y < 1$. (b) Compute $P(X < \frac{1}{2}|Y = \frac{1}{2})$.

5.9 Suppose that the bivariate random vector has the joint probability density function

$$f(x, y) = 2e^{-(x+y)}, \ 0 < x < y$$
$$= 0 \text{ otherwise.}$$

Compute $P(Y < 1|X < 1)$.

5.10 Suppose that the conditional probability density function of Y given $X = x$ and the marginal density function of X are given by

$$f_{Y|X}(y|x) = \frac{2y + 4x}{1 + 4x}, \ 0 < x, y < 1$$
$$= 0 \text{ otherwise}$$

and

$$f_X(x) = \frac{1 + 4x}{3}, \ 0 < x < 1$$
$$= 0 \text{ otherwise}$$

respectively. Determine the marginal density function of Y.

5.11 The joint probability density function of a random vector (X, Y) is given by

$$f(x, y) = c(x^2 - y^2)e^{-x}, \ -x \leq y \leq x, 0 < x < \infty$$
$$= 0 \text{ otherwise.}$$

Find the conditional distribution function of Y given $X = x$.

5.12 Let X denote the percentage of marks obtained by a student in Mathematics and Y denote the percentage in English in the final examinations. Suppose that the random vector (X, Y) has the joint probability density function

$$f(x, y) = \frac{2}{5}(2x + 3y), \ 0 < x, y \leq 1$$
$$= 0 \text{ otherwise.}$$

(a) What percentage of students obtain more than 80% in Mathematics? (b) If a student has obtained 30% in English, what is the probability that he or she gets more than 80% in Mathematics? (c) If a student has obtained 30% in Mathematics, what is the probability that he or she gets more than 80% in English?

5.13 The joint probability density function of a random vector (X, Y) is given by

$$\begin{aligned} f(x,y) &= 12xy(1-y), \;\; 0 < x, y < 1 \\ &= 0 \;\; \text{otherwise.} \end{aligned}$$

Show that X and Y are independent random variables.

5.14 The joint probability density function of a random vector (X, Y) is given by

$$\begin{aligned} f(x,y) &= 4x(1-y), \;\; 0 < x, y < 1 \\ &= 0 \;\; \text{otherwise.} \end{aligned}$$

Determine $P(0 < X < \frac{1}{3}, 0 < Y < \frac{1}{3})$.

5.15 The joint probability density function of a random vector (X, Y) is given by

$$\begin{aligned} f(x,y) &= c\,(x^2 - y^2)e^{-x}, \;\; -x \leq y \leq x, 0 < x < \infty \\ &= 0 \;\; \text{otherwise.} \end{aligned}$$

(a) Determine the constant c and (b) check whether X and Y are independent random variables.

5.16 Suppose that the bivariate random vector has the joint probability density function

$$\begin{aligned} f(x,y) &= xe^{-(x+y)}, \;\; x > 0, y > 0 \\ &= 0 \;\; \text{otherwise.} \end{aligned}$$

Are X and Y independent?

5.17 Determine the correlation coefficient of (X, Y) when the random vector (X, Y) has the joint probability density function given by

$$\begin{aligned} f(x,y) &= 2, \;\; x \geq 0, y \geq 0, x + y \leq 1 \\ &= 0 \;\; \text{otherwise.} \end{aligned}$$

5.18 Show that

$$Cov(aX + b, cY + d) = ac \;\; Cov(X, Y)$$

for any random vector (X, Y) with finite covariance and for arbitrary constants a, b, c and d.

5.19 Suppose that U is a random variable uniformly distributed on the interval $[0, 2\pi]$. Define $X = \cos U$ and $Y = \sin U$. Show that X and Y are uncorrelated. Are X and Y independent?

5.20 If X has the standard normal distribution and $Y = a + bX + cX^2$ with $b \neq 0$ or $c \neq 0$, then show that

$$\rho_{X,Y} = \frac{b}{\sqrt{b^2 + 2c^2}}.$$

5.21 If X and Y are independent random variables and g and h are functions such that $E[g(X)]$ and $E[h(Y)]$ are finite, show that

$$E[g(X)h(Y)] = E[g(X)]E[h(Y)].$$

5.22 Suppose that a bivariate random vector (X, Y) has the joint probability density function

$$\begin{aligned} f(x, y) &= x + y, \quad 0 \leq x, y \leq 1 \\ &= 0 \quad \text{otherwise.} \end{aligned}$$

Find the regression function of Y on X.

5.23 Prove that

$$E[E(Y|X)] = E[Y]$$

for any random vector (X, Y) whenever the expectations exist.

5.24 Suppose that a bivariate random vector (X, Y) has the joint probability density function

$$\begin{aligned} f(x, y) &= y^2 e^{-y(x+1)}, \quad x \geq 0, y \geq 0 \\ &= 0 \quad \text{otherwise.} \end{aligned}$$

Find the regression function of Y on X.

5.25 If (X, Y) is a bivariate normal random vector, then show that

$$E(Y|X = x) = \mu_Y + \frac{\rho\sigma_Y}{\sigma_X}(x - \mu_X)$$

where $\mu_X = E[X], \mu_Y = E[Y], \sigma_X^2 = Var(X), \sigma_Y^2 = Var(Y)$ and ρ is the correlation coefficient between X and Y.

5.26 Show that if X and Y are independent random variables, then

$$E[X|Y = y] = E[X]$$

for all y.

5.27 For any bivariate random vector (X, Y) and for any two functions $g(.)$ and $h(.)$, prove that

$$E[g(X)h(Y)|X = x] = g(x)E[h(Y)|X = x]$$

with probability one whenever the expectations exist.

5.28 Suppose that a random vector (X, Y) has the bivariate normal probability density function with $\mu_X = 5, \mu_Y = 10, \sigma_X = 1, \sigma_Y = 5$ and $\rho > 0$. If $P(4 < Y < 16|X = 5) = 0.954$, find the constant ρ.

5.29 Suppose that a random vector (X, Y) has the bivariate normal density with $\sigma_X = \sigma_Y$. Show that the random variables $X+Y$ and $X-Y$ are independent.

5.30 Suppose that a random vector (X, Y) has the bivariate probability density function defined by

$$f(x, y) = \frac{1}{2\pi} \exp[-\frac{1}{2}(x^2 + y^2)]\{1 + xy \exp[x^2 + y^2 - 2]\}.$$

This function is not a bivariate normal probability density function. Show that the marginal probability density functions of X and Y are normal probability density functions thus establishing the fact that the marginal density functions of X and Y being normal does not imply that the joint distribution is bivariate normal.

5.31 Suppose that a random vector (X, Y) has the bivariate probability density function defined by

$$f(x, y) = c \exp[-(\frac{x^2}{2} + xy + y^2 - x - 2y)].$$

Evaluate the constant c.

5.32 A large lot of items manufactured by a company contains 20% with just one defect, 10% with more than one defect and the rest with no defects. Suppose that n items are randomly selected from the lot. If X_1 denote the number of items with one defect and X_2 denotes the number of items with more than one defect in the sample, the repair costs are $X_1 + 2X_2$. Find the mean and the variance of the repair costs.

5.33 Suppose that a random vector $\mathbf{X}$ has a multivariate normal distribution with the mean vector μ and the covariance matrix $\mathbf{\Sigma}$. Let $Y = \mathbf{c}'\mathbf{X}$ and $Z = \mathbf{d}'\mathbf{X}$. Show that Y and Z are independent if and only if $\mathbf{c}'\mathbf{\Sigma}\mathbf{d} = \mathbf{0}$.

5.34 Suppose that $X_1, \ldots, X_n$ are independent normally distributed random variables with a common variance. Let $Y = \mathbf{c}'\mathbf{X}$ and $Z = \mathbf{d}'\mathbf{X}$. Show that Y and Z are independent if and only if $\mathbf{c}'\mathbf{d} = \mathbf{0}$.

5.35 Suppose a random vector $\boldsymbol{X}' = (X_1, \ldots, X_n)$ has the multivariate normal distribution $N_n(\mu, \mathbf{\Sigma})$ with the mean vector μ and the covariance matrix $\mathbf{\Sigma}$. Let $\mathbf{W} = \mathbf{BX}$ where $\mathbf{B}$ is a matrix of order $k \times n$. Show that the random matrix $\mathbf{W}'$ has the multivariate normal distribution $N_k(\mathbf{B}\mu, \mathbf{B\Sigma B}')$ with the mean vector $\mathbf{B}\mu$ and the covariance matrix $\mathbf{B\Sigma B}'$.

Chapter 6

FUNCTIONS OF RANDOM VECTORS

6.1 Introduction

In the last chapter, we have introduced the concept of multivariate distributions of a random vector $\boldsymbol{X} = (X_1, \ldots, X_n)$. It is of interest and importance to find the probabilistic behaviour of a function $g(\boldsymbol{X}) = g(X_1, \ldots, X_n)$ of the random vector $\boldsymbol{X} = (X_1, \ldots, X_n)$ most of the time. The function $g(.)$ could be either the sum $X_1 + \cdots + X_n$ or the $\max\{X_i, 1 \leq i \leq n\}$ or some other function depending on the phenomenon under study. We will now give three approaches for obtaining the distribution function of the function $g(\boldsymbol{X}) = g(X_1, \ldots, X_n)$ of a random vector $\boldsymbol{X} = (X_1, \ldots, X_n)$.

6.2 Functions of Two Random Variables

Direct approach

Suppose (X, Y) is a bivariate random vector with a probability density function $f(x, y)$. Let $Z = g(X, Y)$ where $g(x, y)$ is a known real-valued function. Define

$$D_z = \{(x, y) : g(x, y) \leq z\}$$

where $-\infty < z < \infty$. Then

$$P(Z \leq z) = \int\int_{D_z} f(x, y) dx dy.$$

The problem reduces to evaluating the double integral given above for all values of z in R.

Example 6.1. Suppose (X, Y) is a bivariate random vector with the uniform distribution on the unit square $[0, 1] \times [0, 1]$. Then the joint density of

(X, Y) is

$$f(x, y) = 1 \text{ if } 0 < x, y < 1$$
$$= 0 \text{ otherwise.}$$

Let us find the distribution function of $Z = g(X, Y) = XY$. It is obvious that

$$F_Z(z) = 0 \text{ if } z \leq 0$$
$$= 1 \text{ if } z \geq 1.$$

Suppose that $0 < z < 1$. Then

$$F_Z(z) = P(XY \leq z)$$
$$= \int\int_{\{(x,y):xy \leq z\}} f(x, y) dx dy$$
$$= \int\int_{\{(x,y):xy \leq z, 0<x<1, 0<y<1\}} dx dy.$$

In order to evaluate the last integral, let us look at the set of all points (x, y) such that $xy \leq z, 0 < x < 1$, and $0 < y < 1$. See the Figure 6.1.

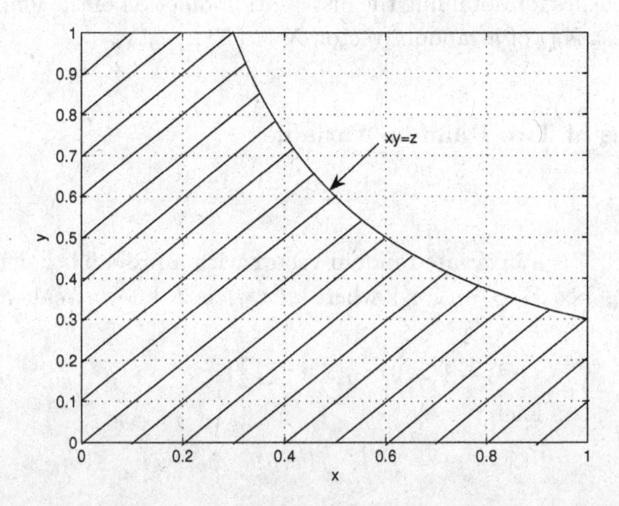

Fig. 6.1

If $0 < x \leq z$, then, for any $0 < y < 1$, the product $xy < z$ and if $x > z$, then $xy \leq z$ only if $0 < y \leq \frac{z}{x}$ (see the area shaded in Figure 6.1). Hence, for $0 < z < 1$,

$$\begin{aligned} F_Z(z) &= P(Z \leq z) \\ &= \int_0^z \{\int_0^1 dy\}\, dx + \int_z^1 \{\int_0^{z/x} dy\}\, dx \\ &= \int_0^z dx + \int_z^1 \frac{z}{x}\, dx \\ &= z + z[\log x]_z^1 \\ &= z - z \log z. \end{aligned}$$

Therefore

$$\begin{aligned} F_Z(z) &= 0 \text{ if } z \leq 0 \\ &= z - z\log z \text{ if } 0 < z < 1 \\ &= 1 \text{ if } z \geq 1 \end{aligned}$$

is the distribution function of the random variable Z. The probability density function $f_Z(z)$ of Z is obtained by differentiating $F_Z(z)$ with respect to z. It is given by

$$\begin{aligned} f_Z(z) &= 0 \text{ if } z \leq 0 \text{ or } z \geq 1 \\ &= -\log z \text{ if } 0 < z < 1. \end{aligned}$$

Example 6.2. Suppose X and Y are independent random variables with the probability density function

$$\begin{aligned} f(x) &= \lambda e^{-\lambda x} \text{ if } x > 0 \\ &= 0 \text{ otherwise.} \end{aligned}$$

Define $Z = X + Y$. Let us find the distribution function of Z. From the definition of the random variable Z,

$$F_Z(z) = P(Z \leq z) = 0 \text{ if } z \leq 0$$

and, for $z > 0$,

$$\begin{aligned} F_Z(z) &= P(Z \leq z) \\ &= \int\int_{\{(x,y):x+y \leq z\}} f(x,y)\ dxdy \end{aligned}$$

where $f(x, y)$ is the joint probability density function of (X, Y). Since X and Y are independent random variables, the joint probability density function of (X, Y) is given by

$$\begin{aligned} f(x,y) &= \lambda^2 e^{-\lambda(x+y)} \text{ if } x > 0, y > 0 \\ &= 0 \text{ otherwise.} \end{aligned}$$

Therefore, for $z > 0$,

$$F_z(z) = \int_0^z \int_0^{z-x} \lambda^2 e^{-\lambda(x+y)} \, dydx.$$

For $z > 0$, the set $\{(x,y) : x + y \leq z, x > 0, y > 0\}$ is the shaded region in the Figure 6.2.

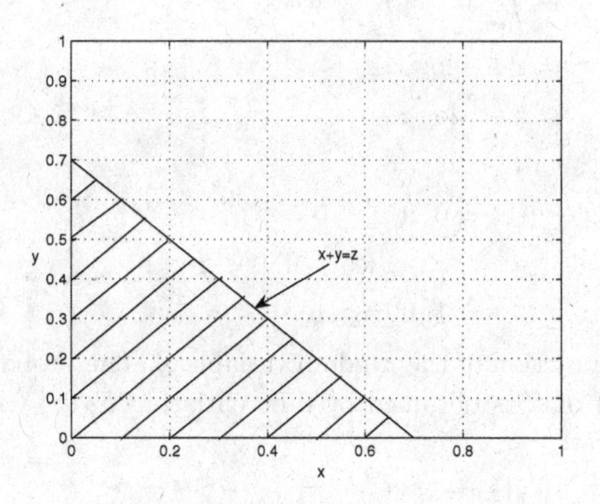

Fig. 6.2

Hence, for $z > 0$,

$$\begin{aligned}
F_z(z) &= \int_0^z [\int_0^{z-x} \lambda e^{-\lambda y} dy] \lambda e^{-\lambda x} dx \\
&= \int_0^z [-e^{-\lambda y}]_0^{z-x} \lambda e^{-\lambda x} dx \\
&= \int_0^z [1 - e^{-\lambda(z-x)}] \lambda e^{-\lambda x} dx \\
&= \int_0^z [\lambda e^{-\lambda x} - \lambda e^{-\lambda z}] dx \\
&= [-e^{-\lambda x}]_0^z - z\lambda e^{-\lambda z} \\
&= 1 - e^{-\lambda z} - \lambda z e^{-\lambda z}.
\end{aligned}$$

It is easy to check that

$$\begin{aligned}
f_Z(z) &= \lambda^2 z e^{-\lambda z} \quad \text{for } z > 0 \\
&= 0 \text{ otherwise.}
\end{aligned}$$

Recall that the above function is the Gamma probability density function discussed in the Chapter 4. Hence the sum of two independent identical exponentially distributed random variables has a gamma distribution.

Let us discuss another example.

Example 6.3. Suppose X and Y are independent random variables with the probability distribution function F. Define $Z = \max(X, Y)$. Then the distribution function of Z is given by

$$\begin{aligned} F_Z(z) &= P(Z \leq z) \\ &= P(\max(X, Y) \leq z) \\ &= P(X \leq z, Y \leq z) \\ &= P(X \leq z)P(Y \leq z) \\ &= [F(z)]^2. \end{aligned}$$

If the random variable X has the probability density function f, then the distribution function F is differentiable almost every where with the derivative equal to f and the random variable $Z = \max(X, Y)$ has the probability density function

$$f_Z(z) = \frac{dF_Z(z)}{dz} = 2F(z)f(z), -\infty < z < \infty.$$

All the three examples discussed above deal with the method of finding the distribution function of a random variable $Z = g(X, Y)$ directly. This method can also be applied even when the random vector (X, Y) has a discrete distribution.

Example 6.4. Suppose X_1 and X_2 are independent random variables with the Binomial distributions with parameters (n_1, p) and (n_2, p) respectively. We will now find the distribution of the random variable $Z = X_1 + X_2$. Note that the possible values for Z are $0, 1, \ldots, n_1 + n_2$. Hence, for any integer r such that $0 \leq r \leq n_1 + n_2$,

$$
\begin{aligned}
P(Z = r) &= P(X_1 + X_2 = r) \\
&= \sum_{j=0}^{r} P(X_1 = j, X_2 = r - j) \\
&= \sum_{j=0}^{r} P(X_1 = j)P(X_2 = r - j) \\
&= \sum_{j=0}^{r} n_1{}_j^C p^j (1-p)^{n_1-j} n_2{}_{r-j}^C p^{r-j} (1-p)^{n_2-(r-j)} \\
&= p^r (1-p)^{n_1+n_2-r} \sum_{j=0}^{r} n_1{}_j^C n_2{}_{r-j}^C \\
&= p^r (1-p)^{n_1+n_2-r} (n_1+n_2)_r^C.
\end{aligned}
$$

Hence the random variable $Z = X_1 + X_2$ has the Binomial distribution with parameters $(n_1 + n_2, p)$.

Example 6.5. Suppose X_1 and X_2 are independent random variables with the Poisson distributions with the parameters λ_1 and λ_2 respectively. We will now find the distribution of the random variable $Z = X_1 + X_2$. Note that the possible values for Z are $0, 1, 2, \ldots$. Hence, for any integer r such that $r \geq 0$,

$$
\begin{aligned}
P(Z = r) &= P(X_1 + X_2 = r) \\
&= \sum_{j=0}^{r} P(X_1 = j, X_2 = r - j) \\
&= \sum_{j=0}^{r} P(X_1 = j)P(X_2 = r - j) \\
&= \sum_{j=0}^{r} e^{-\lambda_1} \frac{\lambda_1^j}{j!} e^{-\lambda_2} \frac{\lambda_2^{r-j}}{(r-j)!} \\
&= e^{-(\lambda_1+\lambda_2)} \sum_{j=0}^{r} \frac{\lambda_1^j}{j!} \frac{\lambda_2^{r-j}}{(r-j)!} \\
&= e^{-(\lambda_1+\lambda_2)} \frac{(\lambda_1+\lambda_2)^r}{r!}.
\end{aligned}
$$

Hence the random variable $Z = X_1 + X_2$ has the Poisson distribution with the parameter $\lambda_1 + \lambda_2$.

Transformation approach

Suppose (X_1, X_2) is a bivariate random vector with the probability density function $f(x_1, x_2)$ and we would like to determine the distribution function of $Z_1 = g_1(X_1, X_2)$ or its probability density function. Suppose that another function $Z_2 = g_2(X_1, X_2)$ can be chosen such that the transformation (x_1, x_2) to (z_1, z_2) is one-to-one. In other words, to every point $(x_1, x_2) \in R^2$ corresponds a point $(z_1, z_2) \in R^2$ via the above transformation and conversely to every point $(z_1, z_2) \in R^2$ there corresponds one and only one point $(x_1, x_2) \in R^2$ such that

$$z_1 = g_1(x_1, x_2) \text{ and } z_2 = g_2(x_1, x_2).$$

For example, suppose $Z_1 = X_1 + X_2$. Choose $Z_2 = X_1 - X_2$. Then the transformation $(x_1, x_2) \to (z_1, z_2) \in R^2$ is one-to-one and

$$X_1 = \frac{Z_1 + Z_2}{2} \text{ and } X_2 = \frac{Z_1 - Z_2}{2}.$$

In general, suppose that we can express (x_1, x_2) in terms of z_1 and z_2 uniquely, that is, there exist real-valued functions h_1 and h_2 defined on R^2 such that

$$X_1 = h_1(Z_1, Z_2) \text{ and } X_2 = h_2(Z_1, Z_2).$$

Furthermore assume that the functions h_1 and h_2 have continuous partial derivatives with respect to z_1 and z_2. Consider the *Jacobian* of the transformation $(z_1, z_2) \to (x_1, x_2)$. It is the determinant J of the matrix

$$\begin{pmatrix} \frac{\partial x_1}{\partial z_1} & \frac{\partial x_1}{\partial z_2} \\ \frac{\partial x_2}{\partial z_1} & \frac{\partial x_2}{\partial z_2} \end{pmatrix}$$

which is

$$\frac{\partial x_1}{\partial z_1}\frac{\partial x_2}{\partial z_2} - \frac{\partial x_1}{\partial z_2}\frac{\partial x_2}{\partial z_1}.$$

We denote the above Jacobian by the symbol

$$J = \frac{\partial(x_1, x_2)}{\partial(z_1, z_2)}.$$

Suppose J is not equal to zero for all (z_1, z_2). Then, it can be shown that the random vector (Z_1, Z_2) has a probability density function and it is given by

$$\begin{aligned} f_{(Z_1,Z_2)}(z_1, z_2) &= f(h_1(z_1, z_2), h_2(z_1, z_2))|J| \text{ if } (z_1, z_2) \in B \\ &= 0 \text{ otherwise} \end{aligned}$$

where

$$B = \{(z_1, z_2) : z_1 = g_1(x_1, x_2), z_2 = g_2(x_1, x_2) \text{ for some } (x_1, x_2) \in R^2\}.$$

This follows from the change of variable formula for double integrals. We do not prove this result here. Here $|J|$ denotes the absolute value of the Jacobian J. From the joint probability density function of (Z_1, Z_2), the marginal probability density function of Z_1 can be derived and it is given by

$$f_{Z_1}(z_1) = \int_{-\infty}^{\infty} f_{(Z_1,Z_2)}(z_1, z_2) dz_2.$$

Example 6.6. Suppose X_1 and X_2 are independent identically distributed standard normal random variables. Let us compute the distribution function of $Z_1 = X_1 + X_2$. Suppose we choose $Z_2 = X_1 - X_2$. Then the transformation $(x_1, x_2) \to (z_1, z_2)$ in R^2 is one-to-one and the set R^2 is mapped onto R^2. Further more

$$X_1 = \frac{Z_1 + Z_2}{2} \text{ and } X_2 = \frac{Z_1 - Z_2}{2}$$

and the Jacobian of the transformation is the determinant J of the matrix

$$\begin{pmatrix} \frac{\partial x_1}{\partial z_1} & \frac{\partial x_1}{\partial z_2} \\ \frac{\partial x_2}{\partial z_1} & \frac{\partial x_2}{\partial z_2} \end{pmatrix}$$

which is equal to the determinant of the matrix

$$\begin{pmatrix} \frac{1}{2} & \frac{1}{2} \\ \frac{1}{2} & -\frac{1}{2} \end{pmatrix}$$

It is easy to see that the determinant is equal to $-\frac{1}{2}$. The joint probability density function of (X_1, X_2) is

$$f(x_1, x_2) = \frac{1}{\sqrt{2\pi}} e^{-\frac{1}{2}x_1^2} \frac{1}{\sqrt{2\pi}} e^{-\frac{1}{2}x_2^2}, -\infty < x_1, x_2 < \infty.$$

Hence the joint probability density function of (Z_1, Z_2) is

$$\begin{aligned} g(z_1, z_2) &= f(\frac{z_1 + z_2}{2}, \frac{z_1 - z_2}{2})|J| \\ &= \frac{1}{2\pi} \exp\{-\frac{1}{2}(\frac{z_1 + z_2}{2})^2) - \frac{1}{2}(\frac{z_1 - z_2}{2})^2\} \frac{1}{2} \\ &= \frac{1}{4\pi} \exp\{-(\frac{z_1^2}{4} + \frac{z_2^2}{4})\} \end{aligned}$$

for $-\infty < x_1, x_2 < \infty$. Check that the marginal density function of Z_1 is

$$f_{Z_1}(z_1) = \frac{1}{\sqrt{4\pi}} e^{-\frac{z_1^2}{4}}, -\infty < z_1 < \infty$$

and that of Z_2 is

$$f_{Z_2}(z_2) = \frac{1}{\sqrt{4\pi}} e^{-\frac{z_2^2}{4}}, -\infty < z_2 < \infty.$$

In other words, the random variables Z_1 and Z_2 have $N(0,2)$ as their probability distributions. In fact the random variables Z_1 and Z_2 are independent since the joint probability density function $g(z_1, z_2)$ of (Z_1, Z_2) is the product of the probability density function $f_{Z_1}(z_1)$ of Z_1 and the probability density function $f_{Z_2}(z_2)$ of Z_2.

We consider another example to illustrate the transformation approach.

Example 6.7. Suppose X_1 and X_2 are independent random variables with a common probability density function

$$\begin{aligned} f(x) &= \frac{1}{2} e^{-\frac{x}{2}}, \text{ if } 0 < x < \infty \\ &= 0 \text{ otherwise.} \end{aligned}$$

Then the joint probability density function of (X_1, X_2) is

$$\begin{aligned} f(x_1, x_2) &= \frac{1}{4} e^{-(\frac{x_1+x_2}{2})}, \text{ if } 0 < x_1, x_2 < \infty \\ &= 0 \text{ otherwise.} \end{aligned}$$

Let us find the distribution function of $Z_1 = \frac{1}{2}(X_1 - X_2)$. Here it is convenient to choose $Z_2 = X_2$. Then the transformation $(x_1, x_2) \to (z_1, z_2)$ is one-to-one from the set $A = \{(x_1, x_2) : 0 < x_1 < \infty, 0 < x_2 < \infty\}$ onto the set $B = \{(z_1, z_2) : z_2 > 0, -\infty < z_1 < \infty$ and $z_2 > -2z_1\}$. This can be seen from the inverse transformation

$$x_1 = 2z_1 + z_2 \text{ and } x_2 = z_2.$$

If $x_1 > 0$, then $2z_1 + z_2 > 0$, and hence $z_2 > -2z_1$. If $x_2 > 0$, then $z_2 > 0$. It is obvious that $-\infty < z_1 < \infty$. Check that the Jacobian of the transformation is equal to 2. Hence the joint probability density function of (Z_1, Z_2) is

$$\begin{aligned} g(z_1, z_2) &= f(2z_1 + z_2, z_2)|J| \text{ if } (z_1, z_2) \in B \\ &= 0 \text{ otherwise.} \end{aligned}$$

Therefore

$$g(z_1, z_2) = \frac{1}{2}e^{-z_1-z_2} \text{ if } (z_1, z_2) \in B$$
$$= 0 \text{ otherwise}$$

and the marginal density function of Z_1 is

$$f_{Z_1}(z_1) = \int_{-2z_1}^{\infty} \frac{1}{2}e^{-z_1-z_2}\; dz_2 \text{ if } -\infty < z_1 < 0$$
$$= \int_{0}^{\infty} \frac{1}{2}e^{-z_1-z_2}\; dz_2 \text{ if } 0 \leq z_1 < \infty.$$

Check that the marginal probability density function of Z_1 is

$$f_{Z_1}(z_1) = \frac{1}{2}e^{-|z_1|}, -\infty < z_1 < \infty.$$

A probability distribution with this density function is called the *double exponential distribution.*

An important application of the transformation approach is to determine the probability distribution of the sum of two *independent* random variables X_1 and X_2 given the probability distributions of the random variables X_1 and X_2. This distribution of the sum $X_1 + X_2$ is said to be the *convolution* of the distributions of X_1 and X_2.

Suppose X_1 and X_2 are two independent random variables with probability density functions $f_1(x_1)$ and $f_2(x_2)$ respectively. Define $Z_1 = X_1+X_2$ and $Z_2 = X_2$. Then

$$X_1 = Z_1 - Z_2 \text{ and } X_2 = Z_2$$

and the transformation $(x_1, x_2) \to (z_1, z_2)$ is one-to-one. In fact the Jacobian of the transformation is equal to one. Since the joint probability density function of (X_1, X_2) is

$$f_{X_1,X_2}(x_1, x_2) = f_1(x_1)f_2(x_2), -\infty < x_1, x_2 < \infty,$$

it follows that the joint probability density function of (Z_1, Z_2) is

$$f_{Z_1,Z_2}(z_1, z_2) = f_1(z_1 - z_2))f_2(z_2), -\infty < z_1, z_2 < \infty.$$

Hence the probability density function of Z_1 is

$$f_{Z_1}(z_1) = \int_{-\infty}^{\infty} f_{Z_1,Z_2}(z_1, z_2)\; dz_2$$
$$= \int_{-\infty}^{\infty} f_1(z_1 - z_2)f_2(z_2)\; dz_2$$

for $-\infty < z_2 < \infty$. This formula, which gives the probability density function of Z_1, is known as the *convolution formula.* The probability density function of the random variable $X_1 + X_2$ is the convolution of the probability density function of the random variable X_1 with the probability density function of the random variable X_2. Note that the convolution of the probability density function of the random variable X_1 with the probability density function of the random variable X_2 is the same as the convolution of the probability density function of the random variable X_2 with the probability density function of the random variable X_1. Both operations lead to the probability density function of $Z = X_1 + X_2$. We now calculate the distribution function of Z_1. Note that

$$\begin{aligned}
F_{Z_1}(z_1) &= \int_{-\infty}^{z_1} f_{Z_1}(u)\, du \\
&= \int_{-\infty}^{z_1} [\int_{-\infty}^{\infty} f_1(u - z_2) f_2(z_2) dz_2]\, du \\
&= \int_{-\infty}^{\infty} [\int_{-\infty}^{z_1} f_1(u - z_2)\, du] f_2(z_2)\, dz_2 \\
&= \int_{-\infty}^{\infty} [\int_{-\infty}^{z_1 - z_2} f_1(u) du] f_2(z_2)\, dz_2 \\
&= \int_{-\infty}^{\infty} F_1(z_1 - z_2) f_2(z_2)\, dz_2
\end{aligned}$$

where $F_1(x_1)$ denotes the distribution function of the random variable X_1. Therefore the distribution function of Z_1 is the convolution of the distribution function of X_1 and the probability density function of X_2. The above relation gives an explicit formula to compute the distribution function of the sum of two independent random variables X_1 and X_2. All the above equations can be justified using the Fubini's theorem. We will not go into the details here. Let us consider an example.

Example 6.8. Suppose X_1 and X_2 are independent random variables with the Gamma distributions having the parameters (α_1, λ) and (α_2, λ) respectively. Let us find the probability density function of the sum $Z = X_1 + X_2$ using the convolution formula. The probability density function of X_i is

$$\begin{aligned}
f_i(x_i) &= \frac{\lambda^{\alpha_i} x_i^{\alpha_i - 1} e^{-\lambda x_i}}{\Gamma(\alpha_i)} \quad \text{if} \quad x_i > 0 \\
&= 0 \text{ otherwise}
\end{aligned}$$

for $i = 1, 2$. Hence the probability density function of Z is, for $z > 0$, given by

$$\begin{aligned} f_Z(z) &= \int_{-\infty}^{\infty} f_1(z-u) f_2(u) du \\ &= \int_0^z f_1(z-u) f_2(u) du \\ &= \int_0^z \frac{\lambda^{\alpha_1}(z-u)^{\alpha_1-1} e^{-\lambda(z-u)}}{\Gamma(\alpha_1)} \frac{\lambda^{\alpha_2} u^{\alpha_2-1} e^{-\lambda u}}{\Gamma(\alpha_2)} du \\ &= \frac{\lambda^{\alpha_1+\alpha_2} e^{-\lambda z}}{\Gamma(\alpha_1)\Gamma(\alpha_2)} \{ \int_0^z (z-u)^{\alpha_1-1} u^{\alpha_2-1} du \} \\ &= \frac{\lambda^{\alpha_1+\alpha_2} e^{-\lambda z}}{\Gamma(\alpha_1)\Gamma(\alpha_2)} z^{\alpha_1+\alpha_2-1} \{ \int_0^1 (1-v)^{\alpha_1-1} v^{\alpha_2-1} dv \} \\ &= \frac{\lambda^{\alpha_1+\alpha_2} e^{-\lambda z}}{\Gamma(\alpha_1)\Gamma(\alpha_2)} z^{\alpha_1+\alpha_2-1} B(\alpha_1, \alpha_2) \\ &= \frac{\lambda^{\alpha_1+\alpha_2} e^{-\lambda z}}{\Gamma(\alpha_1+\alpha_2)} z^{\alpha_1+\alpha_2-1} \end{aligned}$$

by applying the transformation $v = \frac{u}{z}$ at one of the intermediate steps. The last equality follows from the properties of the Beta function and the Gamma function discussed earlier. It is obvious that $f_Z(z) = 0$ for $z \leq 0$. Hence the convolution of the Gamma distribution with parameters (α_1, λ) and (α_2, λ) is the Gamma distribution with parameters $(\alpha_1 + \alpha_2, \lambda)$.

Moment generating function approach

We now consider a third method for determining the distribution function of the function of two random variables. This is by the method of moment generating functions. This is specially useful for finding the distribution functions of the sums or linear combinations of independent random variables. We now illustrate this method through an example.

Example 6.9. Suppose X_1 and X_2 are independent random variables with distributions $N(\mu_1, \sigma_1^2)$ and $N(\mu_2, \sigma_2^2)$ respectively. Define $Z = X_1 + X_2$. Then the moment generating function of Z is

$$\begin{aligned} M_Z(t) &= E[e^{tZ}] \\ &= E[e^{t(X_1+X_2)}] \\ &= E[e^{tX_1}]E[e^{tX_2}] \\ &= M_{X_1}(t) M_{X_2}(t) \end{aligned}$$

from the fact that e^{tX_1} and e^{tX_2} are independent random variables when X_1 and X_2 are independent. We have proved earlier that

$$M_{X_i}(t) = \exp\{\mu_i t + \frac{1}{2}t^2\sigma_i^2\}$$

for $i = 1, 2$. Hence

$$M_Z(t) = \exp\{(\mu_1 + \mu_2)t + \frac{1}{2}t^2(\sigma_1^2 + \sigma_2^2)\}$$

for all $-\infty < t < \infty$. But this function is the moment generating function of the normal distribution with mean $\mu_1 + \mu_2$ and variance $\sigma_1^2 + \sigma_2^2$. Here the moment generating function is finite over the whole real line. From the Theorem 4.1 discussed in the Chapter 4, it follows that the random variable $Z = X_1 + X_2$ has the distribution $N(\mu_1 + \mu_2, \sigma_1^2 + \sigma_2^2)$.

6.3 Functions of Multivariate Random Vectors

Suppose we have n random variables $X_1, \ldots, X_n$ not necessarily independent and we are interested in finding the distribution of a function of the random vector $\boldsymbol{Z} = (Z_1, \ldots, Z_r)$ where $Z_i = g_i(X_1, \ldots, X_n), i = 1, \ldots, r$. The methods described in the previous section can be extended to the general case but we will not go into the detailed description or extension of these methods. We will illustrate by a few examples.

Example 6.10 : Suppose $X_1, \ldots, X_n$ is a random sample of size n , that is $X_1, \ldots, X_n$ are independent and identically distributed (i.i.d.) random variables. Let us assume that the random variable X_1 has a distribution function $F(x)$. Define

$$Z_1 = \min(X_1, \ldots, X_n),$$

and

$$Z_n = \max(X_1, \ldots, X_n).$$

Let us find the joint distribution function $F_{Z_1,Z_n}(z_1, z_n)$ of (Z_1, Z_n). It is obvious that $z_1 \leq z_n$. Then, for any $-\infty < z_1 \leq z_n < \infty$,

$$\begin{aligned}
F_{Z_1,Z_n}(z_1,z_n) &= P(Z_1 \le z_1, Z_n \le z_n) \\
&= P(Z_n \le z_n) - P(Z_1 > z_1, Z_n \le z_n) \\
&= P(Z_n \le z_n) - P(z_1 < Z_1 \le Z_n \le z_n) \\
&= P(Z_n \le z_n) - P(z_1 < X_i \le z_n \text{ for } 1 \le i \le n) \\
&= P(X_i \le z_n \text{ for } 1 \le i \le n) - \Pi_{i=1}^n P(z_1 < X_i \le z_n) \\
&= \Pi_{i=1}^n P(X_i \le z_n) - \Pi_{i=1}^n P(z_1 < X_i \le z_n) \\
&= [F(z_n)]^n - [F(z_n) - F(z_1)]^n.
\end{aligned}$$

Suppose the function $F(x)$ is differentiable with the derivative $f(x)$. In other words, the random variables $X_1, \ldots, X_n$ are i.i.d. with the probability density function $f(x)$. Then the function $F_{Z_1,Z_n}(z_1,z_n)$ is differentiable from the above calculations and the corresponding bivariate probability density function $f_{Z_1,Z_n}(z_1,z_n)$ of (Z_1, Z_n) can be obtained from the relation

$$f_{Z_1,Z_n}(z_1,z_n) = \frac{\partial^2 F_{Z_1,Z_n}(z_1,z_n)}{\partial z_1 \partial z_n}.$$

Check that

$$\begin{aligned}
f_{Z_1,Z_n}(z_1,z_n) &= n(n-1)[F(z_n) - F(z_1)]^{n-2} f(z_1) f(z_n) \\
&\qquad \text{if } -\infty < z_1 < z_n < \infty \\
&= 0 \text{ otherwise.}
\end{aligned}$$

The quantity $Z_n - Z_1$ is called the *sample range* of the random sample $X_1, \ldots, X_n$. It is the difference between the largest and the smallest observations and it is a measure of the spread of the sample. Let us now find the distribution function of the range $W_1 = Z_n - Z_1$. Define $W_2 = Z_1$. Then $Z_1 = W_2$ and $Z_n = W_2 + W_1$ and the transformation $(z_1, z_n) \to (w_1, w_2)$ is one-to -one (check). Further more the Jacobian of the transformation is equal to -1. Hence the joint probability density function of (W_1, W_2) is given by

$$\psi(w_1, w_2) = f_{Z_1,Z_n}(w_2, w_2 + w_1).$$

Hence

$$\begin{aligned}
\psi(w_1, w_2) &= n(n-1)[F(w_2 + w_1) - F(w_2)]^{n-2} f(w_2) f(w_2 + w_1) \\
&\qquad \text{if } 0 < w_1 < \infty, -\infty < w_2 < \infty \\
&= 0 \text{ otherwise.}
\end{aligned}$$

The marginal probability density function of W_1 is given by

$$\psi_{W_1}(w_1) = \int_{-\infty}^{\infty} \psi(w_1, w_2)\ dw_2$$

and hence

$$\begin{aligned}\psi_{W_1}(w_1) &= n(n-1) \int_{-\infty}^{\infty} [F(w_2 + w_1) - F(w_2)]^{n-2} f(w_2) f(w_2 + w_1)\ dw_2 \\ &\quad \text{if } 0 < w_1 < \infty \\ &= 0 \text{ otherwise.}\end{aligned}$$

Let us now consider a special case of the above problem when the random variables $X_1, \ldots, X_n$ are i.i.d. with uniform distribution on the interval $[0, 1]$. Then the distribution funotion of X_1 is

$$\begin{aligned}F(x) &= 0 \text{ for } x < 0 \\ &= x \text{ for } 0 \leq x \leq 1 \\ &= 1 \text{ for } x > 1\end{aligned}$$

and the probability density function of X_1 is

$$\begin{aligned}f(x) &= 0 \text{ for } x < 0 \\ &= 1 \text{ for } 0 \leq x \leq 1 \\ &= 0 \text{ for } x > 1.\end{aligned}$$

In this case, the sample range W_1 has the probability density function

$$\begin{aligned}\psi_{W_1}(w_1) &= n(n-1) \int_{0}^{1-w_1} w_1^{n-2}\ dw_2 = n(n-1) w_1^{n-2}(1 - w_1) \\ &\quad \text{if } 0 < w_1 < 1 \\ &= 0 \text{ otherwise.}\end{aligned}$$

6.4 Sampling Distributions

We will now discuss some important standard probability distributions which are the distributions of some functions of independent standard normal random variables. Such distributions have applications in statistical inference which we will discuss later in this book.

Chi-square distribution

Suppose that a random variable X has the standard normal distribution. Let $Z = X^2$. Then, for any $z \geq 0$,

$$\begin{aligned} F_Z(z) &= P(Z \leq z) \\ &= P(X^2 \leq z) \\ &= P(-\sqrt{z} \leq X \leq \sqrt{z}) \\ &= P(X \leq \sqrt{z}) - P(X < -\sqrt{z}) \\ &= \Phi(\sqrt{z}) - \Phi(-\sqrt{z}) \end{aligned}$$

where $\Phi(x)$ denotes the standard normal distribution function. It is obvious that

$$F_Z(z) = 0 \text{ for } z < 0.$$

Differentiating $F_Z(z)$ with respect to z, we have the probability density function of Z given by

$$\begin{aligned} f_Z(z) &= 0 \text{ for } z < 0 \\ &= \frac{1}{2} z^{-1/2} \phi(\sqrt{z}) + \frac{1}{2} z^{-1/2} \phi(-\sqrt{z}) \text{ for } z \geq 0 \end{aligned}$$

where

$$\phi(u) = \frac{1}{\sqrt{2\pi}} e^{-\frac{u^2}{2}}, \quad -\infty < u < \infty.$$

Hence

$$\begin{aligned} f_Z(z) &= \frac{1}{\sqrt{2\pi}} e^{-\frac{z}{2}} z^{-1/2}, \quad 0 < z < \infty \\ &= 0 \text{ otherwise.} \end{aligned}$$

Another equivalent form of this probability density function is

$$\begin{aligned} f_Z(z) &= \frac{1}{2^{1/2}\Gamma(1/2)} e^{-\frac{z}{2}} z^{(1/2)-1}, \quad 0 < z < \infty \\ &= 0 \text{ otherwise.} \end{aligned}$$

A distribution with the above probability density function is called the *Chi-square distribution* with one degree of freedom and it is denoted by χ_1^2. We have proved that if a random variable X has the standard normal distribution, then the random variable X^2 has the Chi-square distribution with one degree of freedom. Let us now compute the m.g.f. of the random variable $Z = X^2$. By the definition of a m.g.f.,

$$
\begin{aligned}
M_Z(t) &= E[e^{tZ}] \\
&= \int_0^\infty e^{tz} \frac{1}{2^{1/2}\Gamma(1/2)} e^{-\frac{z}{2}} z^{(1/2)-1} dz \\
&= \int_0^\infty \frac{1}{2^{1/2}\Gamma(1/2)} e^{-\frac{z}{2}(1-2t)} z^{(1/2)-1} dz \\
&= \int_0^\infty \frac{1}{2^{1/2}\Gamma(1/2)} e^{-\frac{u}{2}} u^{-(1/2)} \frac{1}{(1-2t)^{1/2}} du
\end{aligned}
$$

provided $t < \frac{1}{2}$. The last equality is obtained by using the transformation $z(1-2t) = u$. Note that the integrals defined above are finite only when $t < \frac{1}{2}$ and hence the m.g.f. exists only on the interval $(-\infty, \frac{1}{2})$ in this case. But

$$\int_0^\infty u^{-1/2} e^{-u/2} du = 2^{1/2}\Gamma(1/2)$$

from the properties of the Gamma function discussed earlier or equivalently from the observation

$$\int_{-\infty}^\infty f_Z(z) = 1.$$

Hence

$$M_Z(t) = (1-2t)^{-1/2} \text{ for } t < \frac{1}{2}.$$

Let us now suppose that $X_1, \ldots, X_n$ is a random sample from the standard normal distribution. Consider the function

$$Z_n = X_1^2 + \cdots + X_n^2.$$

The random variable Z_n is the sum of squares of n i.i.d. standard normal random variables. Let us find the m.g.f. of Z_n. Note that

$$
\begin{aligned}
M_{Z_n}(t) &= E[e^{tZ_n}] \\
&= E[\exp(t\{X_1^2 + \cdots + X_n^2\})] \\
&= \Pi_{i=i}^n E[\exp(tX_i^2)]
\end{aligned}
$$

by the independence of the random variables $X_i, 1 \le i \le n$. Since the random variables $X_i^2, 1 \le i \le n$ are also identically distributed, it follows from the earlier computations that

$$
\begin{aligned}
M_{Z_n}(t) &= \Pi_{i=1}^n (1-2t)^{-1/2} \\
&= (1-2t)^{-n/2}
\end{aligned}
$$

for $t < \frac{1}{2}$. Check that the above function is the m.g.f. of the distribution corresponding to the density function

$$f_Z(z) = \frac{1}{2^{n/2}\Gamma(n/2)} e^{-\frac{z}{2}} z^{(n/2)-1}, \quad 0 < z < \infty$$
$$= 0 \text{ otherwise.}$$

The distribution corresponding to the above probability density function is called the *Chi-square distribution with n degrees of freedom.* An application of the Theorem 4.1 in the Chapter 4 to the m.g.f.'s stated earlier shows that the random variable

$$Z_n = X_1^2 + \cdots + X_n^2$$

has the Chi-square distribution with n degrees of freedom. This distribution is usually denoted by χ_n^2.

Let us now calculate the mean and variance of the Chi-square distribution with n degrees of freedom. Since we can represent the Chi-square distribution with n degrees of freedom as the distribution of the random variable Z_n which is the sum of squares of n i.i.d. standard normal random variables $X_i, 1 \leq i \leq n$, it follows that

$$E[Z_n] = E(X_1^2) + \cdots + E(X_n^2)$$

and

$$Var(Z_n) = Var(X_1^2) + \cdots + Var(X_n^2).$$

But $E(X_1^2) = 1$ and $Var(X_1^2) = E(X_1^4) - [E(X_1^2)]^2 = 2$ for the standard normal random variable X_1. Therefore

$$E(Z_n) = n \text{ and } Var(Z_n) = 2n.$$

Suppose $Y_i, 1 \leq i \leq n$ are independent random variables such that Y_i has the $N(\mu_i, \sigma_i^2)$ as its distribution. Let

$$X_i = \frac{Y_i - \mu_i}{\sigma_i}, 1 \leq i \leq n.$$

Then $X_i, 1 \leq i \leq n$ are independent standard normal random variables and

$$\sum_{i=1}^{n} \frac{(Y_i - \mu_i)^2}{\sigma_i^2} = \sum_{i=1}^{n} X_i^2$$

has the Chi-square distribution with n degrees of freedom. In particular, if $\mu_i = \mu$ and $\sigma_i = \sigma$ for $1 \leq i \leq n$, then the random variables $Y_i, 1 \leq i \leq n$ form a random sample from the $N(\mu, \sigma^2)$, and the random variable

$$Z_n = \sum_{i=1}^{n} \frac{(Y_i - \mu)^2}{\sigma^2}$$

has the Chi-square distribution with n degrees of freedom. We will not go into the reasons why the parameter n is termed degrees of freedom. From the expression for the probability density function of a Chi-square distribution with n degrees of freedom, it is easy to see that it is a special case of the Gamma density function with parameters $\alpha = \frac{n}{2}$ and $\lambda = \frac{1}{2}$. The graphs given in Figure 6.3 show the shape of the Chi-square probability density functions for $n = 1, 2, 3$ and for $n = 4$.

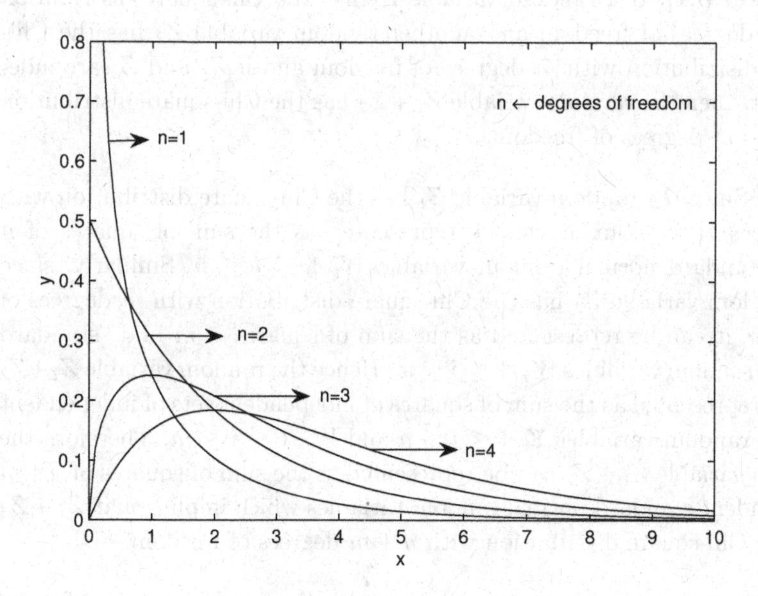

Fig. 6.3 Chi-Square Probability Density Functions

Exact computation of the probabilities of different events under a Chi-square distribution for different values of the degrees of freedom n is difficult. However the tables giving the probabilities under the Chi-square distribution are available (for instance see, *Formulae and Tables for Statistical Work* by C.R.Rao, A. Matthai, S.K. Mitra and K.G. Ramamurti

(1975), Statistical Publishing Society, Calcutta). With the availability of personal computers now, these probabilities can be found out from statistical software packages. Let us compute the $P[3.25 \leq Z \leq 20.5]$ when Z has the Chi-square distribution with 10 degrees of freedom. It is easy to check that

$$\begin{aligned} P(3.25 \leq Z \leq 20.5) &= P(Z \leq 20.5) - P(Z < 3.25) \\ &= 0.975 - 0.025 = 0.95 \end{aligned}$$

from Table C.3.

An important property of the Chi-square distribution is the additivity property. We now discuss this result.

Theorem 6.1. If a random variable Z_1 has the Chi-square distribution with n degrees of freedom and another random variable Z_2 has the Chi-square distribution with m degrees of freedom and if Z_1 and Z_2 are independent, then the random variable $Z_1 + Z_2$ has the Chi-square distribution with $n + m$ degrees of freedom.

Proof. Since the random variable Z_1 has the Chi-square distribution with n degrees of freedom, it can be represented as the sum of squares of n i.i.d. standard normal random variables $Y_i, 1 \leq i \leq n$. Similarly, since the random variable Z_2 has the Chi-square-distribution with m degrees of freedom, it can be represented as the sum of squares of m i.i.d. standard normal random variables $W_j, 1 \leq j \leq m$. Hence the random variable $Z_1 + Z_2$ can be represented as the sum of squares of independent sets of independent normal random variables $Y_i, 1 \leq i \leq n$ and $W_j, 1 \leq j \leq m$. Therefore the random variable $Z_1 + Z_2$ can be represented as the sum of squares of $n + m$ independent standard normal random variables which implies that $Z_1 + Z_2$ has the Chi-square distribution with $n + m$ degrees of freedom.

Another important property dealing with the random samples from a normal distribution is the following. We omit the proof.

Theorem 6.2. Suppose $Y_i, 1 \leq i \leq n$ is a random sample from $N(\mu, \sigma^2)$ where $n > 1$. Let

$$\bar{Y} = \frac{1}{n} \sum_{i=1}^{n} Y_i$$

and

$$S^2 = \frac{1}{n-1}\sum_{i=1}^{n}(Y_i - \bar{Y})^2.$$

Here $\bar{Y}$ is the sample mean and S^2 is the sample variance. Then

(i) $\bar{Y}$ and S^2 are independent random variables;
(ii) $\bar{Y}$ has $N(\mu, \frac{\sigma^2}{n})$ as its probability distribution and
(iii) $\frac{(n-1)S^2}{\sigma^2}$ has the Chi-square distribution with $(n-1)$ degrees of freedom.

The theorem stated above has a large number of applications in statistical inference. We remark that

$$\frac{(n-1)S^2}{\sigma^2} = \sum_{i=1}^{n}\frac{(Y_i - \bar{Y})^2}{\sigma^2}.$$

We have seen earlier that

$$\sum_{i=1}^{n}\frac{(Y_i - \mu)^2}{\sigma^2}$$

has the Chi-square distribution with n degrees of freedom. Applying the previous theorem, observe that, if the population mean μ is replaced by the sample mean $\bar{Y}$ in the expression

$$\sum_{i=1}^{n}\frac{(Y_i - \mu)^2}{\sigma^2},$$

then we can interpret that one degree of freedom is lost and

$$\sum_{i=1}^{n}\frac{(Y_i - \bar{Y})^2}{\sigma^2}$$

has the Chi-square distribution with $(n-1)$ degrees of freedom. A formal proof of this fact can be given using Theorem 6.3 stated later in this chapter. We will not discuss the proof of Theorem 6.2.

t-distribution

Consider two *independent random variables* Y and W where Y has the standard normal distribution and W has the Chi-square distribution with n degrees of freedom. Define

$$U = \frac{Y}{\sqrt{n^{-1}W}}$$

The distribution of the random variable U is called the *t-distribution* with n degrees of freedom. This distribution is also known as the *Student's t-distribution.* The exact derivation of the probability density function of U is beyond the scope of this book. It can be shown that the probability density function of U is given by

$$f(u) = \frac{\Gamma(\frac{n+1}{2})}{\sqrt{(n\pi)}\Gamma(\frac{n}{2})}(1+\frac{u^2}{n})^{-\frac{(n+1)}{2}}, -\infty < u < \infty.$$

It is obvious from the form of the probability density function that it s symmetric about zero. It is bell-shaped and, for large n, it is close to the probability density function for a normal distribution with mean zero. The mean of t-distribution does not exist for $n = 1$. In fact, for $n = 1$, this distribution is the Cauchy distribution. For $n > 1$, the mean of the distribution does exist. Furthermore , for $n > 1, E(|U|^k) < \infty$ for $k < n$ and $E(|U|^k) = \infty$ for $k \geq n$. In other words, the t -distribution with n degrees of freedom , with $n > 1$, has moments up to order $n - 1$ but has no moments of higher order. In particular, it follows that the m.g.f. does not exist for the t-distribution. Typical graphs of the probability density functions of t-distribution are given in Figure 6.4. Probabilities of events under a t-distribution can be computed using the tables of probabilities. See Table C.2. We will discuss some applications of the t-distribution for comparing means of two normal distributions and to other problems of statistical inference in the following chapters.

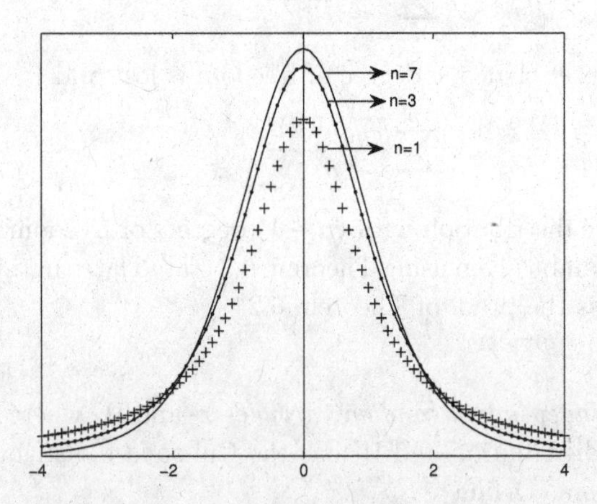

Fig. 6.4 Probability Density Functions of t-Distributions

Let us consider a random sample $X_1, \ldots, X_n$, with $n > 1$, from a normal distribution with mean μ and variance σ^2 and denote the sample mean by $\bar{X}$ and the sample variance by S^2 as defined earlier. We have proved earlier in this chapter that

$$\bar{X} \text{ has } N(\mu, \frac{\sigma^2}{n}).$$

Let us standardize $\bar{X}$ and define

$$Y = \frac{\bar{X} - \mu}{\sigma/\sqrt{n}}.$$

Then the random variable Y has the standard normal distribution. We have seen above that $\bar{X}$ and S^2 are independent and $\frac{(n-1)S^2}{\sigma^2}$ has the Chi-square distribution with $(n-1)$ degrees of freedom. Let

$$W = \frac{(n-1)S^2}{\sigma^2}.$$

Define

$$U = \frac{Y}{\sqrt{\frac{W}{n-1}}}.$$

Then the random variable U has the t-distribution with $(n-1)$ degrees of freedom. Check that

$$U = \frac{\bar{X} - \mu}{S/\sqrt{n}} = \frac{\sqrt{n}(\bar{X} - \mu)}{S}.$$

If μ and σ are known, then we can compute the random variable Y defined above given the sample $X_1, \cdots, X_n$, and the distribution of the random variable is $N(0, 1)$. If σ is not known, then we can estimate σ by S and replace σ by S in the functional form of Y which reduces to the random variable U. The distribution of the random variable U is the t - distribution with $n - 1$ degrees of freedom. This process is known as the *Studentization.* The term is coined after W.S. Gosset who discovered the t-distribution and published it under the name "Student." We will see that S^2 is a *good* estimator of σ^2, in the sense that the higher the sample size is the closer S^2 is to σ^2, with a large probability.

We will discuss more about this later in this book.

F-distribution

The F-distribution is another sampling distribution which plays a major role in statistical inference especially in comparing the means of several normal populations and in comparing variances of normal populations in problems of testing statistical hypotheses. We will discuss applications of this distribution later in this book. Let Y and Z be two *independent* random variables such that Y has the Chi-square distribution with m degrees of freedom and Z has the Chi-square distribution with n degrees of freedom. Define

$$W = \frac{Y/m}{Z/n} = \frac{nY}{mZ}.$$

The distribution of the random variable W is called the *F-distribution* with m and n degrees of freedom. It can be shown that the probability density function of W is

$$\begin{aligned} f(w) &= \frac{\Gamma(\frac{m+n}{2})m^{m/2}n^{n/2}}{\Gamma(\frac{m}{2})\Gamma(\frac{n}{2})} \frac{w^{(m/2)-1}}{(mw+n)^{\frac{m+n}{2}}}, \quad \text{if } w > 0 \\ &= 0 \text{ otherwise.} \end{aligned}$$

Note that the order (m, n) is important in defining the F-distribution. The F-distribution with m and n degrees of freedom is different from the F-distribution with n and m degrees of freedom unless $m = n$. Further more if a random variable W has the F-distribution with m and n degrees of freedom, its reciprocal $\frac{1}{W}$ has the F-distribution with n and m degrees of freedom (check?).

(i) Suppose $X_1, \ldots, X_m$ is a random sample of size m from $N(\mu_1, \sigma_1^2)$ and $Y_1, \ldots, Y_n$ is another independent random sample of size n from $N(\mu_2, \sigma_2^2)$ where μ_1, μ_2, σ_1^2 and σ_2^2 are known. Then

$$\sum_{i=1}^{m} \frac{(X_i - \mu_1)^2}{\sigma_1^2} \text{ is } \chi_m^2$$

and

$$\sum_{j=1}^{n} \frac{(Y_j - \mu_2)^2}{\sigma_2^2} \text{ is } \chi_n^2.$$

Hence

$$W_1 = \frac{\frac{1}{m}\sum_{i=1}^{m} \frac{(X_i - \mu_1)^2}{\sigma_1^2}}{\frac{1}{n}\sum_{j=1}^{n} \frac{(Y_j - \mu_2)^2}{\sigma_2^2}}$$

has the F-distribution with m and n degrees of freedom.

(ii) Let us now suppose that $X_1, \ldots, X_m$ is a random sample of size m from $N(\mu_1, \sigma^2)$ and $Y_1, \ldots, Y_n$ is another independent random sample of size n from $N(\mu_2, \sigma^2)$ where μ_1 and μ_2 are known and the variance σ^2 is possibly unknown. Note that both the samples are from normal populations with the *same variance.* Define

$$W_2 = \frac{\frac{1}{m}\sum_{i=1}^{m}\frac{(X_i-\mu_1)^2}{\sigma^2}}{\frac{1}{n}\sum_{j=1}^{n}\frac{(Y_j-\mu_2)^2}{\sigma^2}}.$$

Then it follows from the above remarks that the random variable W_2 has the F-distribution with m and n degrees of freedom. However note that W_2 does not depend on the value of σ^2 and it can be computed even if the common variance σ^2 of the two normal distributions is unknown. In fact

$$W_2 = \frac{n\sum_{i=1}^{m}(X_i-\mu_1)^2}{m\sum_{j=1}^{n}(Y_j-\mu_2)^2}.$$

In other words the distribution of W_2 is F-distribution with m and n degrees of freedom irrespective of the fact whether the common variance σ^2 of the two normal populations is known or unknown.

(iii) Suppose $X_1, \ldots, X_m$ is a random sample of size m from $N(\mu_1, \sigma_1^2)$ and $Y_1, \ldots, Y_n$ is another independent random sample of size n from $N(\mu_2, \sigma_2^2)$ where $\mu_1, \mu_2, \sigma_1^2, \sigma_2^2$ are all unknown. It is obvious that the random variable W_1 cannot be computed from the observed samples. Let us consider the sample variances

$$S_X^2 = \frac{1}{m-1}\sum_{i=1}^{m}(X_i - \bar{X})^2$$

and

$$S_Y^2 = \frac{1}{n-1}\sum_{j=1}^{n}(Y_j - \bar{Y})^2$$

of the X-sample and Y-sample respectively. From the earlier discussions in this chapter,

$$\frac{(m-1)S_X^2}{\sigma_1^2} \text{ has } \chi^2_{m-1}$$

and

$$\frac{(n-1)S_Y^2}{\sigma_2^2} \text{ has } \chi^2_{n-1}$$

as their distributions. Furthermore these random variables are independent as they are functions of independent samples.

Hence

$$W_3 = \frac{\frac{1}{m-1}\frac{(m-1)S_X^2}{\sigma_1^2}}{\frac{1}{n-1}\frac{(n-1)S_Y^2}{\sigma_2^2}}$$

$$= \frac{\sigma_2^2 S_X^2}{\sigma_1^2 S_Y^2}$$

has the F-distribution with $(m-1)$ and $(n-1)$ degrees of freedom. The expression for the random variable W_3 involves σ_1^2 and σ_2^2. However, if $\sigma_1^2 = \sigma_2^2$, then the random variable W_3 reduces to

$$W = \frac{S_X^2}{S_Y^2}$$

and it can be computed even if the common value of the variance is unknown. Furthermore, in such a case, the random variable W has the F-distribution with $(m-1)$ and $(n-1)$ degrees of freedom.

Before we conclude discussion of this chapter, we will study a result connected with the distributions of quadratic forms involving multivariate normal distributions known as the Cochran's theorem which will have applications in the Chapter 10 on linear regression and correlation.

Distribution theory for quadratic forms

A homogeneous polynomial of degree two in the variables $\mathbf{x} = (x_1, \ldots, x_n)$ is called a quadratic form in $x_i, 1 \leq i \leq n$. Any such quadratic form Q can be represented in the form $Q = \mathbf{x}'\mathbf{A}\mathbf{x}$ where $\mathbf{A}$ is a symmetric matrix. We will consider only those quadratic forms where the variables $x_i, 1 \leq i \leq n$ are real as well as the coefficients of the homogeneous polynomial are real. An example of such a polynomial is

$$s^2 = n^{-1}\sum_{i=1}^{n}(x_i - \bar{x})^2.$$

The following theorem is due to Cochran. We omit the proof.

Theorem 6.3. Let $X_i, 1 \leq i \leq n$ be independent and identically distributed random variables with the normal distribution with mean zero and variance σ^2. Suppose the sum of squares $\sum_{i=1}^{n} X_i^2$ can be written as the sum of k quadratic forms $Q_j, 1 \leq j \leq k$ in the variables $X_i, 1 \leq i \leq n$ with the matrix $\mathbf{A}_j$ as the matrix of the quadratic form Q_j for $1 \leq j \leq k$. Suppose the rank of the matrix $\mathbf{A}_j$ is n_j. Let $\mathbf{X}' = (X_1, \ldots, X_n)$. Then the random variables $Q_j = \mathbf{X}'\mathbf{A}_j\mathbf{X}, 1 \leq j \leq k$ are independent and the random variable Q_j/σ^2 has the Chi-square distribution with n_j degrees of freedom for $1 \leq j \leq k$ if and only if $n = \sum_{j=1}^{k} n_j$.

We omit the proof of this result. There are some other properties of quadratic forms involving i.i.d. normal random variables which are of interest. We shall now state these properties. Let $X_i, 1 \leq i \leq n$ be i.i.d. N(0, σ^2) random variables.

(i) Let $Q = \mathbf{X}'\mathbf{A}\mathbf{X}$. Suppose the rank of the matrix $\mathbf{A}$ is r. Then the random variable Q/σ^2 has the Chi-square distribution with r degrees of freedom if and only if $\mathbf{A}^2 = \mathbf{A}$. Such a matrix $\mathbf{A}$ is said to be *idempotent*.

(ii) Let $Q_j = \mathbf{X}'\mathbf{A}_j\mathbf{X}, j = 1, 2$ be quadratic forms in $\mathbf{X}$ as defined above. Then the random variables $Q_j, j = 1, 2$ are independent if and only if $\mathbf{A}_1\mathbf{A}_2 = \mathbf{0}$.

(iii) Let $Q = \sum_{j=1}^{k} Q_j$ where Q and $Q_j, 1 \leq j \leq k$ are quadratic forms in $\mathbf{X}$ as defined above. Suppose that the random variable Q/σ^2 has the Chi-square distribution with r degrees of freedom and the random variable Q_j/σ^2 has the Chi-square distribution with r_j degrees of freedom for $1 \leq j \leq k-1$. Further suppose that Q_k is nonnegative. Then the random variables $Q_i, 1 \leq i \leq k$ are independent and Q_k/σ^2 has the Chi-square distribution with $r_k = r - \sum_{j=1}^{k-1} r_j$ degrees of freedom.

Suppose $\mathbf{X}$ is a random vector having a multivariate normal distribution. Distributional properties of the quadratic form $Q = \mathbf{X}'\mathbf{A}\mathbf{X}$, are given, for instance, in Montgomery et al. (2001). We do not discuss these results here.

Remarks. We have provided a table for computing the probabilities under the standard normal distribution in Table C.1, a table for the threshold value t_0 such that $P(t \leq t_0) = p$ for the t-distribution with degrees of freedom $n = 1, \ldots, 30$ and $p = 0.90, 0.95, 0.975, 0.99$ and $p = 0.995$ in Table C.2, a table for values u such that $P(\chi^2 \leq u) = p$ for the χ-square distribution with degrees of freedom $n = 1, \ldots, 30$ and $p = 0.005, 0.01, 0.025, 0.975, 0.99$ and $p = 0.995$ in Table C.3 and a table for the threshold value u for the F-distribution with n_1 and n_2 degrees of freedom such that $P(F \leq u) = p$ for $n_1 = 1, \ldots, 15$ and $n_2 = 1, \ldots, 15$ for $p = 0.95$ and $p = 0.975$ in Table C.4.

6.5 Exercises

6.1 Suppose that the bivariate random vector (X, Y) has the joint probability density function

$$\begin{aligned} f(x, y) &= 2(x+y) \text{ for } 0 \leq x \leq y \leq 1 \\ &= 0 \text{ otherwise.} \end{aligned}$$

Determine the probability density function of $Z = X + Y$.

6.2 Suppose that the bivariate random vector (X, Y) has the joint probability density function

$$\begin{aligned} f(x, y) &= x + y \text{ for } 0 \leq x, y \leq 1 \\ &= 0 \text{ otherwise.} \end{aligned}$$

Determine the probability density function of $Z = X + Y$.

6.3 Suppose that the bivariate random vector (X, Y) has the uniform probability density function on the unit square $[0,1] \times [0,1]$. Find the probability density function of $Z = \frac{X}{Y}$.

6.4 Suppose that X_1 and X_2 are independent random variables with a common probability density function $f(x)$. Let $F(x)$ be the corresponding distribution function. Find the probability density function of $Z = \min(X_1, X_2)$.

6.5 Suppose that X_1 and X_2 are independent random variables with the Gamma probability density functions $f_i(x), i = 1, 2$ given by

$$\begin{aligned} f_i(x_i) &= \frac{x_i^{\alpha_i - 1} e^{-x_i}}{\Gamma(\alpha_i)} \text{ if } x_i > 0 \\ &= 0 \text{ otherwise.} \end{aligned}$$

Let $Z_1 = X_1 + X_2$ and $Z_2 = \frac{X_1}{X_1 + X_2}$. Show that Z_1 and Z_2 are independent random variables. Find the distribution functions of Z_1 and Z_2.

6.6 (**Box-Mueller transformation**) Let X_1 and X_2 be independent random variables uniformly distributed on the interval $[0,1]$. Define

$$\begin{aligned} Z_1 &= (-2 \log X_1)^{1/2} \cos(2\pi X_2), \\ Z_2 &= (-2 \log X_1)^{1/2} \sin(2\pi X_2). \end{aligned}$$

Show that the random variables Z_1 and Z_2 are independent standard normal random variables. This transformation gives a method for generating observations from a standard normal distribution from those of a standard uniform distribution.

6.7 Let X_1 and X_2 be independent standard normal random variables. Define $Z_1 = \frac{X_1}{X_2}$. Show that the random variable Z_1 has the standard Cauchy probability density function defined by

$$f_{Z_1}(z_1) = \frac{1}{\pi(1+z_1^2)}, \quad -\infty < z_1 < \infty.$$

(Hint: Define $Z_2 = X_2$ and apply the transformation approach.)

6.8 Suppose that the bivariate random vector (X, Y) has the joint probability density function

$$f(x,y) = 4xy \text{ for } 0 \le x, y \le 1$$
$$= 0 \text{ otherwise.}$$

Define $Z_1 = \frac{X}{Y}$ and $Z_2 = XY$. Determine the joint probability density function of (Z_1, Z_2).

6.9 Suppose that $X_1, \ldots, X_n$ are independent and identically distributed (i.i.d.) random variables with distribution $N(\mu, \sigma^2)$. Define

$$\bar{X} = \frac{1}{n}\sum_{i=1}^{n} X_i.$$

The random variable $\bar{X}$ is called the *sample mean.* Extending the moment generating function approach for more than two random variables, show that the random variable $\bar{X}$ has the distribution $N(\mu, \frac{\sigma^2}{n})$.

6.10 Suppose that $X_1, \ldots, X_n$ are independent random variables and X_i has the distribution $N(\mu_i, \sigma_i^2)$ for $1 \le i \le n$. Find the distribution function of $\sum_{i=1}^{n} c_i X_i$ when c_i are constants not all zero.

6.11 Show that if a random variable Z_1 has the Chi-square distribution with n degrees of freedom and another independent random variable Z_2 has the Chi-square distribution with m degrees of freedom, then the random variable $Z_1 + Z_2$ has the Chi-square distribution with $n + m$ degrees of freedom by using the moment generating function approach.

6.12 Let S^2 be the sample variance corresponding to a random sample $X_i, 1 \le i \le n$ of size $n = 25$ from the normal distribution with mean μ and variance σ^2. Show that

$$E(S^2) = \sigma^2 \text{ and } Var(S^2) = \frac{2\sigma^4}{n-1}.$$

Suppose that $\mu = 3$ and $\sigma^2 = 100$. Find

$$P(0 < \bar{X} < 6, 55.2 < S^2 < 145.6).$$

6.13 Suppose that $X_1, \ldots, X_n$ is a random sample from the exponential distribution with probability density function

$$f(x) = \lambda e^{-\lambda x} \text{ for } x > 0$$
$$= 0 \text{ otherwise.}$$

Show that the random variable $Z = 2\lambda n\bar{X}$ has the Chi-square distribution with $2n$ degrees of freedom.

6.14 Find the mode of the Chi-square distribution with n degrees of freedom for $n > 2$.

6.15 Suppose that $X_1, \ldots, X_n$ are independent random variables and X_i has (absolutely) continuous distribution function F_i. Define

$$Y = -2\sum_{i=1}^{n} \log F_i(X_i).$$

Show that the random variable Y has the Chi-square distribution with $2n$ degrees of freedom. (Hint: Use the fact that the random variable $F_i(X_i)$ has the uniform distribution on the interval $[0, 1]$ for $i = 1, \ldots, n$.)

6.16 If a random variable U has the t-distribution with n degrees of freedom with $n > 2$, show that $Var(U) = n/(n-2)$.

6.17 Suppose a random variable U has the t- distribution with 10 degrees of freedom. Determine $P(|U| > 2.288)$.

6.18 Show that, as $n \to \infty$, the probability density function of the t-distribution with n degrees of freedom converges to the probability density function of the standard normal distribution. (Hint: Use the Stirling's formula: $n! \simeq \sqrt{2\pi n} n^n e^{-n}$ as $n \to \infty$.)

6.19 If a random variable U has the t-distribution with n degrees of freedom, then show that the random variable $Z = U^2$ has the F-distribution with 1 and n degrees of freedom.

6.20 If a random variable W has the F-distribution with m and n degrees of freedom with $n > 4$, show that

$$E(W) = \frac{n}{n-2} \text{ and } Var(W) = \frac{2n^2(m+n-2)}{m(n-2)^2(n-4)}.$$

6.21 If a random variable W has the F-distribution with m and n degrees of freedom, show that the random variable $\frac{1}{W}$ has the F-distribution with n and m degrees of freedom.

6.22 If a random variable W has the F-distribution with 5 and 10 degrees of freedom, find a and b such that $P(W \leq a) = 0.05$ and $P(W \leq b) = 0.95$.

6.23 Suppose that X_1 and X_2 are i.i.d. random variables with the probability density function

$$\begin{aligned} f(x) &= e^{-x} \text{ if } 0 < x < \infty \\ &= 0 \text{ otherwise.} \end{aligned}$$

Show that the random variable $W = \frac{X_1}{X_2}$ has an F-distribution.

6.24 If a random variable W has the F-distribution with m and n degrees of freedom, show that

$$V = \frac{1}{1 + \frac{m}{n} W}$$

has the Beta distribution.

Chapter 7

APPROXIMATIONS TO SOME PROBABILITY DISTRIBUTIONS

7.1 Introduction

We have discussed different methods for obtaining the distribution functions of functions of a random vector in the previous chapter. Even though it is possible to derive the distribution functions explicitly in closed forms in some special cases, this is not true in general. Furthermore the computation of the exact probabilities of some events is difficult and cumbersome at times even when the probability distributions functions are completely known. For instance, it is easy to write down the exact probabilities for the Binomial distribution with parameters $n = 100$ and $p = \frac{1}{50}$. However, computation of the individual probabilities involve factorials for integers of large order which are impossible to handle even with high speed computing facilities. We will now discuss some approximations to probability distributions such as the Binomial by a Poisson or a normal approximation. We will also discuss the *Central limit theorem* which gives the limiting form for the distribution of a sample mean. Central limit theorem essentially states that, whatever be the original distribution from which the random sample is drawn, the sample mean has an approximately normal distribution for large samples (as long as the original distribution has a finite positive variance). The normal and the Poisson distributions can be used for computing approximations to probabilities of certain events when their exact values are not available or difficult to compute even if explicit formulae are available.

7.2 Chebyshev's Inequality

We prove an important result in this section known as the *Chebyshev's inequality* . It gives an upper bound for the probability that a random variable takes values outside any symmetric interval around the mean in terms of the variance and the length of the interval. More precisely, we have the following result.

Theorem 7.1. Suppose X is a random variable with mean μ and finite variance σ^2. Then, for every $\varepsilon > 0$,

$$P(|X-\mu| \geq \varepsilon) \leq \frac{\sigma^2}{\varepsilon^2}.$$

Proof. We prove the result for a random variable X with a probability density function f. Proof in the discrete case is similar. From the definition of $Var(X)$, we have,

$$\sigma^2 = E(X-\mu)^2 = \int_{-\infty}^{\infty} (x-\mu)^2 \; f(x)dx.$$

Then, for every $\varepsilon > 0$,

$$\begin{aligned}
\sigma^2 &= \int_{\{x:|x-\mu|\geq\varepsilon\}} (x-\mu)^2 f(x)dx + \int_{\{x:|x-\mu|<\varepsilon\}} (x-\mu)^2 f(x)dx \\
&= \int_{-\infty}^{\mu-\varepsilon} (x-\mu)^2 f(x)dx + \int_{\mu-\varepsilon}^{\mu+\varepsilon} (x-\mu)^2 f(x)dx + \int_{\mu+\varepsilon}^{\infty} (x-\mu)^2 f(x)dx \\
&\geq \int_{-\infty}^{\mu-\varepsilon} (x-\mu)^2 f(x)dx + \int_{\mu+\varepsilon}^{\infty} (x-\mu)^2 f(x)dx
\end{aligned}$$

since $(x-\mu)^2 f(x) \geq 0$ for all x. Note that if $x \in (-\infty, \mu-\varepsilon]$ or if $x \in [\mu+\varepsilon, \infty)$, then $(x-\mu)^2 \geq \varepsilon^2$ and hence

$$\begin{aligned}
\sigma^2 &\geq \varepsilon^2 \int_{-\infty}^{\mu-\varepsilon} f(x)dx + \varepsilon^2 \int_{\mu+\varepsilon}^{\infty} f(x)dx \\
&= \varepsilon^2 \int_{\{x:|x-\mu|\geq\varepsilon\}} f(x)dx \\
&= \varepsilon^2 P(|X-\mu| \geq \varepsilon).
\end{aligned}$$

Rewriting the above inequality, we have the inequality

$$P(|X-\mu| \geq \varepsilon) \leq \frac{\sigma^2}{\varepsilon^2}$$

for every $\varepsilon > 0$. This inequality also holds when the distribution of the random variable X is neither (absolutely) continuous nor discrete as long as it has finite variance. We will not discuss the proof in the general case here.

Remarks. The above inequality is very general. We need to know nothing about the probability distribution of the random variable X. It could be either a discrete distribution such as the Binomial or the Poisson or a continuous distribution such as the normal, beta or gamma distributions or any other distribution. The only restriction is that the random variable should have finite variance. The upper bound given by the inequality is universal, that is, it is the same for all random variables X with a given mean μ and a given variance σ^2. The price that is paid for such universality is that the upper bound is not sharp in general. If there is more information about the distribution of X, then it might be possible to get a better upper bound than the upper bound given by the Chebyshev's inequality. We will illustrate this observation by the following example.

Example 7.1. Suppose that a random variable X is $N(\mu, \sigma^2)$. Then $E(X) = \mu$ and $Var(X) = \sigma^2$. Let us compute the $P(|X - \mu| \geq 2\sigma)$. Note that

$$P(|X - \mu| \geq 2\sigma) = P(|\frac{X - \mu}{\sigma}| \geq 2)$$
$$= P(|Z| \geq 2)$$

which is equal to 0.0456 since $Z = \frac{X-\mu}{\sigma}$ has the standard normal distribution. This can be checked from the Table C.1 for the standard normal distribution. Direct application of the Chebyshev's inequality leads to the result

$$P(|X - \mu| \geq 2\sigma) \leq \frac{\sigma^2}{4\sigma^2} = 0.25$$

which is substantially large compared to the exact value 0.0456 obtained from the probability distribution of X. Thus we could get a better upper bound in this example by using the exact distribution of X.

Example 7.2. Suppose that X is a random variable taking the values -1 and 1 with $P(X = 1) = P(X = -1) = \frac{1}{2}$. Check that $E(X) = 0$ and $Var(X) = 1$. Hence, by the Chebyshev's inequality, it follows that

$$P(|X - \mu| \geq \sigma) \leq \frac{\sigma^2}{\sigma^2} = 1.$$

On the other hand, direct calculations show that

$$P(|X-\mu| \geq \sigma) = P(|X| \geq 1) = 1.$$

The upper bound obtained by using the Chebyshev's inequality as well as the bound obtained by using the distribution of X are the same in this example.

Example 7.3. Suppose a person makes 100 cheque transactions during a certain period. In balancing the cheque book, suppose he or she rounds off the entries to the nearest integer. Let us find an upper bound to the probability that the total error committed exceeds 5 after 100 transactions. Let X_i denote the roundoff error made for the i-th transaction. Then the total error is $X_1 + \cdots + X_{100}$. We assume that the random variables $X_i, i = 1, \ldots, 100$ are i.i.d. with the uniform distribution on the interval $[-0.5, 0.5]$. We are interested in finding an upper bound for the $P(|S_{100}| > 5)$ where $S_n = X_1 + \cdots + X_n$. It is difficult to compute the exact distribution of S_{100}. However we can use the Chebyshev's inequality to get an upper bound. Check that $E(S_{100}) = 0$ and $Var(S_{100}) = \frac{100}{12}$. Hence

$$P(|S_{100} - E(S_{100})| \geq 5) \leq \frac{Var(S_{100})}{25},$$

that is,

$$P(|S_{100}| \geq 5) \leq \frac{100}{(12)(25)} = \frac{1}{3}.$$

7.3 Weak Law of Large Numbers

We now give an important application of the Chebyshev's inequality. Suppose $X_1, \ldots, X_n$ is a random sample from a distribution with mean μ and finite variance σ^2. Define the sample mean

$$\bar{X}_n = n^{-1} \sum_{i=1}^{n} X_i.$$

Then $E(\bar{X}_n) = \mu$ and $Var(\bar{X}_n) = \frac{\sigma^2}{n}$. Applying the Chebyshev's inequality to the random variable $\bar{X}_n$, we have the inequality

$$P(|\bar{X}_n - \mu| \geq \varepsilon) \leq \frac{\sigma^2}{n\varepsilon^2}$$

for any $\varepsilon > 0$. If $n \to \infty$, then $\frac{\sigma^2}{n\varepsilon^2} \to 0$ and

$$P(|\bar{X}_n - \mu| \geq \varepsilon) \to 0.$$

In other words, the probability that $\bar{X}_n$ differs from μ by more than any given positive number ε becomes smaller and smaller as n increases. An alternate way of stating this result is as follows: for any $\varepsilon > 0$, given any positive number $\delta > 0$, we can choose sufficiently large $N(\varepsilon, \delta) \geq 1$ such that for every $n \geq N(\varepsilon, \delta)$,

$$P(|\bar{X}_n - \mu| \geq \varepsilon) \leq \delta.$$

This result is known as the *Weak law of large numbers* (WLLN). We write that

$$\bar{X}_n \to \mu \text{ as } n \to \infty \text{ in probability}$$

and denote it by

$$\bar{X}_n \xrightarrow{p} \mu \text{ as } n \to \infty.$$

This type of convergence is known as the *convergence in probability.* We now state the theorem.

Theorem 7.2. (Weak Law of Large Numbers) Suppose $X_n, n \geq 1$ is a sequence of i.i.d. random variables with mean μ and finite variance σ^2. Let

$$\bar{X}_n = n^{-1} \sum_{i=1}^{n} X_i.$$

Then

$$P(|\bar{X}_n - \mu| \geq \varepsilon) \to 0 \text{ as } n \to \infty$$

for every $\varepsilon > 0$.

Remarks. The above theorem is true even when the variance is not finite but the mean μ is finite. However this result does not follow as an application of the Chebyshev's inequality. Proof of the WLLN in this general case is beyond the scope of this book. Note that the theorem given above does not indicate that the absolute difference $|\bar{X}_n - \mu|$ is less than given ε for large values of n but specifies that the probability of such an event becomes smaller and smaller as n grows large. There is also a result known as the *Strong law of large numbers* (SLLN) which ensures that, with probability one, the absolute difference $|\bar{X}_n - \mu|$ can be made less than ε for large $n > N$ where N depends on the sample and the value ε. We will not prove this result here.

Example 7.4. Suppose a random experiment has two possible outcomes called a success S and a failure F. Let p be the probability of a success. Suppose the experiment is repeated independently. Let X_i take the value 1 or 0 according as the outcome in the i–th trial of the random experiment is a success or a failure. Then the sequence $X_i, i \geq 1$ is a sequence of i.i.d. random variables with

$$P(X_i = 1) = p \text{ and } P(X_i = 0) = 1 - p, i \geq 1.$$

Check that $E(X_i) = p$ and $Var(X_i) = p(1-p)$ for $i \geq 1$. Let

$$S_n = X_1 + \cdots + X_n.$$

Note that S_n is the number of successes in n trials. Since the mean and the variance of the random variable X_i are finite, it follows by the weak law of large numbers, applied to the sequence $\{X_i, i \geq 1\}$, that

$$P(|\frac{S_n}{n} - p| \geq \varepsilon) \to 0 \text{ as } n \to \infty$$

for every $\varepsilon > 0$. Now, what is $\frac{S_n}{n}$? It is the proportion of successes in n trials. In other words, as the number of trials increases, the proportion of successes tends to stabilize to the probability of a success in a single trial. Of course, one of the basic assumptions behind this conclusion is that the random experiment can be repeated independently. We have now justified the important fact which is basic to the frequency interpretation of the concept of probability discussed in the Chapter 1.

7.4 Poisson Approximation to Binomial Distribution

Suppose X is a random variable with the Binomial distribution with the number of trials n and the probability of success p. We know that

$$P(X = r) = n_r^C p^r (1-p)^{n-r}, r = 0, 1, \ldots, n.$$

Exact computation of these probabilities is complicated when n is large due to the factorial terms involved in the above expression which in turn increase rapidly. We have seen in the Chapter 3 that

$$E(X) = np \text{ and } Var(X) = np(1-p).$$

Let $0 < p_n < 1$ and let us look at the limit of the probability

$$P(X = r) = n_r^C p_n^r (1 - p_n)^{n-r}$$

as $n \to \infty$ such that $np_n \to \lambda$ where $\lambda > 0$ is fixed. Since $n \to \infty$ such that $np_n \to \lambda$, it follows that $p_n \to 0$. In other words, we are considering a situation when n is large and p_n is small such that np_n is of the order λ.

Now

$$\begin{aligned}
P(X = r) &= n_r^C p_n^r (1 - p_n)^{n-r} \\
&= n_r^C (\frac{np_n}{n})^r (1 - \frac{np_n}{n})^{n-r} \\
&= \frac{n!}{r!(n-r)!}(np_n)^r \frac{1}{n^r}(1 - \frac{np_n}{n})^n (1 - \frac{np_n}{n})^{-r} \\
&= \frac{(np_n)^r}{r!} \frac{n!}{(n-r)!} \frac{1}{(n - np_n)^r}(1 - \frac{np_n}{n})^n.
\end{aligned}$$

Applying the elementary result

$$\lim_{n\to\infty} (1 - \frac{\lambda}{n})^n = e^{-\lambda}$$

and the fact that $np_n \to \lambda$ and observing that

$$\frac{n!}{(n-r)!(n-np_n)^r} = \frac{n(n-1)\dots(n-r+1)}{(n-np_n)^r}$$

tends to one as $n \to \infty$, it follows that

$$n_r^C p_n^r (1 - p_n)^{n-r} \to e^{-\lambda}\frac{\lambda^r}{r!} \text{ as } n \to \infty.$$

In other words, the probability for r successes out of n Bernoulli trials, with probability of a success on a single trial equal to p_n, can be approximated by the corresponding Poisson probability for r events to occur with $\lambda = np_n$ as the mean. This approximation is good when n is *large* and p_n is *small.*

In order to get an idea of what we mean by n is large and p_n is small, let us consider the case when $\lambda = 1$. Since $\lambda \simeq np_n$, it follows that $p_n \simeq \frac{1}{n}$. For illustrative purpose, consider the cases $n = 5$ and $p_n = \frac{1}{5}$ and $n = 100$ and $p_n = \frac{1}{100}$. See Table 7.1 and Table 7.2 for comparison of the exact probabilities under the Binomial distribution and the corresponding approximate probabilities computed under the Poisson distribution with $\lambda \simeq np_n = 1$.

Table 7.1
$n = 5, \; p_n = \frac{1}{5}, \; \lambda = 1$

r	Binomial(n, p_n)	Poisson(λ)
0	0.328	0.368
1	0.410	0.368
2	0.205	0.184
3	0.051	0.061
4	0.006	0.015
5	0.000	0.003
6	0.000	0.001

Table 7.2
$n = 100, \; p_n = \frac{1}{100}, \; \lambda = 1$

r	Binomial(n, p_n)	Poisson(λ)
0	0.366032	0.367879
1	0.369730	0.367879
2	0.184865	0.183940
3	0.060999	0.061313
4	0.014942	0.015328
5	0.002898	0.003066
6	0.000463	0.000511
7	0.000063	0.000073
8	0.000007	0.000009
9	0.000001	0.000001
10	0.000000	0.000001

The entries are rounded off to the nearest three decimal places in Table 7.1 and to six decimal places in Table 7.2. The entries corresponding to the values of r greater than 10 are zero when rounded off and hence are not presented. As we can see from the Tables 7.1 and 7.2, the approximation of the Binomial probabilities by the Poisson probabilities is not good when n is small but it is very good for large n provided of course that p_n is small. The corresponding entries in the second and the third column of Table 7.2 agree even up to the third decimal place.

Example 7.5. Suppose a company produces a large number of items of a certain type and the probability that an item produced by it is defective is 0.1. Let us compute the probability that a random sample of 10 items produced by the same company contains at most one defective. Since the lot size is large, if X denotes the number of defective items in a random sample of size $n = 10$ from the lot, then we can assume that X has the Binomial distribution with the parameters $n = 10$ and $p = 0.1$. We want to find the $P(X \leq 1)$. The exact probability is given by

$$\begin{aligned} P(X \leq 1) &= P(X = 0) + P(X = 1) \\ &= 10^C_0(0.1)^0(0.9)^{10} + 10^C_1(0.1)^1(0.9)^9 \\ &= 0.7361. \end{aligned}$$

On the other hand, suppose we approximate the distribution of X by the Poisson distribution with $\lambda = np = (10)(0.1) = 1$. Then $P(X \leq 1)$ is approximately equal to

$$\begin{aligned} P(Y \leq 1) &= P(Y = 0) + P(Y = 1) \\ &= e^{-1} + e^{-1} \\ &= 0.7358 \end{aligned}$$

where Y has the Poisson distribution with mean $\lambda = 1$. Note the close approximation between the exact probability computed using the Binomial distribution and an approximate probability computed using the Poisson distribution.

7.5 Central Limit Theorem

The central limit theorem (CLT) is one of the most important and useful results in probability theory. We have seen earlier that the sum of a finite number of independent normally distributed random variables has a normal distribution. However it is easy to see that the sum of a finite number of independent random variables which do not have a normal distribution need not have a normal distribution. Even in such a case, the sum of a *large* number of independent random variables has a distribution that is approximately normal under some general conditions according to the central limit theorem. The CLT provides a method of computing the approximate probabilities for events concerning sums of large number of independent random variables. This theorem also suggests the possible reasons why most of the data observed in practice leads to bell-shaped curves after a suitable transformation. We now state the central limit theorem.

Theorem 7.3. (Central limit theorem) Let $X_n, n \geq 1$ be a sequence of independent and identically distributed (i.i.d.) random variables with mean μ and finite positive variance σ^2. Then, for any x,

$$P(\frac{X_1 + \cdots + X_n - n\mu}{\sigma\sqrt{n}} \leq x) \to \Phi(x) \text{ as } n \to \infty$$

where $\Phi(x)$ is the standard normal distribution function.

A rigourous proof of this result involves the notion of the *characteristic function* of a random variable and some complex analysis beyond the scope of this book. Let us understand the statement of the theorem. Let

$$S_n = X_1 + \cdots + X_n.$$

Then we know that

$$E(S_n) = n\mu \text{ and } Var(S_n) = n\sigma^2.$$

Suppose we standardize the random variable S_n. Then the standardized random variable

$$\frac{S_n - n\mu}{\sigma\sqrt{n}}$$

has mean zero and variance one. The CLT says that the distribution function of the standardized random variable

$$\frac{S_n - n\mu}{\sigma\sqrt{n}}$$

can be approximated by the standard normal distribution function for large n. In other words, the distribution function of the sum

$$S_n = X_1 + \cdots + X_n$$

will be approximately normal for large n as long as the random variables $X_i, 1 \leq i \leq n$ are i.i.d. with finite positive variance irrespective of the fact whether the underlying distribution of X_1 is continuous or discrete or a mixture of the two types or any other distribution. The theorem can be extended to independent random variables which are not necessarily identically distributed or to dependent random variables under some conditions. We will not go into these aspects here.

The central limit theorem has many applications in statistical inference. An important application is to a sequence of independent Bernoulli random variables. This application of the theorem leads to an approximation of a Binomial distribution by a suitable normal distribution as we will see in the next section.

7.6 Normal Approximation to the Binomial Distribution

Let $X_i, i \geq 1$ be a sequence of i.i.d. random variables such that $P(X_i = 1) = p = 1 - P(X_i = 0)$ where $0 < p < 1$. Such a sequence is called a Bernoulli sequence of independent random variables. Observe that $S_n = X_1 + \cdots + X_n$ has the Binomial distribution with parameters n and p. Check that $E(X_1) = p$ and $Var(X_1) = p(1-p)$ which is finite and positive. An application of the CLT gives the following result: for every x,

$$P(\frac{S_n - np}{\sqrt{n}\sqrt{p(1-p)}} \leq x) \to \Phi(x) \text{ as } n \to \infty.$$

In other words, for large n,

$$P(S_n \leq np + x\sqrt{np(1-p)}) \asymp \Phi(x)$$

where $\asymp$ denotes that the quantities on both sides are approximately equal to each other.

An alternate way of interpreting the above approximation is that any Binomial distribution with parameters n and p tends to be close to a suitable normal distribution for large n. Let us explain this in a more detailed way.

Suppose a random variable S_n has the Binomial distribution with parameters n and p where $0 < p < 1$. Then, for any integer r such that $1 \leq r \leq n$,

$$\begin{aligned} P(S_n \leq r) &= P(\frac{S_n - np}{\sqrt{np(1-p)}} \leq \frac{r - np}{\sqrt{np(1-p)}}) \\ &\asymp \Phi(\frac{r - np}{\sqrt{np(1-p)}}) \end{aligned}$$

for large n. It is difficult to compute the exact probability

$$P(S_n \leq r) = \sum_{j=0}^{r} n_j^C p^j (1-p)^{n-j}$$

when n is large. A close approximation to this probability can be obtained by computing

$$\Phi(\frac{r - np}{\sqrt{np(1-p)}}).$$

It has been found from empirical studies that the approximation is good whenever $n \geq 30$ and a better approximation can be obtained by applying a slight correction, namely, by choosing

$$\Phi(\frac{r + \frac{1}{2} - np}{\sqrt{np(1-p)}})$$

as the normal approximation for $P(S_n \leq r)$. We will now illustrate this method of approximation through an example.

Example 7.6. The ideal size of the first year class of students in a college is 150. It is known from an earlier data that on the average only 30% of those accepted for admission to the college will actually join. Suppose the college accepts admission of 450 students. What is the probability that more than 150 students will actually join in the first year class?

Let S_n denote the number of students who actually join in the first year class when admission is offered to n students. Assuming that all the students take independent decisions of either joining or not joining the college, we can suppose that S_n has the Binomial distribution with parameters n and $p = 0.3$. Here $n = 450$ and we have to find $P(S_n \geq 150)$. Note that

$$E(S_n) = np = 450(0.3) = 135 \quad \text{and} \quad Var(S_n) = 450(0.3)(0.7).$$

Further more

$$\begin{aligned} P(S_n \geq 150) &= 1 - P(S_n < 150) \\ &= 1 - P(S_n \leq 149) \end{aligned}$$

and

$$\begin{aligned} P(S_n \leq 149) &\asymp \Phi(\frac{149 + \frac{1}{2} - 135}{\sqrt{450(0.3)(0.7)}}) \\ &= \Phi(1.59). \end{aligned}$$

Hence

$$\begin{aligned} P(S_n \geq 150) &\asymp 1 - \Phi(1.59) \\ &= 0.0559. \end{aligned}$$

This shows that the probability that more than 150 students will join the first year class among the 450 offered admission is less than 6%.

7.7 Approximation of a Chi-square Distribution by a Normal Distribution

Suppose $X_i, i \geq 1$ is a sequence of i.i.d. standard normal random variables. Then $X_i^2, i \geq 1$ is a sequence of i.i.d. random variables each with a Chi-square distribution with one degree of freedom. Note that

$$E(X_1^2) = 1 \text{ and } Var(X_1^2) = 2.$$

Let us apply the central limit theorem to the sequence $X_i^2, i \geq 1$. It follows that

$$P(\frac{X_i^2 + \cdots + X_n^2 - n}{\sqrt{n}\sqrt{2}}) \to \Phi(x) \text{ as } n \to \infty$$

for every x. But $S_n = X_1^2 + \cdots + X_n^2$ has the Chi-square distribution with n degrees of freedom. What we have shown just now is that, if a random variable S_n has the Chi-square distribution with n degrees of freedom, then $\frac{S_n - n}{\sqrt{2n}}$ has an approximate standard normal distribution. In other words, for every x,

$$P(\frac{S_n - n}{\sqrt{2n}} \leq x) \asymp \Phi(x)$$

for large n whenever S_n has the Chi-square distribution with n degrees of freedom.

7.8 Convergence of Sequences of Random Variables

We have proved the Weak law of large numbers earlier in the Section 3. This result deals with a special type of convergence, known as convergence in probability, for a sequence of averages of i.i.d. random variables. We will now briefly discuss two types of convergence notions for general sequences of random variables. A sequence of random variables X_n is said to *converge in probability* to a random variable X if

$$P(|X_n - X| > \varepsilon) \to 0 \text{ as } n \to \infty$$

for every $\varepsilon > 0$. We denote this convergence by $X_n \xrightarrow{p} X$ as $n \to \infty$. Let $F_n(.)$ be the distribution function of X_n and $F(.)$ be the distribution function of X. A sequence of distribution functions F_n is said to *converge weakly* if

$$F_n(x) \to F(x) \text{ as } n \to \infty$$

for all x at which the distribution function F is continuous. We call this type of convergence as *convergence in law* or *convergence in distribution* and denote this convergence by $X_n \xrightarrow{\mathcal{L}} X$ as $n \to \infty$. It can be proved that if a sequence of random variables X_n converges in probability to a random variable X, then it also converges in law to X. However, the converse is not true, that is, if a sequence of random variables X_n converges in law to a random variable X, then it is not necessary that $X_n \xrightarrow{p} X$ as $n \to \infty$. The two types of convergence are equivalent, that is, each one implies the other if the limiting random variable X is degenerate, that is, there exists a constant c such that $P(X = c) = 1$.

The following fundamental result, which is useful later in this book, is due to Slutsky.

Theorem 7.4. **(Slutsky's lemma)** Let $\{X_n\}$ and $\{Y_n\}$ be sequences of random variables such that $X_n \xrightarrow{\mathcal{L}} X$ and $Y_n \xrightarrow{p} c$ as $n \to \infty$ where c is a constant. Then

(i) $X_n + Y_n \xrightarrow{\mathcal{L}} X + c$;
(ii) $X_n Y_n \xrightarrow{\mathcal{L}} cX$;
(iii) if $c \neq 0$, then $X_n/Y_n \xrightarrow{\mathcal{L}} X/c$ as $n \to \infty$.

Another result, which will be used later in this book, is the following.

Theorem 7.5. Let $\{X_n\}$ and $\{Y_n\}$ be sequences of random variables such that $X_n - Y_n \xrightarrow{p} 0$ and $Y_n \xrightarrow{\mathcal{L}} Y$ as $n \to \infty$. Then $X_n \xrightarrow{\mathcal{L}} Y$ as $n \to \infty$.

We do not go into the proofs of these results. As an application of these results, we now discuss a method for obtaining the limiting distribution of functions of random variables.

Delta method

Suppose T_n is a sequence of random variables such that

$$\sqrt{n}(T_n - \theta) \xrightarrow{\mathcal{L}} N(0, \sigma^2) \quad \text{as } n \to \infty.$$

Let us now consider a function $g(T_n)$ of T_n. How do we determine the limiting distribution of the sequence of random variables $g(T_n)$ if any? For instance, the function $g(x)$ could be x^2. Answers to such questions are useful in the study of the interval estimation or in constructing the confidence intervals discussed in the Chapter 9.

Suppose the function $g(.)$ is twice differentiable. Expanding the function $g(.)$ around the point θ, we get that

$$g(x) \simeq g(\theta) + (x - \theta)g'(\theta)$$

and hence

$$g(x) - g(\theta) \simeq (x - \theta)g'(\theta).$$

Under some conditions, it can now be checked that

$$\sqrt{n}(g(T_n) - g(\theta)) - \sqrt{n}(T_n - \theta)g'(\theta) \xrightarrow{p} 0$$

as $n \to \infty$ by using Theorem 7.4 and Theorem 7.5 stated above. Applying Theorems 7.4 and 7.5 again, we get that

$$\sqrt{n}(g(T_n) - g(\theta)) \xrightarrow{\mathcal{L}} N(0, (g'(\theta))^2\sigma^2)$$

as $n \to \infty$. In particular, if the function $g(.)$ is chosen so that $(g'(\theta))^2\sigma^2$ is a constant c for all θ, then such a transformation is called a *variance stabilizing transformation* as it stabilizes the variance. Such transformations are useful in constructing confidence intervals as discussed in the Chapter 9.

7.9 Exercises

7.1 If X is a random variable with $E(X) = \mu$ and $Var(X) = 0$, show that $P(X = \mu) = 1$.(Hint: Use the Chebyshev's inequality.)

7.2 If X is a random variable with $E(X) = \mu$ and $Var(X) = \sigma^2$, find an upper bound for $P(|X - \mu| \geq 3\sigma)$.

7.3 If X is a random variable with $E(X) = 3$ and $E(X^2) = 13$, find a lower bound for $P(-2 \leq X \leq 8)$.

7.4 Suppose that X is a random variable with the exponential probability density function given by

$$f(x) = e^{-x} \text{ for } x > 0$$
$$= 0 \text{ otherwise.}$$

Compute $E(X) = \mu$ and $Var(X) = \sigma^2$. Compute an upper bound for $P(|X - \mu| \geq 2\sigma)$ using the Chebyshev's inequality and compare it with the exact probability obtained from the distribution of X.

7.5 How large the size of a random sample should be, from a population with mean μ and finite variance σ^2, in order that the probability that the sample mean will be within 2σ limits of the population mean μ is at least 0.99?

7.6 Defects in a particular kind of a metal sheet occur at an average rate of one per 100 square metres. Find the probability that two or more defects occur in a sheet of size 40 square metres.

7.7 In a particular book of 520 pages, 390 printing errors were there. What is the probability that a page selected from this book at random will contain no errors?

7.8 In a large population, the proportion of people having a certain disease is 0.01. Find the probability that at least four will have the disease in a random group of 200 people.

7.9 If ten fair dice are rolled, find the approximate probability that the sum of the numbers observed is between 30 and 40.

7.10 Suppose $X_i, 1 \leq i \leq 10$ are independent random variables each uniform on $[0, 1]$. Determine an approximation to $P(X_1 + \cdots + X_{10} > 6)$.

7.11 Show that if a random variable Y has the Poisson distribution with parameter λ, then $\frac{Y-\lambda}{\sqrt{\lambda}}$ has approximately a standard normal distribution when λ is large. (Hint: Use the fact that if Y_1 and Y_2 are independent random variables with Poisson distributions with the parameters λ_1 and λ_2 respectively, then the random variable $Y_1 + Y_2$ has the Poisson distribution with the parameter $\lambda_1 + \lambda_2$.)

7.12 Let $X_1, \ldots, X_{15}$ be a random sample of size 15 from a population with the probability density function

$$\begin{aligned} f(x) &= 3(1-x)^2 \text{ if } 0 < x < 1 \\ &= 0 \text{ otherwise.} \end{aligned}$$

Find the approximate probability for the event $\frac{1}{8} < \bar{X} < \frac{3}{8}$.

7.13 If a random variable X has the Binomial distribution with $n = 100$ and $p = \frac{1}{2}$, find an approximation for $P(X = 50)$.

7.14 If a random variable X has the Binomial distribution with parameters n and $p = 0.55$, determine the smallest value of n for which

$$P(X > \frac{n}{2}) \geq 0.95$$

approximately.

7.15 If a random variable X has the Binomial distribution with parameters $n = 100$ and $p = 0.1$, find the approximate value of $P(12 < X < 14)$ using (a) the normal approximation, and (b) the Poisson approximation.

Chapter 8

ESTIMATION

8.1 Introduction

We have studied the concept of probability and its various applications to model building for different random phenomena in the previous chapters. In probability theory, we proceed from a *known* distribution assumed to be modelling some data and derive the probabilities of different events associated with the phenomenon underlying the data. Let us illustrate this by an example. Suppose a manufacturer of automobiles is interested in the purchase of a wagon load of a particular type of items, say, the car tyres from a vendor. As it is too expensive to check for all the defective items in the lot, he or she chooses a random sample for examination. Suppose the lot contains N tyres of which r are defective and a sample of size n is chosen at random from among the N tyres available. Let D denote the number of defective tyres found in the sample chosen. From the results in the Chapter 3, we know that the random variable D has the Hypergeometric distribution and

$$P(D = d) = \frac{r_d^C \quad (N-r)_{(n-d)}^C}{N_n^C}, d = 0, 1, \ldots, \min(n, r, N).$$

This type of result is not useful in practice since the number r is unknown. If the value of r is known, then there is no necessity of obtaining a sample! We are interested to know whether we can draw information about the unknown value of r based on the observed value of D. It is obvious that the higher the sample size is, the better the information we can obtain about r. If we examine all the N items, then we hope to have the exact information about D. This involves a huge expenditure due to the cost of sampling and involves large amount of time and at times it is not even possible to examine all the items, for instance, if the item for examination happens to be a (say) light bulb instead of a tyre and our interest is in its life time. Examining whether a light bulb is defective or not entails its use and full examination of all the light bulbs will reduce the total lot to zero and there won't be any left for the actual use of the customer! Complete checking of the lot of N items will also not give the number of defective items due to non sampling errors. For instance, errors might also creep up in recording whether an item is defective or not by the investigator in view of the large amount of checking. In the light of these remarks, it is necessary to draw some conclusion about the value of the unknown constant r based on a random sample chosen from the total lot. Hence the basic problem can be thought of as how to draw inference about a population on the basis of a random sample from the same population. Methods dealing with problems of this nature constitute the subject of *Statistical Inference*. *Probability Theory* deals with probabilities of various events based on a sample from a *known* population where as the subject of *Statistical Inference* is concerned with the methods of drawing inferences about *unknown* features of a population based on a random sample chosen from the same population. Let us get back to the problem of estimating the number of defective in the wagon load of car tyres discussed above. Here we do not have information about the total number r of defective tyres in the lot but the exact number N of of items in the lot is known as well as the sample size n chosen. In other words, the probability distribution of the random variable D is completely known except for the value of r. If r is known, then we have the full information about the lot. Such a constant r is called a *parameter*. The problem of statistical inference consists in obtaining information about the unknown parameters of a population based on the observed samples drawn from the same population according to some sampling scheme. We shall study different methods of estimation of parameters and illustrate these methods by several examples. We will center our interest around two methods which are widely used in practice, namely, the method of moments and the method of maximum likelihood. The former is easy to implement in practice while the latter gives estimators with "good" properties. A natural question that arises immediately is how does one say whether a particular estimator is *good* or not as compared to another estimator. In other words, a criterion or criteria have to be specified to judge the performance of the estimators before classifying an estimator as good in comparison with another estimator. All these questions will be studied later in this chapter.

8.2 Estimation

The problem of estimation of a parameter can be explained in the following way. Suppose X is a random variable with the distribution function $F(x;\theta)$ where θ is a parameter. The parameter θ could be a scalar parameter or a vector parameter. It is a scalar if it is real-valued and a vector if it is of the form $(\theta_1, \theta_2, \ldots, \theta_k)$ for some finite k where $\theta_i, 1 \leq i \leq k$ are real-valued. The set of possible values of the parameter θ is called the *parameter space* Θ. The parameter space Θ is a subset of the real line R or the k-dimensional Euclidean space R^k for some k in most of our discussions. We assume that the functional form of the distribution function F is known except for the parameter θ. For instance, F might be the distribution function for a Binomial distribution where n, the number of trials is known but p, the probability of success is unknown or F is the distribution function for a normal distribution with mean μ and variance σ^2 both unknown. One is an example of a discrete distribution function and the other is an example of an (absolutely) continuous distribution function. It is possible that F could be a mixture of both continuous and discrete distributions. However, in our discussions later in this chapter, we will consider the case where F corresponds to either a discrete distribution or a continuous distribution. In order to estimate the parameter θ, a random sample of observations $X_1, \ldots, X_n$ is chosen from the population with the distribution function $F(x;\theta)$. Let $x_1, \ldots, x_n$ be the actual values or observations of a set of independent identically distributed (i.i.d.) random variables $X_1, \ldots, X_n$. It is natural that the information that can be retrieved about the parameter θ should depend on the observed data $x_1, \ldots, x_n$. Any function T depending only on the random variables $X_1, \ldots, X_n$ is called a *Statistic.* Note that a statistic T does not depend on the parameter θ by definition. In the context of the problem of estimation, if the statistic T evaluated at the observations $x_1, \ldots, x_n$ is chosen as an estimate for the parameter θ, then the statistic $T(X_1, \ldots, X_n)$ is called an *estimator.* If we substitute the actual observations $x_1, \ldots, x_n$ in the functional form of T and compute the value $T(x_1, \cdots, x_n)$, then this value is called an *estimate* for θ. Here after we do not distinguish between an estimator and an estimate. The meaning should be clear from the context. Now that we have explained what we mean by an estimator, the next question is how to find one. Obviously there are many different functions T of $(X_1, \ldots, X_n)$ which can be chosen as estimators for θ. Not all of them are suitable. We first study a couple of criteria which can be used to judge the performance of an estimator.

Unbiased estimator

Let $g(.)$ be a known function defined on the parameter space Θ. An estimator $T(X_1, \ldots, X_n)$ is said to be an *unbiased* estimator for the function $g(\theta)$ of the parameter θ if the expectation of T exists and is equal to $g(\theta)$ for all θ in the parameter space Θ, that is,

$$E_\theta(T) = g(\theta)$$

for all $\theta \in \Theta$. Here E_θ denotes the expectation taken when θ is the true parameter. The intuitive implication is that even though the value of T evaluated at the actual observations $x_1, \ldots, x_n$ might not be near $g(\theta)$ for a particular realization $(x_1, \ldots, x_n)$, on the average, T is close to $g(\theta)$. The likelihood of over estimation of $g(\theta)$ by T is the same as the likelihood of under estimation of $g(\theta)$ by T. Let us illustrate this concept by some examples.

Example 8.1. Suppose $X_i, 1 \leq i \leq n$ constitute n independent Bernoulli trials each with the same probability of success p. Our problem is to find an unbiased estimator of p, if there exists any, based on the observations, $X_i, 1 \leq i \leq n$. From the definition of a Bernoulli trial,

$$P(X_i = 1) = p \text{ and } P(X_i = 0) = 1 - p, i = 1, \ldots, n.$$

Hence

$$E_p(X_i) = p$$

for any $1 \leq i \leq n$. In other words every one of the observations can be considered as an unbiased estimator of p. These are not the only unbiased estimators. There are others. For instance, if

$$\bar{X} = \frac{X_1 + \cdots + X_n}{n},$$

then

$$E_p(\bar{X}) = p$$

and $\bar{X}$ is an unbiased estimator for p.

Example 8.2. Suppose $X_1, \ldots, X_n$ are independent and identically distributed (i.i.d.) random variables with the normal distribution with mean μ and variance σ^2. It is easy to check that the sample mean $\bar{X} = \frac{1}{n}\sum_{i=1}^{n} X_i$ is an unbiased estimator for μ. We leave the detailed checking to the reader. Let us concentrate on the parameter σ^2. If μ is known, then the function

$(X_i - \mu)^2$ will be an unbiased estimator for σ^2 since $E(X_i - \mu)^2 = \sigma^2$. Since this equality holds for all the observations $X_i, 1 \leq i \leq n$, it can be seen that

$$E\{\frac{1}{n}\sum_{i=1}^{n}(X_i - \mu)^2\} = \sigma^2$$

and hence

$$\tilde{\sigma}^2 = \frac{1}{n}\sum_{i=1}^{n}(X_i - \mu)^2$$

is an unbiased estimator for σ^2. However, if μ is not known, then $\tilde{\sigma}^2$ is not a statistic as it depends on the unknown parameter μ. Let us estimate μ by $\bar{X}$ and consider

$$\hat{\sigma}^2 = \frac{1}{n}\sum_{i=1}^{n}(X_i - \bar{X})^2.$$

It is known that

$$\sum_{i=1}^{n}\frac{(X_i - \bar{X})^2}{\sigma^2}$$

has the Chi-square distribution with $(n-1)$ degrees of freedom and hence

$$E\left[\sum_{i=1}^{n}\frac{(X_i - \bar{X})^2}{\sigma^2}\right] = n - 1$$

from the properties of the Chi-square distribution. Therefore

$$E[\hat{\sigma}^2] = \frac{n-1}{n}\sigma^2$$

and $\hat{\sigma}^2$ is not an unbiased estimator of σ^2. A slightly perturbed version of the statistic $\tilde{\sigma}^2$, namely,

$$(\sigma^2)^* \equiv S^2 = \frac{1}{n-1}\sum_{i=1}^{n}(X_i - \bar{X})^2$$

can now be seen to be an unbiased estimator of σ^2. The statistic S^2 is some times called the *sample variance.* Observe that S^2 is defined only when the sample size n is at least two.

We will discuss other examples later in this chapter.

If an estimator T is not unbiased for a scalar parameter θ in the sense that $E_\theta(T)$ exists but it might be different from θ, then the quantity $E_\theta(T) -$

θ is called the *bias* of the estimator T with respect to θ. If T is unbiased, then its bias is zero for the parameter θ. The question now arises: how to compare two different unbiased estimators of the same parameter whenever they exist? We consider only those unbiased estimators which have finite variances. For instance X_1 and $\bar{X}$ are both unbiased estimators in the Example 8.1. Suppose T is an unbiased estimator of a scalar parameter θ. Then $T - \theta$ measures the deviation of the estimator T from θ. Since the random variable $T - \theta$ is a function of the observations $X_1, \ldots, X_n$ and the parameter θ, a measure of the deviation of T from θ may be chosen as $E_\theta(T - \theta)$. However this measure is not useful since $E_\theta(T - \theta) = 0$ for all unbiased estimators T for the parameter θ and there is no way that this measure can be used to compare two unbiased estimators. The fact that the estimator T is unbiased for the parameter θ essentially means that the likelihood that T under estimates θ is the same as the likelihood that it over estimates θ. In the light of this observation, a better measure of the deviation can be obtained by considering $|T - \theta|$ or in general $|T - \theta|^r$ where $r \geq 1$. It turns out that the value $r = 2$ is a convenient choice and the precision of T can be measured by the quantity $E_\theta(T - \theta)^2$. Since T is unbiased for θ, this quantity is the variance of the estimator T. Let us write $V_\theta(T)$ for the variance of the unbiased estimator T when θ is the true parameter. The larger the variance of T is, the higher the deviation of T from θ on the average. For obvious reasons, one prefers an estimator whose spread is not too large. In other words an unbiased estimator T_1 for a parameter θ is preferred to another unbiased estimator T_2 for the same parameter θ if

$$V_\theta(T_1) \leq V_\theta(T_2)$$

for all θ with the strict inequality for at least one value of θ. It is quite possible that $V_\theta(T_1)$ and $V_\theta(T_2)$ might not be comparable in the sense that $V_\theta(T_1)$ might be strictly larger than $V_\theta(T_2)$ for some values of θ and $V_\theta(T_2)$ might be strictly larger than $V_\theta(T_1)$ for some other values of θ as in the Figure 8.1.

In this case, no judgement can be made regarding whether T_1 is preferable to T_2 or T_2 is preferable to T_1 for estimating the parameter θ. The choice between T_1 and T_2 depends on the value of θ which is unknown in the first instance.

Let us now consider the class of all unbiased estimators for the parameter θ with finite variances. Can we find an estimator which is "best" in this

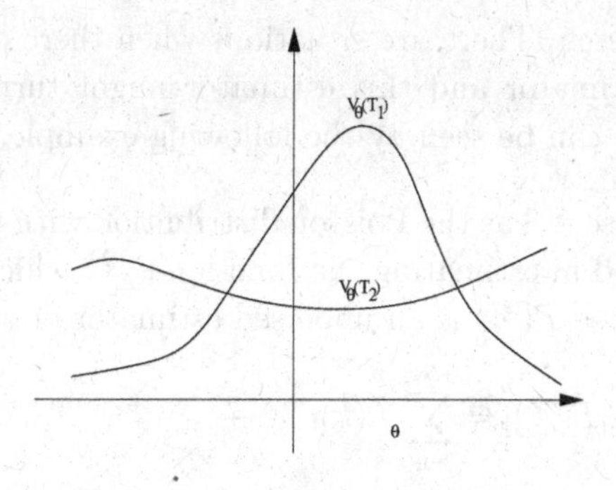

Fig. 8.1

case? That is, is there a *minimum variance unbiased estimator* (MVUE) for θ and if it does exist, how does one find it? Another question which comes up here is, given an unbiased estimator T, is it possible to improve this estimator? We will come back to these and related questions later in this chapter. It should be pointed out that it might be preferable at times to use a slightly biased estimator than an unbiased estimator even though the latter might exist. For instance, suppose T_1 and T_2 are two estimators of θ the former being unbiased and the latter biased. Let us look at the sampling distributions of T_1 and T_2. Suppose the probability density functions of T_1 and T_2 exist and are given as in the Figure 8.2.

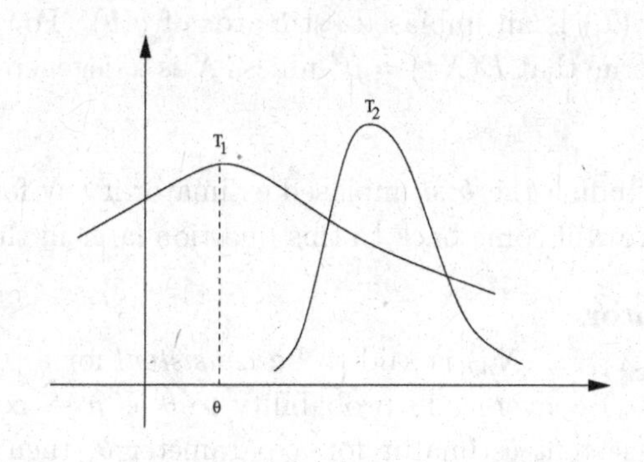

Fig. 8.2

Obviously the statistic T_2 takes values close to θ more often than T_1 does. But T_2 is obviously biased. From the graph, a rational choice, among the statistics T_1 and T_2 as estimators for θ, is T_2 even though T_2 is not an unbiased estimator. Some times unbiased estimators might not exist in some particular problems. There are situations when there exists one and only one unbiased estimator and this estimator might turn out to be a meaningless one. This can be seen by the following example.

Example 8.3. Suppose X has the Poisson distribution with the parameter λ and we are interested in estimating the function $e^{-2\lambda}$ which is obviously a function of λ. Suppose $T(X)$ is an unbiased estimator of $e^{-2\lambda}$. Then

$$E_\lambda[T(X)] = \sum_{k=0}^{\infty} T(k) e^{-\lambda} \frac{\lambda^k}{k!} = e^{-2\lambda}$$

for all $\lambda > 0$. Hence

$$\sum_{k=0}^{\infty} T(k) \frac{\lambda^k}{k!} = e^{-\lambda} = \sum_{k=0}^{\infty} \frac{(-\lambda)^k}{k!}.$$

Since this identity holds for all $\lambda > 0$, comparing the coefficients of λ^k and equating them for all $k \geq 0$, we find that

$$T(k) = (-1)^k, k \geq 0.$$

It is clear that the function $e^{-2\lambda} > 0$ for all $\lambda > 0$. However the unique unbiased estimator $T(X)$, of the positive valued function $e^{-2\lambda}$ whose range is $(0, 1)$, takes the values 1 or -1 which is absurd.

Finally, we should observe that, if T is an unbiased estimator of θ, it is not necessary that $g(T)$ is an unbiased estimator of $g(\theta)$. For instance if $E(X) = \mu$, it is not true that $E(X^2) = \mu^2$ unless X is a degenerate random variable at μ..

The question of finding the *best* unbiased estimator if any for a parameter still remains. We will come back to this question later in this chapter.

Consistent estimator

An estimator $T(X_1, \ldots, X_n)$ is said to be *consistent* for a parameter θ if $T_n \equiv T(X_1, \ldots, X_n)$ converges in probability to θ as $n \to \infty$. In other words, if T is a consistent estimator for a parameter θ, then the larger the sample size n is, the closer the values of T are to θ with probability approaching one. In mathematical terms,

$$P_\theta(|T_n - \theta| > \varepsilon) \to 0 \text{ as } n \to \infty$$

for every $\varepsilon > 0$. An estimator which is *not consistent* is *not useful* as there is no guarantee that the values of the estimator will be close to θ even if the sample size is large. Here $P_\theta(E)$ denotes the probability of the event E when θ is the true parameter.

Example 8.4. (Example 8.1 continued) Observe that the sample mean $\bar{X}$ is a consistent estimator for p. In fact

$$P_p(|\bar{X} - p| > \varepsilon) \to 0 \text{ as } n \to \infty$$

for every $\varepsilon > 0$. This can be seen by using the Chebyshev's inequality since

$$P_p(|\bar{X} - p| > \varepsilon) \leq \frac{1}{\varepsilon^2} E_p(\bar{X} - p)^2 = \frac{1}{\varepsilon^2} \text{Var}_p(\bar{X}).$$

But $\sum_{i=1}^{n} X_i$ has the Binomial distribution with the parameters n and p and hence $\text{Var}_p(\bar{X}) = \frac{p(1-p)}{n}$. The last term in the above inequality tends to zero as $n \to \infty$. This proves that $\bar{X}$ converges in probability to p as $n \to \infty$. Therefore $\bar{X}$ is a consistent estimator for p.

Example 8.5. (Example 8.2 continued) Let us again suppose that we have a random sample $X_1, \ldots, X_n$ from the normal distribution with mean μ and variance σ^2. It can be checked that the sample mean $\bar{X}$ is a consistent estimator for μ. This can be proved either by using the Chebyshev's inequality or by an application of the WLLN. We leave the details to the reader. Let us now consider the sample variance and check whether the sample variance S^2 is a consistent estimator for σ^2. Note that

$$P(|S^2 - \sigma^2| > \varepsilon) \leq \frac{1}{\varepsilon^2} E(S^2 - \sigma^2)^2$$

and $E(S^2) = \sigma^2$. Hence

$$\begin{aligned}
E(S^2 - \sigma^2)^2 &= \text{Var}(S^2) \\
&= \text{Var}\left(\frac{1}{n-1} \sum_{i=1}^{n} (X_i - \bar{X})^2\right) \\
&= \frac{1}{(n-1)^2} \text{Var}\left(\sum_{i=1}^{n} (X_i - \bar{X})^2\right) \\
&= \frac{\sigma^4}{(n-1)^2} \text{Var}\left(\sum_{i=1}^{n} \frac{(X_i - \bar{X})^2}{\sigma^2}\right) \\
&= \frac{2\sigma^4 (n-1)}{(n-1)^2} = \frac{2\sigma^4}{n-1}
\end{aligned}$$

and the last equality follows from observing that

$$\frac{\sum_{i=1}^{n}(X_i - \bar{X})^2}{\sigma^2}$$

has the Chi-square distribution with $(n-1)$ degrees of freedom and that the variance for such a distribution is $2(n-1)$. Hence

$$E(S^2 - \sigma^2)^2 \to 0 \text{ as } n \to \infty$$

which implies that

$$P(|S^2 - \sigma^2| > \varepsilon) \to 0 \text{ as } n \to \infty$$

for every $\varepsilon > 0$ by the Chebyshev's inequality. Therefore S^2 is a consistent estimator for σ^2.

From the examples given above, it can be seen that a set of sufficient conditions for an estimator T_n to be consistent for a parameter θ are

$$\text{(i)} \quad E_\theta(T_n) \to \theta \text{ as } n \to \infty$$

and

$$\text{(ii)} \quad \text{Var}_\theta(T_n) \to 0 \text{ as } n \to \infty.$$

Details are left to the reader. We should point out that a consistent estimator need not be an unbiased estimator. If it is unbiased, then the condition (i) holds automatically and the condition (ii) needs to be checked only for consistency. For instance, in the Example 8.4, it is easy to check that

$$\frac{1}{n+1}\sum_{i=1}^{n} X_i$$

is a consistent estimator for p but it is not unbiased for the parameter p.

Example 8.6. Let $X_i, 1 \le i \le n$ be i.i.d. random variables with the uniform distribution on $[0, \theta]$. It can be checked that$E_\theta(X_1) = \theta/2$ and $\text{Var}_\theta(X_1) = \theta^2/12$. Therefore $E_\theta(2\bar{X}) = \theta$ and $\text{Var}_\theta(2\bar{X}) = \theta^2/3n$. This proves that $2\bar{X}$ is an unbiased estimator of θ and it is also consistent since $\text{Var}_\theta(2\bar{X}) \to 0$ as $n \to \infty$. On the other hand, let $X_{(n)}$ be the n-th order statistic based on $X_i, 1 \le i \le n$, that is, $X_{(n)}$ is the maximum of the observations $X_i, 1 \le i \le n$. Check that the distribution function of $X_{(n)}$ is given by

$$\begin{aligned} P(X_{(n)} \le x) &= \left(\frac{x}{\theta}\right)^n \text{ for } 0 \le x \le \theta \\ &= 0 \text{ for } x < \theta \\ &= 1 \text{ for } x > \theta. \end{aligned}$$

Therefore the probability density function of $Y = X_{(n)}$ is

$$f_n(y) = n\left(\frac{y}{\theta}\right)^{n-1}\frac{1}{\theta} \text{ for } 0 \le y \le \theta$$
$$= 0 \quad \text{otherwise.}$$

Let us now compute the mean and the variance of Y. Check that

$$\begin{aligned} E_\theta(Y^r) &= \int_0^\theta y^r n\left(\frac{y}{\theta}\right)^{n-1}\frac{1}{\theta}dy \\ &= \frac{n}{\theta^n}\int_0^\theta y^{n-1+r}dy \\ &= \frac{n}{\theta^n}\left[\frac{y^{n+r}}{n+r}\right]_0^\theta \\ &= \frac{n\theta^r}{n+r}. \end{aligned}$$

Hence $E_\theta(Y) = \frac{n\theta}{n+1}$ and $E_\theta(Y^2) = \frac{n\theta^2}{n+2}$. Therefore

$$\begin{aligned} \text{Var}_\theta(Y) &= \frac{n\theta^2}{n+2} - \left(\frac{n\theta}{n+1}\right)^2 \\ &= n\theta^2\left[\frac{1}{n+2} - \frac{n}{(n+1)^2}\right] \\ &= n\theta^2\left[\frac{(n+1)^2 - n(n+2)}{(n+1)^2(n+2)}\right] \\ &= \frac{n\theta^2}{(n+1)^2(n+2)}. \end{aligned}$$

Note that $E_\theta\left(\frac{n+1}{n}Y\right) = \theta$ and $\text{Var}_\theta\left(\frac{n+1}{n}Y\right) = \frac{\theta^2}{n(n+2)}$. Here we have an instance where $\hat{\theta}_n = 2\bar{X}$ and $\tilde{\theta}_n = \frac{n+1}{n}\max(X_1, \ldots, X_n)$ are both unbiased estimators of θ with

$$\text{Var}_\theta(\hat{\theta}_n) = \frac{\theta^2}{3n} \text{ and } \text{Var}_\theta(\tilde{\theta}_n) = \frac{\theta^2}{n(n+2)}.$$

It is clear that both are consistent estimator too. However $\tilde{\theta}_n$ is preferable to $\hat{\theta}_n$ since $\text{Var}(\tilde{\theta}_n) < \text{Var}(\hat{\theta}_n)$ for all $n > 1$. In fact $\text{Var}(\tilde{\theta}_n)$ converges to zero at a faster rate than the rate of convergence of $\text{Var}(\hat{\theta}_n)$ to zero. The estimator $\hat{\theta}_n$ depends on the observations $X_1, \ldots, X_n$ through their sum where as the estimator $\tilde{\theta}_n$ depends through their maximum.

8.3 Some Methods of Estimation

We shall now discuss different methods of finding estimators. It needs to be checked in every case whether the estimators obtained by using these methods have the properties of unbiasedness and consistency discussed in the previous section. Even if an estimator is not unbiased, it does not lead to erroneous inference but the consistency of an estimator is more important and one should consider only consistent estimators in practice.

Method of moments

This method is quite easy to use in practice in view of its simplicity. It consists in equating the sample moments to the population moments and solving the resultant equations. More specifically, suppose $X_1, X_2, \ldots, X_n$ is a random sample from a population with the distribution function depending on a k-dimensional parameter $\theta = (\theta_1, \ldots, \theta_k)$. The problem is to find a suitable estimator for θ based on the observed sample. Let

$$m_{n,r} = \frac{1}{n}\sum_{i=1}^{n} X_i^r.$$

The quantity $m_{n,r}$ is called the r-th *sample moment* of the random sample $X_1, X_2, \ldots, X_n$. Suppose the r-th moment $\mu_r = E(X^r)$ exists for $1 \leq r \leq k$. Note that μ_r is a function of the parameter $\theta = (\theta_1, \ldots, \theta_k)$. The method of moments involves in solving the equations

$$m_{n,r} = \mu_r(\theta_1, \ldots, \theta_k), 1 \leq r \leq k.$$

It is obvious that one needs to equate at least k moments to the corresponding k sample moments in order to solve for the k components of θ. However it is not specified by the method that which of the k moments are to be equated. In practice, the first k population moments are equated to the first k sample moments. We illustrate the method by some examples.

Example 8.7. Suppose $X_1, \ldots, X_n$ is a random sample from the Poisson distribution with mean λ. Suppose the parameter λ is unknown. It is known from the earlier results that $E_\lambda(X_1) = \lambda$ and $\text{Var}_\lambda(X_1) = \lambda$. Following the principle underlying the method of moments, we equate the sample mean to the population mean and solve for λ. Since the parameter in this example is one-dimensional, we need to use one and only one equation. Here the equation can be written in the form

$$\frac{1}{n}\sum_{i=1}^{n} X_i = \lambda.$$

Solving this equation, we get that $\hat{\lambda} = \bar{X}$ is a moment estimator for λ. It is also possible to use the second moment to obtain another estimator of λ by solving the equation

$$\frac{1}{n}\sum_{i=1}^{n} X_i^2 = E_\lambda(X_1^2) = \lambda + \lambda^2.$$

Since $\lambda > 0$, we prefer the estimator to be positive and it can be seen that a unique positive solution to the above equation is

$$\tilde{\lambda} = \frac{-1 + \sqrt{1 + \frac{4}{n}\sum_{i=1}^{n} X_i^2}}{2}.$$

However, for simplicity of computation, one still prefers the estimator $\hat{\lambda}$ to the estimator $\tilde{\lambda}$ as an estimator for λ. There are other reasons too why $\hat{\lambda}$ is preferable to $\tilde{\lambda}$. We will come back to these questions later in this chapter.

The next example deals with case when the parameter θ is 2-dimensional.

Example 8.8. Suppose $X_1, \ldots, X_n$ is a random sample from the normal distribution with the mean μ and the variance σ^2 both unknown. Applying the method of moments, we equate the first two sample moments to the corresponding population moments:

$$\frac{1}{n}\sum_{i=1}^{n} X_i = E(X) = \mu$$

and

$$\frac{1}{n}\sum_{i=1}^{n} X_i^2 = E(X^2) = \sigma^2 + \mu^2.$$

Solving these two equations, we get the estimators

$$\hat{\mu} = \frac{1}{n}\sum_{i=1}^{n} X_i = \bar{X}$$

and

$$\hat{\sigma}^2 = \frac{1}{n}\sum_{i=1}^{n} X_i^2 - \bar{X}^2 = \frac{1}{n}\sum_{i=1}^{n}(X_i - \bar{X})^2$$

as the moment estimators of μ and σ^2 respectively. It is easy to check that the estimator $\hat{\mu} = \bar{X}$ is an unbiased estimator of μ where as the estimator

$\hat{\sigma}^2$ is not an unbiased estimator of σ^2. However both the estimators $\hat{\mu}$ and $\hat{\sigma}^2$ are consistent estimators of μ and σ^2 respectively.

As has been pointed out earlier, the method of moments is useful for its simplicity but the properties of the estimators obtained by this method have to be investigated separately for each estimator. It is not clear whether the estimators obtained this way are the *best* in any sense. An alternate method which will give *best* or *efficient* estimators for large samples under some conditions is the so called *method of maximum likelihood.* We shall now study this method of estimation.

Method of maximum likelihood

In order to explain this method, let us start with an example. Suppose we have a random sample of observations $x_1, \ldots, x_n$ from the Poisson distribution with the mean λ. The problem is to estimate λ. It is easy to see that the probability, that the sample observed is $(x_1, \ldots, x_n)$, is given by

$$\Pi_{i=1}^{n} e^{-\lambda} \frac{\lambda^{x_i}}{x_i!},$$

if λ is the true value of the parameter. The method of maximum likelihood consists in choosing as an estimator of λ, that value of λ, say λ_0, among all the possible values of λ, which *maximizes* the probability of occurrence of the sample $(x_1, \ldots, x_n)$. The value λ_0 is called a maximum likelihood estimator of λ. Obviously λ_0 depends on the observed sample $x_1, \ldots, x_n$. Here we have illustrated the likelihood principle when the sample is from a discrete probability distribution. Likelihood principle essentially states that we should choose those values of the parameters as estimators of the unknown parameters under which the sample observed is most likely to have occurred.

Before we introduce the general principle underlying this method, consider a random sample $X_1, \ldots, X_n$, that is, a sequence of independent and identically distributed random variables from a population with the parameter θ. The parameter θ could be either one-dimensional or multi-dimensional. Let us denote by $f(x, \theta)$ the probability function of X_1 if we are dealing with a discrete random variable or the probability density function if we are concerned with a continuous random variable. The joint probability function or the joint probability density function of $(X_1, \ldots, X_n)$ can be written in the form

$$\Pi_{i=1}^{n} f(x_i, \theta).$$

Denote this function by $L_n(\theta)$. We have suppressed the variables $x_1, \ldots, x_n$ in our definition of $L_n(\theta)$. The function $L_n(\theta)$ is called the *likelihood function* based on the observations $x_1, \ldots, x_n$ when considered as a function of the parameter θ. Suppose that $\theta \in \Theta$ and the set Θ is the *parameter space.* An estimator $\hat{\theta}_n(x_1, \ldots, x_n)$ is called a *maximum likelihood estimator* (MLE) of θ if

$$L_n(\hat{\theta}_n) = \sup_{\theta \in \Theta} L_n(\theta).$$

As was pointed out earlier, it is possible that there might not exist an MLE or even if it does exist, it might not be unique. In other words the supremum of the function $L_n(\theta)$ over the set Θ may not be attained or may be attained at more than one value in Θ. We will present examples indicating these problems later.

Example 8.9. Let us get back to the problem of estimation of the mean λ of the Poisson distribution based on the random sample $X_1, \ldots, X_n$. Observe that the parameter space is $(0, \infty)$ here and the likelihood function is given by

$$L_n(\lambda) = \Pi_{i=1}^n e^{-\lambda} \frac{\lambda^{x_i}}{x_i!}$$

for $0 < \lambda < \infty$. Suppose that at least one of the observations $x_i, 1 \leq i \leq n$ is not equal to zero. We have to find the value $\hat{\lambda}_n$ at which $L_n(\lambda)$ is maximum over $(0, \infty)$ if there exists one. We observe that

$$\log L_n(\lambda) = -n\lambda + \left(\sum_{i=1}^n x_i\right) \log \lambda - \sum_{i=1}^n \log(x_i!)$$

where $\log x$ denote the natural logarithm of x. The function $\log L_n(\lambda)$ attains its maximum at a point λ_0 if and only if $L_n(\lambda)$ attains its maximum at λ_0. Here the function $L_n(\lambda)$ is differentiable with respect to λ and

$$\frac{d \log L_n}{d\lambda} = -n + \frac{\sum_{i=1}^n x_i}{\lambda}.$$

Therefore $\frac{d \log L_n}{d\lambda}|_{\lambda=\hat{\lambda}} = 0$ provided $\hat{\lambda} = \frac{1}{n}\sum_{i=1}^n x_i$. In order to find out whether $L_n(\lambda)$ is maximum at $\lambda = \hat{\lambda}$, we have to compute the second derivative of $L_n(\lambda)$ with respect to λ and check that $\frac{d^2 \log L_n}{d\lambda^2}|_{\lambda=\hat{\lambda}} < 0$. Here

$$\frac{d^2 \log L_n}{d\lambda^2} = -\frac{1}{\lambda^2}\left[\sum_{i=1}^n x_i\right].$$

It is obvious that $\frac{d^2 \log L_n}{d\lambda^2} < 0$ when evaluated at $\lambda = \hat{\lambda}$. This proves that $L_n(\lambda)$ is maximum at $\hat{\lambda}$. There is a unique maximum for the function $L_n(\lambda)$ over the interval $(0, \infty)$ and the maximum is achieved at $\lambda = \hat{\lambda}$. Therefore $\hat{\lambda}$ is the maximum likelihood estimator of λ.

Let us now consider an example where the parameter θ is a vector parameter.

Example 8.10. Let us now consider the problem of estimation of $\theta = (\mu, \sigma^2)$ where both the components μ and σ^2 for a normal distribution with the mean μ and the variance σ^2 are unknown. It is easy to write down the likelihood function $L_n(\mu, \sigma^2)$. It is given by

$$L_n(\mu, \sigma^2) = \Pi_{i=1}^n (2\pi\sigma^2)^{-\frac{1}{2}} \exp\{-\frac{1}{2\sigma^2}(x_i - \mu)^2\}.$$

The problem is to maximize $L_n(\mu, \sigma^2)$ over the set of all (μ, σ^2) such that $-\infty < \mu < \infty, 0 < \sigma^2 < \infty$. We have to maximize the function $L_n(\mu, \sigma^2)$ of two variables over $R \times R_+$. Here

$$\log L_n(\mu, \sigma^2) = \sum_{i=1}^n \{-\frac{1}{2\sigma^2}(x_i - \mu)^2\} - \frac{n}{2}\log(2\pi\sigma^2).$$

Furthermore

$$\frac{\partial \log L_n(\mu, \sigma^2)}{\partial \mu} = \frac{1}{\sigma^2}\sum_{i=1}^n (x_i - \mu)$$

and

$$\frac{\partial \log L_n(\mu, \sigma^2)}{\partial \sigma^2} = \frac{1}{2\sigma^4}\sum_{i=1}^n (x_i - \mu)^2 - \frac{n}{2\sigma^2}$$

are the partial derivatives of $\log L_n(\mu, \sigma^2)$ with respect to μ and σ^2 respectively. Equating the partial derivatives to zero, we get the likelihood equations. These equations have the unique solutions

$$\hat{\mu} = \frac{1}{n}\sum_{i=1}^n x_i = \bar{x} \text{ and } \hat{\sigma}^2 = \frac{1}{n}\sum_{i=1}^n (x_i - \bar{x})^2.$$

It can be shown that these solutions do maximize the likelihood function and hence $\hat{\mu}$ and $\hat{\sigma}^2$ are the maximum likelihood estimators of μ and σ^2 respectively. We leave it to the reader to check that the estimator $\hat{\mu}$ is unbiased and consistent for the parameter μ where as the estimator $\hat{\sigma}^2$ is not an unbiased estimator for σ^2 but a consistent estimator for σ^2. We have noticed in the above example that the likelihood function is a function of two variables. Locating points of maxima of such functions is in general difficult and it is necessary to use special techniques and methods depending on the problem on hand. In the case of a scalar parameter, the likelihood function is a function of one variable and if this function is differentiable twice in the domain of its definition, then one can use the techniques from the calculus of functions of one variable to find the maxima. It is also possible that the likelihood function itself may not be differentiable. The next example illustrates this situation.

Example 8.11. Suppose that $X_i, 1 \leq i \leq n$ is a random sample selected from the uniform distribution on $[0, \theta], \theta > 0$. The problem is to find a MLE of θ based on $X_i, 1 \leq i \leq n$. It is easy to see that the likelihood function is

$$L_n(\theta) = \begin{cases} \theta^{-n} & \text{if } 0 \leq x_i \leq \theta, 1 \leq i \leq n \\ 0 & \text{otherwise} \end{cases}$$

An alternate way of writing this likelihood function is

$$\begin{aligned} L_n(\theta) &= \theta^{-n} \text{ if } 0 \leq x_{(1)} \leq x_{(n)} \leq \theta \\ &= 0 \text{ otherwise} \end{aligned}$$

where $x_{(1)} = \min(x_1, \ldots, x_n)$ and $x_{(n)} = \max(x_1, \ldots, x_n)$. The function $L_n(\theta)$ is not differentiable as a function of θ. In fact the function $L_n(\theta)$ has a discontinuity at $\theta = x_{(n)}$. The graph of the function is sketched in Figure 8.3.

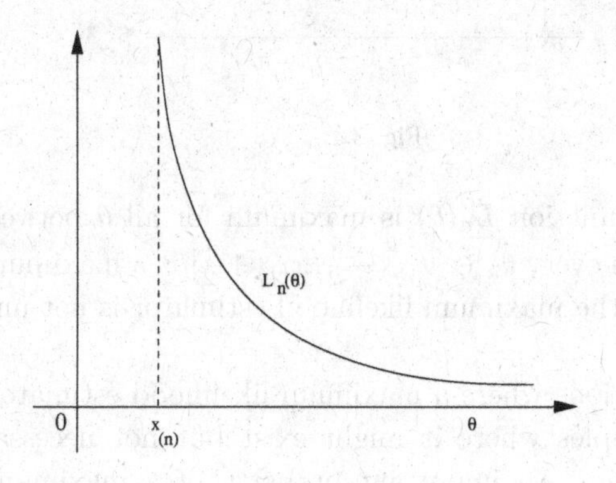

Fig. 8.3

Note that there is a jump at the point $\theta = x_{(n)}$ for the function $L_n(\theta)$ and $L_n(\theta) \downarrow 0$ as $\theta \to \infty$. Hence the function $L_n(\theta)$ attains its maximum at

$\hat{\theta}_n = x_{(n)}$ and the statistic $X_{(n)}$ is the unique maximum likelihood estimator for θ.

Example 8.12. Suppose that $X_i, 1 \le i \le n$ is a random sample selected from the uniform distribution on $[\theta - \frac{1}{2}, \theta + \frac{1}{2}], -\infty < \theta < \infty$. We would like to find a MLE for θ based on the sample $X_1, \ldots, X_n$. The likelihood function here can be written as

$$\begin{aligned} L_n(\theta) &= 1 \text{ if } \theta - \frac{1}{2} \le x_i \le \theta + \frac{1}{2} \text{ for } 1 \le i \le n \\ &= 0 \text{ otherwise} \end{aligned}$$

or equivalently

$$\begin{aligned} L_n(\theta) &= 1 \text{ if } x_{(n)} - \frac{1}{2} \le \theta \le x_{(1)} + \frac{1}{2} \\ &= 0 \text{ otherwise} \end{aligned}$$

where $x_{(1)} = \min(x_1, \ldots, x_n)$ and $x_{(n)} = \max(x_1, \ldots, x_n)$. The graph of the function is sketched in Figure 8.4.

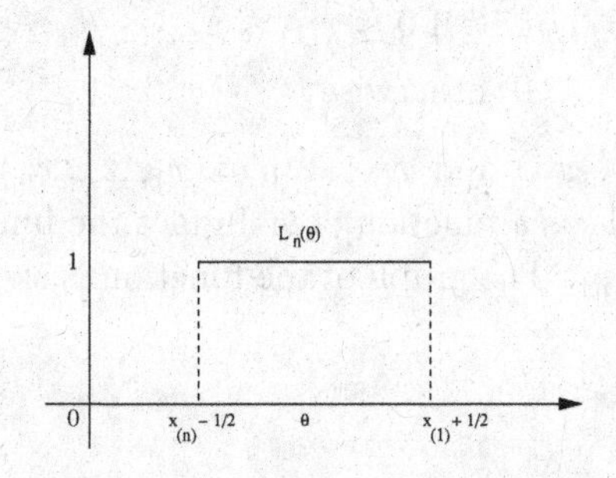

Fig. 8.4

We note that the function $L_n(\theta)$ is maximum for all θ between $x_{(n)} - \frac{1}{2}$ and $x_{(1)} + \frac{1}{2}$. Hence every $\hat{\theta}_n \in [x_{(n)} - \frac{1}{2}, x_{(1)} + \frac{1}{2}]$ is a maximum likelihood estimator of θ and the maximum likelihood estimator is not unique in this example.

There are situations where a maximum likelihood estimator might not exist. Other examples where it might exist but not necessarily unique are presented above. An important property of a maximum likelihood estimator is its invariance.

Theorem 8.1. Let $\hat{\theta}$ be a maximum likelihood estimator of a parameter $\theta \in \Theta \subset R^k$. Suppose $\hat{\theta} \in \Theta$. Let $g : \Theta \to \Omega$ be a function mapping Θ into a set $\Omega \subset R^r$ for some $1 \leq r \leq k$. Then the estimator $g(\hat{\theta})$ is a maximum likelihood estimator of the parametric function $g(\theta)$ in the sense that it maximizes the induced likelihood.

Proof. For each $\omega \in \Omega$, define

$$G(\omega) = \{\theta : \theta \in \Theta, g(\theta) = \omega\}$$

and

$$M(\boldsymbol{x}; \omega) = \sup_{\theta \in G(\omega)} L(\boldsymbol{x}; \theta)$$

where $L(\boldsymbol{x}; \theta)$ is the likelihood function when the observed sample is $\boldsymbol{x}$ and the parameter is θ. The function $M(\boldsymbol{x}; \omega)$ is the induced likelihood function. Note that the collection of subsets $\{G(\omega) : \omega \in \Omega\}$ forms a partition of the set Θ. Since the maximum likelihood estimator $\hat{\theta} \in \Theta$, it belongs to one and only one member of the partition, say, $G(\hat{\omega})$. Further more

$$\begin{aligned} L(\boldsymbol{x}; \hat{\theta}) &\leq \sup_{\theta \in G(\hat{\omega})} L(\boldsymbol{x}; \theta) \\ &= M(\boldsymbol{x}; \hat{\omega}) \\ &\leq \sup_{\omega \in \Omega} M(\boldsymbol{x}; \omega) \\ &= \sup_{\theta \in \Theta} L(\boldsymbol{x}; \theta) \\ &= L(\boldsymbol{x}; \hat{\theta}). \end{aligned}$$

Hence

$$M(\boldsymbol{x}; \hat{\omega}) = \sup_{\omega \in \Omega} M(\boldsymbol{x}; \omega) = \sup_{\theta \in \Theta} L(\boldsymbol{x}; \theta) = L(\boldsymbol{x}; \hat{\theta})$$

which implies that $\hat{\omega}$ is a maximum likelihood estimator of $\omega = g(\theta)$. Since $\hat{\theta} \in G(\hat{\omega})$, it follows that

$$g(\hat{\theta}) = \hat{\omega}.$$

Therefore $g(\hat{\theta})$ is a maximum likelihood estimator of $g(\theta)$.

As an application of the above theorem, it follows that

$$\hat{\sigma} = \sqrt{\frac{1}{n} \sum_{i=1}^{n} (x_i - \bar{x})^2}$$

is a maximum likelihood estimator of the standard deviation σ in the Example 8.10.

Remarks. Note that the invariance property of a maximum likelihood estimator of a parameter θ is an important property as it gives a technique to obtain the maximum likelihood estimator of any function $g(\theta)$ of the parameter θ once the maximum likelihood estimator of the parameter θ is known. However such a property does not hold for an unbiased estimator, that is, if an estimator $\tilde{\theta}$ is an unbiased estimator of θ, it is not necessary that $g(\tilde{\theta})$ is an unbiased estimator of the function $g(\theta)$. For instance, $E_\theta(X) = \theta$ does not imply that $E_\theta(X^2) = \theta^2$ unless X is a degenerate random variable at *theta*.

The point now is to find whether the maximum likelihood estimators are *good* or *efficient* with respect to any specified criterion. This is the topic of our discussion later in this chapter.

We will discuss another method of estimation which is useful when the data is obtained in groups or the data is classified into different groups.

Method of minimum Chi-square

Suppose we have classified a set of independent observations into k different mutually exclusive groups $G_i, 1 \leq i \leq k$. Suppose the groups G_i for classification have been specified by the statistician before the data was obtained. It might also be possible that the actual observations are not recorded completely except that it is known whether a particular observation belongs to a particular group G_i or not. In other words the frequencies of the number of observations in the sample in the different groups are known. Let $p_i > 0$ be the probability that an observation falls into the group G_i for $1 \leq i \leq k$. Since the groups are mutually exclusive, we have $p_1 + \cdots + p_k = 1$. Suppose that the total number of independent observations is n. Let n_i denote the number of observations in the sample belonging to the group G_i. It is clear that the value n can be interpreted as the number of trials and the value n_i can be interpreted as the frequency of the outcomes of type i, that is, the number of observations belonging to the group G_i for $1 \leq i \leq k$. Hence the random vector $(n_i, n_2, \ldots, n_k)$ has the Multinomial distribution with n as the number of trials and $\boldsymbol{p} = (p_1, p_2, \ldots, p_k)$ as the probability vector. For a discussion of the Multinomial distribution, see Chapter 5. Our problem is to estimate the vector $\boldsymbol{p}$. Since the random vector $(n_1, n_2, \ldots, n_k)$ has the Multinomial distribution, it follows that $E(n_i) = np_i$, for $1 \leq i \leq k$. The quantity n_i is the observed frequency in the group G_i and np_i is the expected frequency for the group G_i since p_i is the probability for an observation to belong to the group G_i. If p_i is known, then one expects the difference $n_i - np_i$ to be small. Motivated by this reasoning, we define

$$\chi^2 = \sum_{i=1}^{k} \frac{(n_i - np_i)^2}{np_i}$$

and choose $\hat{p}_i, 1 \leq i \leq k$, as estimators of $p_i, 1 \leq i \leq k$, as those values of p_i, $1 \leq i \leq k$ which minimize χ^2 and hence this method is known as the *method of minimum Chi-square.*

Let us illustrate the method by an example. Suppose k=2. In other words we have two groups G_1 and G_2. Let p be the probability that an observation belongs to the group G_1 and suppose that $0 < p < 1$. Then the probability that an observation belongs to the group G_2 is $1 - p = q$. Here

$$\chi^2 = \frac{(n_1 - np)^2}{np} + \frac{(n_2 - nq)^2}{nq}$$

where n_i is the frequency of observations that belong to the group G_i for $i = 1, 2$ and n is the total number of independent observations that were classified into the two groups. Then $n = n_1 + n_2$ and we can rewrite the expression for χ^2 in the form

$$\begin{aligned} \chi^2 &= \frac{(n_1 - np)^2}{np} + \frac{((n - n_1) - n(1 - p))^2}{n(1 - p)} \\ &= \frac{(n_1 - np)^2}{np} + \frac{(n_1 - np)^2}{n(1 - p)} \\ &= \frac{(n_1 - np)^2}{np(1 - p)}. \end{aligned}$$

Observe that the function χ^2 is a function of the parameter p in this case. Differentiating χ^2 with respect to p and solving the resulting equation, it can be checked that $\hat{p} = \frac{n_1}{n}$ is a solution of the equation (check?). In fact the function χ^2 is minimum at this point (why?). Hence $\hat{p}$ is the estimator of p obtained through the method of minimum chi-square. Note that the classification of n independent observations, into two groups with p as the probability for an observation to fall into the group G_1, implies that the frequency n_1 has the Binomial distribution with the parameters n and p.

Suppose that the probabilities $p_i, 1 \leq i \leq k$ are in turn known functions $p_i = p_i(\theta)$ of a parameter $\theta = (\theta_1, \theta_2, \ldots, \theta_r)$. In this case, the method of estimating the parameter θ consists in choosing the estimator $\hat{\theta}$ as that value of θ which minimizes

$$\chi^2 = \sum_{i=1}^{k} \frac{(n_i - np_i(\theta))^2}{np_i(\theta)}.$$

Example 8.13. Suppose that a sample of n independent observations obtained from a population are classified into three groups $G_i, i = 1, 2, 3$ with $p_1(\theta) = \theta^2$, $p_2(\theta) = 2\theta(1-\theta)$ and $p_3(\theta) = (1-\theta)^2$ where $0 < \theta < 1$. The problem is to estimate θ by the method of minimum Chi-square. Let n_i denote the number of observations belonging to the group G_i for $i = 1, 2, 3$. Here

$$\begin{aligned}\chi^2 &= \sum_{i=1}^{3} \frac{(n_i - np_i(\theta))^2}{np_i(\theta)} \\ &= \frac{(n_1 - n\theta^2)^2}{n\theta^2} + \frac{(n_2 - 2n\theta(1-\theta))^2}{2n\theta(1-\theta)} + \frac{(n_3 - n(1-\theta)^2)^2}{n(1-\theta)^2}.\end{aligned}$$

We leave it to the reader to check that $\hat{\theta} = \frac{n_1+n_2}{2n}$ minimizes the function χ^2 over all $0 < \theta < 1$.

There are other variations of this method. We will not go into further discussion here. Under some regularity conditions, it can be shown that the estimators obtained by this method have nice properties for large samples. Detailed discussion on this topic is beyond the scope of this book.

8.4 Cramer-Rao Inequality and Efficient Estimation

In order to find out whether the estimators derived by the method of maximum likelihood or by any other method of estimation have *good* or *optimum* properties in some sense, it is necessary to specify the criterion under which the optimality of an estimator can be judged. We now discuss one such criterion and indicate its application. Let us now consider a population with a probability density function or probability function $f(x, \theta)$ where $\theta \in \Theta \subset R$ is a scalar parameter. *We assume that the support of the probability function or the probability density function, that is the set of all x where $f(x, \theta) > 0$ does not depend on the parameter $\theta \in \Theta$.* For instance, the following discussion is not applicable to the uniform distribution on $(0, \theta)$ since the support $(0, \theta)$ of the distribution depends on the parameter θ. The problem is to estimate the parameter θ based on a random sample $X_1, \ldots, X_n$ and obtain an efficient estimator of θ, if any, efficiency to be defined later. Let $\boldsymbol{X} = (X_1, \ldots, X_n)$. For definiteness, let us first assume that the function $f(x, \theta)$ is a probability density function. The discussion in the case of discrete distributions follows on similar lines. The likelihood function based on the random sample $X_1, \ldots, X_n$ is

$$L_n(\boldsymbol{X}; \theta) = \Pi_{i=1}^n f(X_i, \theta). \tag{8.1}$$

Suppose the statistic $g(\boldsymbol{X})$ is an estimator of the parameter θ with $E_\theta[g^2(\boldsymbol{X})] < \infty$. Let

$$B(\theta) = E_\theta[g(\boldsymbol{X})] - \theta. \tag{8.2}$$

Then $B(\theta)$ is the *bias* of the estimator $g(\boldsymbol{X})$ with respect to θ. If the estimator $g(\boldsymbol{X})$ is unbiased for the parameter θ, then $B(\theta) = 0$. Observe that

$$\log L_n(\boldsymbol{X}; \theta) = \sum_{i=1}^n \log f(X_i, \theta) \tag{8.3}$$

and

$$\frac{d \log L_n(\boldsymbol{X}; \theta)}{d\theta} = \sum_{i=1}^n \frac{d \log f(X_i, \theta)}{d\theta} \tag{8.4}$$

assuming that $f(x, \theta)$ is differentiable with respect to θ. The function

$$\frac{d \log L_n(\boldsymbol{X}, \theta)}{d\theta}$$

is called the *score function* based on the observations $X_1, \ldots, X_n$. Since $f(x, \theta)$ is a probability density function, it follows that

$$\underbrace{\int_{-\infty}^{\infty} \cdots \int_{-\infty}^{\infty}}_{n \text{ times}} f(x_1, \theta) \ldots f(x_n, \theta) \; dx_1 \ldots dx_n = 1$$

for all θ. For simplicity of notation, we shall write the above equation in the form

$$\int_{R^n} \Pi_{i=1}^n f(x_i, \theta) \; d\boldsymbol{x} = 1. \tag{8.5}$$

Since $E_\theta[g(\boldsymbol{X})] = \theta + B(\theta)$, we have

$$\int_{R^n} g(\boldsymbol{x}) \Pi_{i=1}^n f(x_i, \theta) \; d\boldsymbol{x} = \theta + B(\theta). \tag{8.6}$$

The basic assumption in the present discussion is that the relations (8.5) and (8.6) can be differentiated and that differentiation under the integral sign with respect to θ is permissible. This assumption implies that

$$\frac{d}{d\theta}\left[\int_{R^n} L_n(\boldsymbol{x};\theta)d\boldsymbol{x}\right] = \int_{R^n} \frac{dL_n(\boldsymbol{x};\theta)}{d\theta}d\boldsymbol{x} = 0 \tag{8.7}$$

and

$$\frac{d}{d\theta}\left[\int_{R^n} g(\boldsymbol{x})L_n(\boldsymbol{x};\theta)d\boldsymbol{x}\right] = \int_{R^n} g(\boldsymbol{x})\frac{dL_n(\boldsymbol{x};\theta)}{d\theta}d\boldsymbol{x} = 1 + B'(\theta) \tag{8.8}$$

where $B'(\theta)$ denotes the derivative of $B(\theta)$ with respect to θ. Let us write the equations (8.7) and (8.8) in a slightly different form using the relations

$$\frac{dL_n(\boldsymbol{x};\theta)}{d\theta} = \frac{d\log L_n(\boldsymbol{x};\theta)}{d\theta}L_n(\boldsymbol{x};\theta).$$

We can rewrite (8.7) and (8.8) respectively in the form

$$\int_{R^n} \frac{d\log L_n(\boldsymbol{x},\theta)}{d\theta}L_n(\boldsymbol{x},\theta)d\boldsymbol{x} = 0 \tag{8.9}$$

and

$$\int_{R^n} g(\boldsymbol{x})\frac{d\log L_n(\boldsymbol{x};\theta)}{d\theta}L_n(\boldsymbol{x};\theta)d\boldsymbol{x} = 1 + B'(\theta). \tag{8.10}$$

Since $L_n(\boldsymbol{x};\theta)$ is the joint probability density function of $\boldsymbol{X} = (X_1, \ldots, X_n)$ when θ is the true parameter, the above relations can be written in term of the expectations as

$$E_\theta\left[\frac{d\log L_n(\boldsymbol{X};\theta)}{d\theta}\right] = 0 \tag{8.11}$$

and

$$E_\theta\left[g(\boldsymbol{X})\frac{d\log L_n(\boldsymbol{X};\theta)}{d\theta}\right] = 1 + B'(\theta). \tag{8.12}$$

Combining the above relations, we have

$$E_\theta\left[(g(\boldsymbol{X}) - \theta)\frac{d\log L_n(\boldsymbol{X};\theta)}{d\theta}\right] = 1 + B'(\theta). \tag{8.13}$$

We now use the Cauchy-Schwartz inequality proved earlier in the Chapter 5 which can be stated in the following form. For any two random variables K and H with $E(K^2) < \infty$ and $E(H^2) < \infty$,

$$(E_\theta[KH])^2 \leq E_\theta(K^2)E_\theta(H^2) \tag{8.14}$$

equality occurring if and only if K and H are linearly related. Let

$$K = g(\boldsymbol{X}) - \theta \text{ and } H = \frac{d \log L_n(\boldsymbol{X};\theta)}{d\theta}.$$

Then, it follows from (8.13), that

$$\begin{aligned}(1+B'(\theta))^2 &= \left(E_\theta[(g(\boldsymbol{X})-\theta)\frac{d\log L_n(\boldsymbol{X};\theta)}{d\theta}]\right)^2 \\ &\leq E_\theta[g(\boldsymbol{X})-\theta]^2 E_\theta\left[\frac{d\log L_n(\boldsymbol{X};\theta)}{d\theta}\right]^2 \qquad (8.15)\end{aligned}$$

from (8.14). Let us denote

$$I_n(\theta) = E_\theta\left(\frac{d\log L_n(\boldsymbol{X};\theta)}{d\theta}\right)^2. \qquad (8.16)$$

The function $I_n(\theta)$ is called the *Fisher information* in the sample $(X_1,\ldots,X_n)$. Note that $I_n(\theta) \geq 0$ for all θ. Suppose that $0 < I_n(\theta) < \infty$ for all $\theta \in \Theta$. We have the following inequality from (8.15). For any estimator $g(\boldsymbol{X})$ of θ,

$$E_\theta[g(\boldsymbol{X})-\theta]^2 \geq \frac{[1+B'(\theta)]^2}{I_n(\theta)} \qquad (8.17)$$

where $B(\theta)$ is the bias of the estimator $g(\boldsymbol{X})$ and $I_n(\theta)$ is the Fisher information contained in the observations $X_1,\ldots,X_n$. This inequality is known as the *Cramer-Rao inequality* and as the *Information inequality* in some of the recent literature. As was shown above, it is an easy consequence of the Cauchy-Schwartz inequality. We should caution that the inequality is valid only under various assumptions stated above and it is not applicable, for instance, when the support of the probability density function depends on the parameter θ. Modified versions of the Cramer-Rao inequality are available when this assumption is violated but a discussion of these results is beyond the scope of this book. We are not going to discuss also the multivariate version of the inequality (8.17). Let us again get back to the notion of Fisher information. Note that

$$\begin{aligned}I_n(\theta) &= E_\theta\left[\frac{d\log L_n(\boldsymbol{X};\theta)}{d\theta}\right]^2 \\ &= \mathrm{Var}_\theta\left[\frac{d\log L_n(\boldsymbol{X};\theta)}{d\theta}\right] \qquad (8.18)\end{aligned}$$

since the expectation of the random variable $\frac{d \log L_n(\boldsymbol{X};\theta)}{d\theta}$ is zero when θ is the true parameter as shown in the equation (8.11). Hence

$$\begin{aligned}
I_n(\theta) &= \mathrm{Var}_\theta \left[\sum_{i=1}^{n} \frac{d \log f(X_i, \theta)}{d\theta} \right] \\
&= \sum_{i=1}^{n} \mathrm{Var}_\theta \left[\frac{d \log f(X_i, \theta)}{d\theta} \right] \quad \text{(from the independence of } X_i, 1 \le i \le n) \\
&= n \ \mathrm{Var}_\theta \left[\frac{d \log f(X_1, \theta)}{d\theta} \right] \quad \text{(since } X_i, 1 \le i \le n \\
&\qquad \text{are identically distributed)} \\
&= n \ E_\theta \left[\frac{d \log f(X_1, \theta)}{d\theta} \right]^2 \quad \text{(since } E_\theta \left(\frac{d \log f(X_1, \theta)}{d\theta} \right) = 0) \qquad (8.19)
\end{aligned}$$

which shows that the Fisher-information in n observations is n times the Fisher information contained in one observation. In other words the Fisher-information is *additive*. Cramer-Rao inequality can now be written as

$$E_\theta[g(\boldsymbol{X}) - \theta]^2 \ge \frac{[1 + B'(\theta)]^2}{n \ I(\theta)} \qquad (8.20)$$

where we write $I(\theta)$ for $I_1(\theta)$ for simplicity. The function $I(\theta)$ is the Fisher information contained in one observation. In particular, if $g(\boldsymbol{X})$ is an unbiased estimator for θ, then $B(\theta) = 0$ for all $\theta \in \Theta$ and hence $B'(\theta) = 0$ and $E_\theta[g(\boldsymbol{X}) - \theta]^2 = \mathrm{Var}_\theta[g(\boldsymbol{X})]$. The inequality (8.20) reduces to

$$\mathrm{Var}_\theta[g(\boldsymbol{X})] \ge \frac{1}{n \ I(\theta)}. \qquad (8.21)$$

The lower bound $\frac{1}{n \ I(\theta)}$ is called the *Cramer-Rao lower bound.* The inequality (8.21) can be interpreted in the following manner. Whatever be the unbiased estimator $g(\boldsymbol{X})$ of θ is, its variance cannot be less than $\frac{1}{n \ I(\theta)}$, subject to, of course, the regularity conditions assumed earlier. Hence, if one finds an unbiased estimator $g_0(\boldsymbol{X})$ of θ whose variance attains the lower bound $\frac{1}{n \ I(\theta)}$, then the estimator $g_0(\boldsymbol{X})$ is the *best* unbiased estimator as every other unbiased estimator with finite variance has its variance greater than or equal to that of $g_0(\boldsymbol{X})$ for all $\theta \in \Theta$. Recall that, among two unbiased estimators $g_0(\boldsymbol{X})$ and $g_1(\boldsymbol{X})$ of θ , the estimator $g_0(\boldsymbol{X})$ is said to be as good as $g_1(\boldsymbol{X})$ if the variance of $g_0(\boldsymbol{X})$ is less than or equal to that of the estimator $g_1(\boldsymbol{X})$ for all values of the parameter θ. On the basis of this discussion, we can now define an *efficient unbiased estimator.* An unbiased estimator $g(\boldsymbol{X})$ of θ based on a random sample $\boldsymbol{X} = (X_1, \ldots, X_n)$

is said to be *efficient* in the Cramer-Rao sense if its variance is equal to $\frac{1}{n\ I(\theta)}$ where n is the sample size and $I(\theta)$ is the Fisher information in one observation. It is also a *minimum variance unbiased estimator* (MVUE) in the sense that it has the smallest variance uniformly over the class of all unbiased estimators with finite variances for all $\theta \in \Theta$. It should be pointed out that there could be a MVUE for θ yet its variance might not attain the Cramer-Rao lower bound. We will illustrate our discussion by some examples later.

In order to compute the Fisher information $I(\theta)$ in a simpler manner, let us get an alternate expression for $I(\theta)$. Since

$$\int_R f(x,\theta)\, dx = 1, \tag{8.22}$$

differentiating under the integral sign both sides with respect to θ, we have

$$\int_R \frac{df(x,\theta)}{d\theta} dx = 0$$

or equivalently

$$\int_R \frac{d \log f(x,\theta)}{d\theta} f(x,\theta) dx = 0. \tag{8.23}$$

Assuming that we can differentiate under the integral sign with respect to θ on both sides of the above equation again, we conclude that

$$\int_R \left[\frac{d^2 \log f(x,\theta)}{d\theta^2} f(x,\theta) + \frac{d \log f(x,\theta)}{d\theta} \frac{df(x,\theta)}{d\theta} \right] dx = 0$$

or equivalently

$$\int_R \left[\frac{d^2 \log f(x,\theta)}{d\theta^2} + \frac{d \log f(x,\theta)}{d\theta} \frac{d \log f(x,\theta)}{d\theta} \right] f(x,\theta) dx = 0.$$

In other words

$$E_\theta \left[\frac{d^2 \log f(X,\theta)}{d\theta^2} + \left(\frac{d \log f(X,\theta)}{d\theta} \right)^2 \right] = 0. \tag{8.24}$$

Hence

$$\begin{aligned} I(\theta) &= E_\theta \left[\frac{d \log f(X,\theta)}{d\theta} \right]^2 \\ &= -E_\theta \left[\frac{d^2 \log f(X,\theta)}{d\theta^2} \right] \end{aligned} \tag{8.25}$$

from (8.24). This formula gives an alternate way for computing the Fisher information.

Example 8.14. Suppose $X_1, X_2, \ldots, X_n$ is a random sample from the normal distribution with mean μ and known variance σ^2. The problem of interest is to derive the Cramer-Rao lower bound corresponding to the parameter μ and examine whether there exists an unbiased estimator of μ which attains this lower bound. The probability density function of X_1 is

$$f(x,\mu) = \frac{1}{\sqrt{2\pi\sigma^2}} \exp\left\{-\frac{1}{2\sigma^2}(x-\mu)^2\right\}$$

and

$$\log f(x,\mu) = -\frac{1}{2}\log(2\pi\sigma^2) - \frac{1}{2\sigma^2}(x-\mu)^2.$$

Therefore

$$\frac{d\log f(x,\mu)}{d\mu} = \frac{1}{\sigma^2}(x-\mu)$$

and

$$-\frac{d^2\log f(x,\mu)}{d\mu^2} = \frac{1}{\sigma^2}.$$

Hence

$$I(\mu) = E_\mu\left[\frac{d\log f(X,\mu)}{d\mu}\right]^2 = \frac{1}{\sigma^4}E_\mu[X-\mu]^2 = \frac{1}{\sigma^2}.$$

It can also be derived using the relation (8.25). Observe that

$$I(\mu) = -E_\mu\left[\frac{d^2\log f(X,\mu)}{d\mu^2}\right] = \frac{1}{\sigma^2}$$

which proves that the Cramer-Rao lower bound is $\frac{1}{n[1/\sigma^2]} = \frac{\sigma^2}{n}$ from (8.21). It is easy to see that the estimator $\bar{X}$ is an unbiased estimator of μ and $\mathrm{Var}_\mu(\bar{X}) = \sigma^2/n$. Hence $\mathrm{Var}_\mu(\bar{X})$ is the same as the Cramer-Rao lower bound in this example. Therefore the estimator $\bar{X}$ is a minimum variance unbiased estimator for the parameter μ. It can be shown that there is one and only one such estimator.

Let us now consider another example to illustrate the results obtained earlier.

Example 8.15. Suppose $X_1, X_2, \ldots, X_n$ are independent Bernoulli trials, that is, $X_i, 1 \le i \le n$ are independent random variables with

$$P(X_i = 1) = p \text{ and } P(X_i = 0) = 1 - p.$$

The problem is to obtain the Cramer-Rao lower bound for unbiased estimators of p. It is easy to see that the probability function of X_1 can be written in the form

$$f(x,p) = p^x(1-p)^{1-x}, x = 0, 1.$$

Therefore

$$\frac{d \log f(x,p)}{dp} = \frac{x}{p} - \frac{1-x}{1-p}$$

and

$$\frac{d^2 \log f(x,p)}{dp^2} = -\frac{x}{p^2} - \frac{(1-x)}{(1-p)^2}.$$

Hence

$$\begin{aligned} I(p) &= E_p\left[-\frac{d^2 \log f(X,p)}{dp^2}\right] \\ &= E_p\left[\frac{X}{p^2} + \frac{(1-X)}{(1-p)^2}\right] \\ &= \frac{1}{p} + \frac{1}{1-p} = \frac{1}{p(1-p)} \end{aligned}$$

and the Cramer-Rao lower bound is $\frac{p(1-p)}{n}$. It is easy to see that, if $T = \sum_{i=1}^n X_i$, then the statistic $\frac{T}{n}$ is an unbiased estimator for p and

$$\text{Var}_p\left(\frac{T}{n}\right) = \frac{p(1-p)}{n}.$$

Hence the estimator $\frac{T}{n}$ is a minimum variance unbiased estimator for p. Observe that the estimator $\frac{T}{n}$ is the proportion of successes in n trials.

The next example deals with a case where a minimum variance unbiased estimator for a parameter exists but the Cramer-Rao lower bound is not attained.

Example 8.16. Suppose $X_1, X_2, \ldots, X_n$ is a random sample of size $n > 2$ from a population with the probability density function

$$\begin{aligned} f(x,\lambda) &= \lambda e^{-\lambda x}, x > 0 \\ &= 0 \quad x \leq 0. \end{aligned}$$

The problem is to find the Cramer-Rao lower bound for estimating the parameter λ and find a minimum variance unbiased estimator for λ if any. Note that, for $x > 0$,

$$\log f(x,\lambda) = \log \lambda - \lambda x,$$

$$\frac{d \log f(x, \lambda)}{d\lambda} = \frac{1}{\lambda} - x,$$

and

$$\frac{d^2 \log f(x, \lambda)}{d\lambda^2} = -\frac{1}{\lambda^2}.$$

Hence

$$I(\lambda) = -E_\lambda \left[\frac{d^2 \log f(X, \lambda)}{d\lambda^2} \right] = \frac{1}{\lambda^2}$$

and the Cramer-Rao lower bound is $\frac{\lambda^2}{n}$. It is easy to see that $E_\lambda(\bar{X}) = \frac{1}{\lambda}$. One can consider the statistic $\frac{1}{\bar{X}}$ as an estimator of λ. However it is not an unbiased estimator of λ. In fact the random variable $\sum_{i=1}^{n} X_i$ has the Gamma distribution with the parameters n and λ. We leave it to the reader to show that

$$E_\lambda \left(\frac{1}{\sum_{i=1}^n X_i} \right) = \frac{\lambda}{n-1}.$$

Hence

$$T = \frac{n-1}{\sum_{i=1}^n X_i}$$

is an unbiased estimator of λ. Check that

$$\mathrm{Var}_\lambda(T) = \frac{\lambda^2}{n-2}.$$

Note that $\mathrm{Var}_\lambda(T)$ is strictly bigger than the Cramer-Rao lower bound $\frac{\lambda^2}{n}$. Using other considerations, it can be shown that T is in fact a minimum variance unbiased estimator for the parameter λ.

8.5 Sufficient Statistics

As we have mentioned earlier, any statistic is, by definition, a function of the observations $X_i, 1 \leq i \leq n$ only. Suppose the joint distribution function of $(X_1, \ldots, X_n)$ depends on a parameter $\theta \in \Theta$ which might be a scalar or a vector. The question arises whether it is possible to find a statistic, that is a real-valued or vector-valued function of the observations $X_i, 1 \leq i \leq n$, which contains all the *information* about the parameter θ. Such a question is important when we want to summarize the data on hand as the storing of a large amount of data is expensive and liable to recording errors. It is also unnecessary to store all the data if we can summarize the data without loosing any relevant *information*. We now give a precise mathematical definition leading to this idea.

Definition. A statistic $T = T(X_1, \ldots, X_n)$ is said to be a *sufficient statistic* for the parameter θ if the conditional distribution of the random vector $(X_1, \ldots, X_n)$ given T does not depend on the parameter $\theta \in \Theta$.

Here the statistic T could be one-dimensional or multi-dimensional. It is beyond the scope of this book to discuss all the properties and the results concerning sufficient statistics. We first discuss an example to explain the concept.

Example 8.17. Let $X_1, \ldots, X_n$ be a random sample from the Poisson distribution with the parameter λ. Then the joint probability function of $(X_1, \ldots, X_n)$ is

$$P(X_1 = x_1, \ldots, X_n = x_n) = \Pi_{i=1}^n e^{-\lambda} \frac{\lambda^{x_i}}{x_i!}, \quad x_i = 0, 1, 2, \ldots; i \geq 1.$$

We now compute the conditional distribution of $(X_1, \ldots, X_n)$ given $T = X_1 + \ldots + X_n$. From our earlier discussions in the Chapter 5, check that the random variable T has the Poisson distribution with parameter $n\lambda$. Hence, for any integer $t \geq 0$,

$$P[X_1 = x_1, \ldots, X_n = x_n | T = t] = \frac{P[X_1 = x_1, \ldots, X_n = x_n; T = t]}{P[T = t]}.$$

The numerator in the expression on the right side of the above equation is zero if $x_1 + \cdots + x_n \neq t$. For $x_1 + \cdots + x_n = t$,

$$\begin{aligned} P[X_1 = x_1, \ldots, X_n = x_n | T = t] &= \frac{P[X_1 = x_1, \ldots, X_n = x_n]}{P[T = t]} \\ &= \left(\Pi_{i=1}^n e^{-\lambda} \frac{\lambda^{x_i}}{x_i!} \right) / \left(e^{-n\lambda} \frac{(n\lambda)^t}{t!} \right) \\ &= \frac{t!}{x_1! \ldots x_n!} \left(\frac{1}{n} \right)^t. \end{aligned}$$

Observe that the last term does not depend on λ. In fact the conditional distribution of $(X_1, \ldots, X_n)$ given $T = t$ is the Multinomial distribution with t as the number of trials and $\frac{1}{n}$ as the probability for each type of the n possible outcomes and this distribution does not depend on the parameter λ. Hence T is a sufficient statistic for the parameter λ.

In general, it is difficult to find a sufficient statistic directly from the definition as it involves the computation of the conditional distributions. We can use an alternate method to find a sufficient statistic. This is done by using the Neyman factorization theorem.

Theorem 8.2. **(Neyman factorization theorem)** Suppose $f(\boldsymbol{x};\theta)$ is the joint probability function (or the joint probability density function in the continuous case) of a random vector $\boldsymbol{X} = (X_1, \ldots, X_n)$ where $\theta \in \Theta$. A statistic $T = T(X_1, \ldots, X_n)$ is sufficient for θ if and only if there exists a nonnegative function g depending only on T and θ and a nonnegative function h depending on $\boldsymbol{X} = (X_1, \ldots, X_n)$ only but not on $\theta \in \Theta$ such that

$$f(\boldsymbol{x};\theta) = g(T(\boldsymbol{x});\theta)h(\boldsymbol{x})$$

for all $\boldsymbol{x} = (x_1, \ldots, x_n)$ in the support of f and for all $\theta \in \Theta$.

Proof. We will prove the result when the random vector $\boldsymbol{X}$ has a discrete distribution. The proof in the continuous case involves many technicalities and we omit the proof in this case.

Suppose that a random vector $\boldsymbol{X}$ has the probability function $f(\boldsymbol{x};\theta) = P_\theta(\boldsymbol{X} = \boldsymbol{x})$ where P_θ denotes the probability when θ is the true parameter. Suppose that there exists a factorization

$$f(\boldsymbol{x};\theta) = g(T(\boldsymbol{x}),\theta)h(\boldsymbol{x})$$

as stated in the theorem. For any possible value t of T, let

$$A_t = \{\boldsymbol{x} : T(\boldsymbol{x}) = t\}.$$

Then, for any $\boldsymbol{x} \in A_t$,

$$\begin{aligned}
P_\theta(\boldsymbol{X} = \boldsymbol{x} | T = t) &= \frac{P_\theta(\boldsymbol{X} = \boldsymbol{x}; T = t)}{P_\theta(T = t)} \\
&= \frac{P_\theta(\boldsymbol{X} = \boldsymbol{x})}{P_\theta(T = t)} \\
&= \frac{f(\boldsymbol{x};\theta)}{\sum_{\boldsymbol{y} \in A_t} f(\boldsymbol{y};\theta)} \\
&= \frac{g(T(\boldsymbol{x});\theta)h(\boldsymbol{x})}{\sum_{\boldsymbol{y} \in A_t} g(T(\boldsymbol{y});\theta)h(\boldsymbol{y})} \\
&= \frac{h(\boldsymbol{x})}{\sum_{\boldsymbol{y} \in A_t} h(\boldsymbol{y})}
\end{aligned}$$

since $T(\boldsymbol{y}) = t = T(\boldsymbol{x})$ for any $\boldsymbol{y} \in A_t$. Note that the expression on the right side of the above equation does not depend on $\theta \in \Theta$. Furthermore, if $\boldsymbol{x}$ is not in A_t, then

$$P_\theta(\boldsymbol{X} = \boldsymbol{x}|T = t) = 0$$

which holds for every $\theta \in \Theta$. Hence the conditional distribution of $\boldsymbol{X}$ given the statistic T does not depend on θ. Therefore T is a sufficient statistic for the parameter $\theta \in \Theta$.

Conversely suppose that T is a sufficient statistic for the parameter $\theta \in \Theta$. Then, for any fixed value t of T and any $\boldsymbol{x} \in A_t$ and for any $\theta \in \Theta$, the conditional probability

$$P_\theta(\boldsymbol{X} = \boldsymbol{x}|T = t)$$

should not depend on $\theta \in \Theta$. Hence it is of the form $h(\boldsymbol{x})$ for some function $h(.)$. Let $g(t;\theta) = P_\theta(T = t)$. Then

$$\begin{aligned} f(\boldsymbol{x};\theta) &= P_\theta(\boldsymbol{X} = \boldsymbol{x}) \\ &= P_\theta(\boldsymbol{X} = \boldsymbol{x}; T = t) \\ &= P_\theta(\boldsymbol{X} = \boldsymbol{x}|T = t)P_\theta(T = t) \\ &= h(\boldsymbol{x})g(t;\theta) \\ &= h(\boldsymbol{x})g(T(\boldsymbol{x});\theta). \end{aligned}$$

Hence the probability function $f(\boldsymbol{x};\theta)$ can be factored in the form specified in the theorem.

Let us see how we can apply this theorem to find a sufficient statistic for the parameter λ in the Example 8.17 discussed earlier. Here

$$\begin{aligned} f(x:\lambda) &= \Pi_{i=1}^n e^{-\lambda}\frac{\lambda^{x_i}}{x_i!} = e^{-n\lambda}\frac{\lambda^{x_1+\cdots+x_n}}{x_1!\ldots x_n!} \\ &= g(x_1+\cdots+x_n,\lambda)\ h(x_1,\ldots,x_n) \end{aligned}$$

where

$$g(t,\lambda) = e^{-n\lambda}\lambda^t \text{ and } h(x) = \frac{1}{x_1!\ldots x_n!}$$

for $t \geq 0$ and $x_i \geq 0, 1 \leq i \leq n$. This proves that the statistic $T = X_1 + \cdots + X_n$ is a sufficient statistic for the parameter λ.

Example 8.18. Suppose that $X_1,\ldots,X_n$ is a random sample from the uniform distribution on $(0,\theta)$. Then the joint probability density function of $(X_1,\ldots,X_n)$ is

$$\begin{aligned} f(\boldsymbol{x};\theta) &= \frac{1}{\theta^n} \text{ if } 0 \leq x_i \leq \theta, 1 \leq i \leq n \\ &= 0 \qquad \text{otherwise.} \end{aligned}$$

This can also be written in the form

$$f(\boldsymbol{x};\theta) = \frac{1}{\theta^n} \text{ if } 0 \leq \max(x_i, 1 \leq i \leq n) \leq \theta$$
$$= 0 \text{ otherwise,}$$

or equivalently

$$f(\boldsymbol{x};\theta) = \frac{1}{\theta^n} R(\max_{1 \leq i \leq n} x_i; \theta)$$

where $R(a,b) = 1$ if $0 \leq a \leq b$ and $R(a,b) = 0$ otherwise. Define $g(t,y) = \frac{1}{y^n} R(t,y)$ and $h(\boldsymbol{x}) \equiv 1$. Then

$$f(\boldsymbol{x};\theta) = g(\max_{1 \leq i \leq n} x_i, \theta) h(\boldsymbol{x})$$

and it follows that $\max_{1 \leq i \leq n} X_i$ is a sufficient statistic for the parameter θ by the Neyman factorization theorem.

An interesting and important use of a sufficient statistic is in improving an estimator. The following theorem is due to Rao and Blackwell.

Theorem 8.3. **(Rao-Blackwell theorem)** Suppose T is a sufficient statistic for a parameter $\theta \in \Theta$. Let $\delta(\boldsymbol{X})$ be an unbiased estimator for a parametric function $g(\theta)$. Then $h(T) = E_\theta[\delta(\boldsymbol{X})|T]$, the conditional expectation of $\delta(\boldsymbol{X})$ given T, is an unbiased estimator of $g(\theta)$ and

$$\text{Var}_\theta[h(T)] \leq \text{Var}_\theta[\delta(\boldsymbol{X})]$$

for all $\theta \in \theta$ equality occurring if and only if $\delta(\boldsymbol{X})$ is a function of T with probability one.

Proof. We observe that the random variable $h(T)$ is the conditional expectation of the random variable $\delta(X)$ given the sufficient statistic T when θ is the true parameter. Another way of considering the same is that the random variable $h(T)$ is the expectation of the conditional distribution of the random variable $\delta(X)$ given the sufficient statistic T when θ is the true parameter. Since T is a sufficient statistic for the parameter θ, it follows that the conditional distribution and hence the conditional expectation $h(T) = E_\theta[\delta(\boldsymbol{X})|T]$ does not depend on the parameter θ. Hence the function $h(T)$ is a statistic. Furthermore, it follows from the result stated in the Exercise 5.23 that

$$\begin{aligned} E_\theta[h(T)] &= E_\theta(E_\theta[\delta(\boldsymbol{X})|T]) \\ &= E_\theta(\delta(\boldsymbol{X})) \\ &= g(\theta). \end{aligned}$$

Hence the estimator $h(T)$ is an unbiased estimator for $g(\theta)$. It is easy to see that, for any bivariate random vector (X, Y),

$$E(X^2|Y) \geq [E(X|Y)]^2 \tag{8.26}$$

provided the conditional expectations are well defined. This inequality follows from the fact that the conditional variance of the random variable X given Y is greater than or equal to zero. Applying this inequality, we get that

$$E_\theta[(\delta(\boldsymbol{X}) - g(\theta))^2|T] \geq [E_\theta(\delta(\boldsymbol{X})|T) - g(\theta)]^2 = (h(T) - g(\theta))^2.$$

Taking the expectations on both the sides of the above equation with respect to the distribution of T and again applying the result in the Exercise 5.23, we get that

$$E_\theta[(\delta(\boldsymbol{X}) - g(\theta))^2] \geq E[(h(T) - g(\theta))^2].$$

This result proves that

$$\text{Var}_\theta[h(T)] \leq \text{Var}_\theta[\delta(\boldsymbol{X})]$$

for every $\theta \in \Theta$. It is easy to see that the equality occurs in the relation (8.26) if and only if the random variable X is a function of Y with probability one. Applying this remark, we obtain that the equality occurs in the above inequality if and only if $\delta(\boldsymbol{X})$ is a function of T with probability one.

Remarks. Rao-Blackwell theorem essentially implies that an unbiased estimator $\delta(\boldsymbol{X})$ for a parameter θ can be improved, in the sense of obtaining another unbiased estimator with a smaller variance than that of the estimator $\delta(\boldsymbol{X})$, if there is a sufficient statistic T for the parameter. The improvement is strict if $\delta(\boldsymbol{X})$ is not a function of T with probability one.

Example 8.19. Let $X_i, 1 \leq i \leq n$ be n independent Bernoulli random variables with

$$P(X_i = 1) = 1 - P(X_i = 0) = p, 1 \leq i \leq n.$$

It is easy to see that X_1 is an unbiased estimator of p and $T = \sum_{i=1}^{n} X_i$ is a sufficient statistic for p. An application of the Rao-Blackwell theorem shows that $E[X_1|T]$ is an unbiased estimator of p. Let us compute $E[X_1|T)$. We note that

$$\begin{aligned}
E[X_1|T=t] &= 0.P[X_1=0|T=t]+1.P[X_1=1|T=t]\\
&= \frac{P[X_1=1,T=t]}{P[T=t]}\\
&= \frac{P[X_1=1,X_2+\cdots+X_n=t-1]}{P[T=t]}\\
&= \frac{P[X_1=1]P[X_2+\cdots+X_n=t-1]}{P[T=t]}\\
&= \frac{p\,(n-1)^C_{(t-1)}\;p^{t-1}(1-p)^{n-t}}{n^C_t\,p^t(1-p)^{n-t}}\\
&= \frac{(n-1)^C_{(t-1)}}{n^C_t}=\frac{t}{n}
\end{aligned}$$

and hence the estimator $\frac{T}{n}$ is an unbiased estimator of p. In fact,

$$\mathrm{Var}_p\left(\frac{T}{n}\right)=\frac{p(1-p)}{n}$$

and it is strictly smaller than $\mathrm{Var}_p(X_1)=p(1-p)$ for $n>1$. Hence $h(T)=\frac{T}{n}$ is a "better" estimator than X_1 for estimating p. The estimator $\frac{T}{n}$ cannot be improved further as it is a function of the sufficient statistic T.

The question now arises whether there could be two unbiased estimators of a parametric function $g(\theta)$ which are functions of T and which are distinct almost surely. Suppose $h_i(T), i=1,2$ are two such unbiased estimators. Then $E_\theta[h_1(T)-h_2(T)]=0$ for all θ. A statistic T is said to be *complete* for $\theta\in\Theta$ if $E_\theta[r(T)]=0$ for every $\theta\in\Theta$ implies $r(T)=0$ with probability one. Hence, if T is a complete and sufficient statistic, then $h_1(T)=h_2(T)$ with probability one and there exists a unique unbiased estimator (with probability one) and it is *the minimum variance unbiased estimator* for the parametric function $g(\theta)$. This gives us another method of finding a minimum variance unbiased estimator for a parametric function $g(\theta)$ of a parameter θ whenever it exists. The method can be described as follows:

Step 1: Find a sufficient statistic T for θ if any;
Step 2: find an unbiased estimator $\delta(\boldsymbol{X})$ for $g(\theta)$ if any;
Step 3: compute the conditional expectation of $\delta(\boldsymbol{X})$ given T;and
Step 4: check whether the statistic T is complete.

If we can execute the Step 1 and the Step 2 and if we can prove that T is complete in Step 4, then the Step 3 gives the unique minimum variance unbiased estimator for $g(\theta)$. Otherwise, the procedure, following Step 1 to 3, gives an unbiased estimator of $g(\theta)$ which is as good as or an improvement over the unbiased estimator $\delta(\boldsymbol{X})$ in the sense of smaller variance and it will be a minimum variance unbiased estimator of $g(\theta)$.

8.6 Properties of a Maximum Likelihood Estimator

We have seen, in the previous sections, that the minimum variance unbiased estimators might not exist for a given parameter. If they do exist, it might be possible to find them either by finding an unbiased estimator whose variance attains the Cramer-Rao Lower bound or by finding a complete and sufficient statistic and then applying the Rao-Blackwell theorem. It is possible that the Cramer-Rao lower bound might not be attainable by an unbiased estimator for all the sample sizes but it might hold asymptotically in the sense explained below. An important property of a maximum likelihood estimator (MLE) $\hat{\theta}_n$ of a parameter θ, is that, under some regularity conditions, the MLE is consistent, asymptotically normal and asymptotically efficient estimator. We will discuss the case when the parameter θ is a scalar parameter. The estimator $\hat{\theta}_n$ is *asymptotically efficient* in the sense that the variance of the limiting distribution of $\hat{\theta}_n$ is the Cramer-Rao lower bound $\frac{1}{nI(\theta)}$; more precisely, if $\hat{\theta}_n$ denotes a MLE for θ based on a random sample $X_i, 1 \leq i \leq n$, of size n, then

(i) $\hat{\theta}_n \xrightarrow{p} \theta$ as $n \to \infty$, that is, for every $\varepsilon > 0$

$$P_\theta(|\hat{\theta}_n - \theta| > \varepsilon| \to 0 \quad \text{as } n \to \infty;$$

(ii) $\sqrt{n}(\hat{\theta}_n - \theta) \xrightarrow{\mathcal{L}} N(0, \frac{1}{I(\theta)})$ as $n \to \infty$, that is, for every real x,

$$P_\theta(\sqrt{n}(\hat{\theta}_n - \theta) \leq x) \longrightarrow \sqrt{\frac{I(\theta)}{2\pi}} \int_{-\infty}^{x} e^{-\frac{I(\theta)}{2}y^2} dy \text{ as } n \to \infty$$

under some regularity conditions. Here the measure P_θ indicates that the probability is computed when θ is the true parameter. The first property shows that, if the sample size is sufficiently large, then the estimator $\hat{\theta}_n$ is close to the true parameter θ with probability approaching one as $n \to \infty$. The estimator $\hat{\theta}_n$ is said to be a consistent estimator in such a case. The second property indicates that, for large samples, the distribution of $\hat{\theta}_n$ is approximately normal with mean θ and the variance $\frac{1}{nI(\theta)}$. Let us now see why this is the case.

Let X be a random variable with the probability density function (or probability function in the discrete case) $f(x, \theta), \theta \in \Theta \subset I\!R$. Suppose that the identifiability condition holds, that is two different values of the parameter θ correspond two different distributions, and the function $f(x, \theta)$ is differentiable with respect to θ. Further suppose that the support of the probability density function (probability function in the discrete case) $f(x, \theta)$, that is the set $\{x : f(x, \theta) > 0\}$ does not depend on the parameter θ. For instance, the uniform probability density function on the interval $[0, \theta]$ does not satisfy this condition. Let $X_1, \ldots, X_n$ be i.i.d. random variables distributed as the random variable X. The likelihood function of the random sample $X_1, \ldots, X_n$ is

$$L_n(\boldsymbol{X}; \theta) = \Pi_{i=1}^n f(X_i, \theta).$$

Taking logarithms on both sides of the above equation we get that

$$\log L_n(\boldsymbol{X}; \theta) = \sum_{i=1}^n \log f(X_i, \theta).$$

The equation

$$\frac{\partial \log L_n(\boldsymbol{X}; \theta)}{\partial \theta} = 0$$

is called the *likelihood equation.*

Consistency of MLE

Let θ_0 be the true parameter. Suppose further that there is an open neighbourhood of θ_0 contained in the set Θ. Let $\delta > 0$. We will now show that the probability that there exists a solution of the likelihood equation in the interval $[\theta_0 - \delta, \theta_0 + \delta)]$ tends to one as $n \to \infty$. It follows by the Jensen's inequality (see Exercise 8.11) that

$$E_{\theta_0}\{\log \frac{f(X, \theta_0 + \delta)}{f(X, \theta_0)}\} < 0$$

and

$$E_{\theta_0}\{\log \frac{f(X, \theta_0 - \delta)}{f(X, \theta_0)}\} < 0$$

for any $\delta > 0$. Hence, by the Weak law of large numbers (see Chapter 7) applied to the sequence of random variables $\log \frac{f(X_i, \theta_0 - \delta)}{f(X_i, \theta_0)}, 1 \le i \le n$, we have

$$n^{-1} \sum_{i=1}^n \log \frac{f(X_i, \theta_0 - \delta)}{f(X_i, \theta_0)} \xrightarrow{p} E_{\theta_0}\{\log \frac{f(X, \theta_0 - \delta)}{f(X, \theta_0)}\} < 0$$

as $n \to \infty$. Therefore

$$n^{-1}[\log L_n(\boldsymbol{X};\theta_0-\delta)) - \log L_n(\boldsymbol{X};\theta_0)] \xrightarrow{p} E_{\theta_0}\{\log \frac{f(X,\theta_0-\delta)}{f(X,\theta_0)}\} < 0$$

which implies that the probability that

$$\log L_n(\boldsymbol{X};\theta_0-\delta) - \log L_n(\boldsymbol{X};\theta_0) < 0 \tag{8.27}$$

tends to one as $n \to \infty$. Similarly the probability that

$$\log L_n(\boldsymbol{X};\theta_0+\delta) - \log L_n(\boldsymbol{X};\theta_0) < 0 \tag{8.28}$$

tends to one as $n \to \infty$. Since the function $\log L_n(\boldsymbol{X};\theta)$ is continuous in the closed interval $[\theta_0-\delta,\theta_0+\delta)]$, it attains its local maximum in the interval $[\theta_0-\delta,\theta_0+\delta]$. But this local maximum is not attained at the boundary points $\theta_0+\delta$ and $\theta_0-\delta$ in view of (8.27) and (8.28) and it is attained in the open interval $(\theta_0-\delta,\theta_0+\delta)$. Since the function $\log L_n(\boldsymbol{X};\theta)$ is differentiable with respect to θ, it follows that

$$\frac{\partial \log L_n(\boldsymbol{X};\theta)}{\partial\theta} = 0$$

at any local maximum. Since δ is an arbitrary positive number, it follows that the probability that there exists a solution of the likelihood equation

$$\frac{\partial \log L_n(\boldsymbol{X};\theta)}{\partial\theta} = 0$$

in the interval $(\theta_0-\delta,\theta_0+\delta)$ tends to one as $n \to \infty$. In other words there exists a consistent solution of the likelihood equation.

Asymptotic normality of MLE

Suppose that the probability density function (or the probability function) $f(x,\theta)$ is differentiable thrice in a neighbourhood of the true parameter θ_0 and the Fisher information $I(\theta)$ is finite and positive. Let $\hat{\theta}_n$ be a consistent solution of the likelihood equation

$$\frac{\partial \log L_n(\boldsymbol{X};\theta)}{\partial\theta} = 0.$$

Then

$$[\frac{\partial \log L_n(\boldsymbol{X};\theta)}{\partial\theta}]_{\theta=\hat{\theta_n}} = 0. \tag{8.29}$$

Let us expand the function $\frac{\partial \log L_n(\boldsymbol{X};\theta)}{\partial\theta}$ in a neighbourhood of the true parameter θ_0 by the Taylor's theorem. Then

$$\frac{\partial \log L_n(\boldsymbol{X};\theta)}{\partial\theta} = [\frac{\partial \log L_n(\boldsymbol{X};\theta)}{\partial\theta}]_{\theta=\theta_0} + (\theta-\theta_0[\frac{\partial^2 \log L_n(\boldsymbol{X};\theta)}{\partial\theta^2}]_{\theta=\theta_0}$$
$$+\frac{1}{2}(\theta-\theta_0)^2[\frac{\partial^3 \log L_n(\boldsymbol{X};\theta)}{\partial\theta^3}]_{\theta=\theta'}$$

where $|\theta - \theta'| \leq |\theta - \theta_0|$. Since $\hat{\theta}_n$ is a consistent estimator of the true parameter θ_0, it follows that $P(|\hat{\theta}_n - \theta_0| > \delta)$ tends to zero as $n \to \infty$. Substituting $\theta = \hat{\theta}_n$ on the left side of the above equation, we get that, with probability tending to one,

$$0 = [\frac{\partial \log L_n(\boldsymbol{X};\theta)}{\partial\theta}]_{\theta=\hat{\theta}_n} = [\frac{\partial \log L_n(\boldsymbol{X};\theta)}{\partial\theta}]_{\theta=\theta_0} + (\hat{\theta}_n - \theta_0)[\frac{\partial^2 \log L_n(\boldsymbol{X};\theta)}{\partial\theta^2}]_{\theta=\theta_0} + \frac{1}{2}(\hat{\theta}_n - \theta_0)^2[\frac{\partial^3 \log L_n(\boldsymbol{X};\theta)}{\partial\theta^3}]_{\theta=\tilde{\theta}_n}$$

where $|\tilde{\theta}_n - \theta_0| \leq |\hat{\theta}_n - \theta_0|$. In particular, we have

$$\hat{\theta}_n - \theta_0 = \frac{-[\frac{\partial \log L_n(\boldsymbol{X};\theta)}{\partial\theta}]_{\theta=\theta_0}}{[\frac{\partial^2 \log L_n(\boldsymbol{X};\theta)}{\partial\theta^2}]_{\theta=\theta_0} + \frac{1}{2}(\hat{\theta}_n - \theta_0)[\frac{\partial^3 \log L_n(\boldsymbol{X};\theta)}{\partial\theta^3}]_{\theta=\tilde{\theta}_n}}.$$

Therefore

$$\sqrt{n}(\hat{\theta}_n - \theta_0) = \frac{-\frac{1}{\sqrt{n}}[\frac{\partial \log L_n(\boldsymbol{X};\theta)}{\partial\theta}]_{\theta=\theta_0}}{\frac{1}{n}([\frac{\partial^2 \log L_n(\boldsymbol{X};\theta)}{\partial\theta^2}]_{\theta=\theta_0} + \frac{1}{2}(\hat{\theta}_n - \theta_0)[\frac{\partial^3 \log L_n(\boldsymbol{X};\theta)}{\partial\theta^3}]_{\theta=\tilde{\theta}_n})}.$$

Numerator of the expression on the right side can be written in the form

$$-\frac{1}{\sqrt{n}}\sum_{i=1}^{n}[\frac{\partial \log f(X_i;\theta)}{\partial\theta}]_{\theta=\theta_0}.$$

The random variables

$$[\frac{\partial \log f(X_i;\theta)}{\partial\theta}]_{\theta=\theta_0}, 1 \leq i \leq n,$$

are i.i.d. with the mean

$$E_{\theta_0}([\frac{\partial \log f(X_i;\theta)}{\partial\theta}]_{\theta=\theta_0}) = 0$$

and the variance

$$E_{\theta_0}([\frac{\partial \log f(X_1;\theta)}{\partial\theta}]^2_{\theta=\theta_0}]) = I(\theta_0)$$

which we assume to be positive and finite. Applying the Central limit theorem, we get that

$$-\frac{1}{\sqrt{n}}[\frac{\partial \log L_n(\boldsymbol{X};\theta)}{\partial\theta}]_{\theta=\theta_0} \xrightarrow{\mathcal{L}} N(0, I(\theta_0)) \text{ as } n \to \infty.$$

Applying the Weak law of large numbers and the consistency properties of the estimator $\hat{\theta}_n$, check that the denominator converges in probability to

$$E_{\theta_0}([\frac{\partial^2 \log f(X_1;\theta)}{\partial\theta}]_{\theta=\theta_0}]) = -I(\theta_0)$$

under some further regularity conditions. Using these facts, it can be shown that

$$\sqrt{n}(\hat{\theta}_n - \theta_0) \xrightarrow{\mathcal{L}} N(0, \frac{1}{I(\theta_0)}) \text{ as } n \to \infty.$$

We will not go into the detailed proof here.

Remarks. It is important to note that we have only given some sufficient conditions for the existence of a consistent solution of the likelihood equation and showed that maximum likelihood estimator is a solution of the likelihood equation. Of course these facts do not prove that the maximum likelihood estimator is consistent unless there is a unique solution to the likelihood equation. In a few examples, one can explicitly compute the maximum likelihood estimator and check the consistency and asymptotic normality of the estimator directly. We now illustrate this result by a couple of such examples.

Example 8.20. Suppose that $X_1, \ldots, X_n$ is a random sample from the normal distribution with mean μ and known variance σ^2. It can be checked that $\bar{X}_n$ is the MLE for μ based on $X_1, \ldots, X_n$. From the Chebyshev's inequality stated in Chapter 7, it follows that

$$P_\mu(|\bar{X}_n - \mu| > \varepsilon) \leq \frac{1}{\varepsilon^2} E_\mu(\bar{X}_n - \mu)^2 = \frac{\sigma^2}{n\varepsilon^2}$$

and the last term tends to zero as $n \to \infty$. Hence $\bar{X}_n \xrightarrow{p} \mu$ as $n \to \infty$, that is, $\bar{X}_n$ is a consistent estimator for μ. Let us now study its distribution. Since $\bar{X}_n$ is a linear combination of independent normal random variables, the distribution of $\bar{X}$ is normal and it is easy to see that $E_\mu(\bar{X}_n) = \mu$ and $\text{Var}_\mu(\bar{X}_n) = \frac{\sigma^2}{n}$. Hence $\bar{X}_n$ has $N(\mu, \frac{\sigma^2}{n})$ as its distribution. It can be checked that the Cramer-Rao lower bound for estimation of μ is $\frac{\sigma^2}{n}$. Hence MLE $\bar{X}_n$ for μ is a consistent and an efficient estimator of μ.

Example 8.21. Suppose that $X_i, 1 \leq i \leq n$ are independent Bernoulli trials with p as the probability for a success. Then

$$P(X_i = 1) = 1 - P(X_i = 0) = p, 1 \leq i \leq n.$$

It is easy to see that

$$\hat{p} = \frac{X_1 + \cdots + X_n}{n}$$

is the MLE for p. By the Weak law of large numbers from Chapter 7,

$$\hat{p} = \frac{X_1 + \cdots + X_n}{n} \to E_p(X_1) = p \text{ in probability as } n \to \infty$$

and an application of the Central limit theorem from Chapter 7 implies that

$$\frac{X_1 + \cdots + X_n - np}{\sqrt{np(1-p)}} \xrightarrow{\mathcal{L}} N(0,1) \text{ as } n \to \infty$$

or equivalently

$$\frac{\hat{p} - p}{\sqrt{\frac{p(1-p)}{n}}} \xrightarrow{\mathcal{L}} N(0,1) \text{ as } n \to \infty.$$

8.7 Bayes Estimation

In all the previous sections, we assumed that the distribution function F or the probability density function f depends on a fixed but unknown parameter θ and the problem was to estimate the parameter θ based on a set of observations with the probability density f.

In many problems, the experimenter or the statistician may have some information about the parameter θ. For instance, if we are estimating the average height of a person, it is obvious that it is very unlikely to find a person who is 8 feet tall or less than 6 inches in height. We call such an information as prior information and quantify it through a prior distribution function or a prior probability density function.

Suppose that we know that the true value of the parameter θ lies in an interval Θ of the real line. The interval Θ could be a bounded interval of the real line or the whole real line. Suppose we have a sample of i.i.d. observations $X_1, \ldots, X_n$. Let us denote the sample $(X_1, \ldots, X_n)$ by the random vector $\boldsymbol{X}$. The problem is to estimate the parameter θ. Let us suppose that we estimate the parameter θ by the estimator $\delta(\boldsymbol{X})$. Such a choice will possibly involve a loss. Let us denote the loss so incurred by $L(\theta, \delta(\boldsymbol{X}))$. Let $\lambda(\theta)$ be the *prior* probability density function of θ reflecting the prior information on the parameter θ. For any fixed value of θ, the expected loss from using the estimator $\delta(\boldsymbol{X})$, is

$$E_\theta[L(\theta, \delta(\boldsymbol{X}))] = \int_{\mathcal{X}} L(\theta, \delta(\boldsymbol{x})) f(\boldsymbol{x}, \theta) d\boldsymbol{x}.$$

Here $\mathcal{X}$ is the sample space corresponding to the random vector $\boldsymbol{X}$ and $f(\boldsymbol{x}, \theta)$ is the conditional probability density function of the random vector

$\boldsymbol{X}$ given that θ is the true parameter. The risk $R(\lambda, \delta)$ of the estimator $\delta(\boldsymbol{X})$ is defined to be

$$\int_{\Theta} E_\theta[L(\theta, \delta(\boldsymbol{X}))]\lambda(\theta)d\theta.$$

The basic idea is to choose the estimator $\delta(\boldsymbol{X})$ so that this risk is minimum. Such an estimator is called a *Bayes estimator* with respect to the loss function $L(.,.)$ and the prior probability density function λ. Typically, a loss function $L(\theta, a)$ is chosen to be nonnegative for all $\theta \in \Theta$ and for all $a \in \Theta$. Examples of such loss functions are $L(\theta, a) = (\theta - a)^2$ which is termed as the squared error loss function and $L(\theta, a) = |\theta - a|^p, p \geq 1$. For further discussion, we assume that the loss function is squared error loss function. Then the risk of the estimator $\delta(\boldsymbol{X})$ can be written in the form

$$\begin{aligned}
R(\lambda, \delta) &= \int_{\Theta} E_\theta[L(\theta, \delta(\boldsymbol{X}))]\lambda(\theta) \\
&= \int_{\Theta}\{\int_{\mathcal{X}} [\delta(\boldsymbol{x}) - \theta]^2 f(\boldsymbol{x}, \theta)d\boldsymbol{x}\}\lambda(\theta)d\theta \\
&= \int_{\mathcal{X}}\{\int_{\Theta} [\delta(\boldsymbol{x}) - \theta]^2 f(\boldsymbol{x}, \theta)\lambda(\theta)d\theta\}d\boldsymbol{x} \\
&= \int_{\mathcal{X}}\{\int_{\Theta} [\delta(\boldsymbol{x}) - \theta]^2 \frac{f(\boldsymbol{x}, \theta)\lambda(\theta)}{\int_\Theta f(\boldsymbol{x}, u)\lambda(u)du}[\int_{\Theta} f(\boldsymbol{x}, u)\lambda(u)du]d\theta\}d\boldsymbol{x} \\
&= \int_{\mathcal{X}}\{\int_{\Theta} [\delta(\boldsymbol{x}) - \theta]^2 \frac{f(\boldsymbol{x}, \theta)\lambda(\theta)}{\int_\Theta f(\boldsymbol{x}, u)\lambda(u)du}d\theta\}f_{\boldsymbol{X}}(\boldsymbol{x})d\boldsymbol{x}
\end{aligned}$$

where $f_{\boldsymbol{X}}(\boldsymbol{x})$ denotes the marginal density of the random vector $\boldsymbol{X}$. Since the integrand is nonnegative, the interchange of order of integration done above and the last equality can be justified by using a result known as the Fubini's theorem. We do not go into the details here.

In order to choose an estimator $\delta(\boldsymbol{X})$ which minimizes $R(\lambda, \delta)$, it is sufficient to choose the estimate $\delta(\boldsymbol{x})$ which minimizes the inner integral

$$\int_{\Theta} [\delta(\boldsymbol{x}) - \theta]^2 \frac{f(\boldsymbol{x}, \theta)\lambda(\theta)}{\int_\Theta f(\boldsymbol{x}, u)\lambda(u)du}d\theta \tag{$*$}$$

for each $\boldsymbol{x} \in \mathcal{X}$. Let

$$f(\theta|\boldsymbol{x}) = \frac{f(\boldsymbol{x}, \theta)\lambda(\theta)}{\int_\Theta f(\boldsymbol{x}, u)\lambda(u)du}.$$

Note that the function $f(\theta|\boldsymbol{x})$ is the conditional probability density function of θ given the observation $\boldsymbol{x}$. This probability density function is called the *posterior density* of θ given the observation $\boldsymbol{x}$. The integral expressed in (*) can be written in the form

$$\int_{\Theta} [\delta(\boldsymbol{x}) - \theta]^2 f(\theta|\boldsymbol{x}) d\theta.$$

It is the *posterior expected loss* given the observation $\boldsymbol{x}$, that is, the expected loss with respect to the posterior distribution. Since the loss function is the squared error loss function, it can be checked that it is minimum when $\delta(\boldsymbol{x})$ is the mean of the posterior density(see Exercise 8.12). Hence the Bayes estimator of θ given the squared loss function and the prior λ is the conditional mean of the posterior distribution of θ given the observation $\boldsymbol{x}$, that is $E(\theta|\boldsymbol{X} = \boldsymbol{x})$, which can also be written in the form

$$\delta(\boldsymbol{x}) = \int_{\Theta} \theta f(\theta|\boldsymbol{x}) d\theta.$$

Furthermore, it easy to check that the Bayes estimator of a function $g(\theta)$, given the observation $\boldsymbol{x}$, is given by $E(g(\theta)|\boldsymbol{X} = \boldsymbol{x})$ for a squared error loss function.

Remarks. We have explained the concept of Bayes estimation when the parameter θ is a a scalar parameter and the random vector $\boldsymbol{X}$ has a probability density functions. However the concept can be extended to vector parameter and for discrete probability distributions as well by replacing the integrals by appropriate sums and the probability density functions by probability functions. We do not go into more discussion here.

We will now discuss an example.

Example 8.22. Suppose that a random variable X has the Binomial distribution with parameters n and θ. Suppose that the prior probability density function of the parameter θ is the Beta probability density function given by

$$\begin{aligned} \lambda(\theta) &= \frac{\Gamma(\alpha+\beta)}{\Gamma(\alpha)\Gamma(\beta)} \theta^{\alpha-1}(1-\theta)^{\beta-1}, \quad 0 < \theta < 1 \\ &= 0 \ \text{ otherwise.} \end{aligned}$$

Check that the posterior density $f(\theta|X = x)$ of the parameter θ given $X = x$ is again a Beta probability density function with parameters $x + \alpha$ and $n - x + \beta$. If we consider the loss function as the squared error loss function, then the Bayes estimator of θ given the observation x is the conditional mean $E(\theta|X = x)$, that is the mean of the Beta probability density function with parameters $x+\alpha$ and $n-x+\beta$. It can be checked that the mean can be expressed in the form

$$\delta(x) = \frac{\alpha + x}{\alpha + \beta + n}.$$

8.8 Estimation of a Probability Density Function

In all the previous sections of this chapter, we assumed that the data is obtained from a parametric family of distributions and the problem was to obtain *good* point estimators for the unknown parameters. The assumption that the data was obtained from a specified parametric model might be too strong because there might not be a parametric model that is suitable for the data or the parametric model assumed to be the correct model need not be the *actual* model, even if one hopes there is one such, in other words the model may be misspecified. Statistical methods developed for a particular parametric model might lead to erroneous conclusions if they are applied either to a completely different model or even to a slightly perturbed model. These problems lead to the recent developments in non-parametric methods of estimation. One of the fundamental problems is the estimation of the distribution function or the probability density function of a random variable or random vector whenever it exists as these functions capture the information on the probabilities of different events concerning the phenomenon under consideration. Suppose $X_1, \ldots, X_n$ are i.i.d. random variables with the distribution function F and the probability density function f which are unknown. We would like to estimate the function $f(x)$ based on the random sample $X_1, \ldots, X_n$. It can be shown that there exists no unbiased estimator of the probability density $f(x)$ based on the random sample $X_1, \ldots, X_n$ under some conditions. Since the function $f(x)$ is the derivative of the function $F(x)$ almost every where, it is natural to expect that the quantity

$$\frac{F(x+h) - F(x-h)}{2h}$$

is close to $f(x)$ for sufficiently small $h > 0$. Let

$$F_n(x) = \frac{\text{Number of observations less than or equal to } x}{n}$$

for any $-\infty < x < \infty$. The function $F_n(x)$ is called the *empirical distribution function* based on the random sample $X_1, \ldots, X_n$. Check that, for any fixed x, the function $nF_n(x)$ is a random variable with the Binomial distribution with the parameters n and $F(x)$. In particular

$$E(nF_n(x)) = nF(x) \text{ and } Var(nF_n(x)) = nF(x)(1 - F(x))$$

or equivalently

$$E(F_n(x)) = F(x) \text{ and } Var(F_n(x)) = \frac{F(x)(1 - F(x))}{n}.$$

These relations show that the function $F_n(x)$ is an unbiased estimator of $F(x)$ for every x and applying the Chebyshev's inequality, check that

$$F_n(x) \xrightarrow{p} F(x) \text{ as } n \to \infty.$$

In view of this observation, we can consider

$$f_n^*(x) = \frac{F_n(x + h_n) - F_n(x - h_n)}{2h_n}$$

as an estimator of $f(x)$ where h_n is any positive sequence of real numbers converging to zero. This estimator can also be written in the form

$$f_n^*(x) = \frac{1}{nh_n} \sum_{i=1}^{n} K_0(\frac{x - X_i}{h_n})$$

where

$$K_0(x) = \frac{1}{2} \text{ if } x \in [-1, 1)$$
$$= 0 \text{ otherwise.}$$

In general, let $K(.)$ be a probability density function and define

$$f_n(x) = \frac{1}{nh_n} \sum_{i=1}^{n} K(\frac{x - X_i}{h_n})$$

where $0 < h_n \to 0$ as $n \to \infty$. The function $f_n(x)$ can be considered as an estimator for the probability density function $f(x)$. The function $K(.)$ is called the *kernel* and the sequence h_n is called the *bandwidth* of the estimator $f_n(x)$. It is known that the asymptotic properties of the estimator $f_n(x)$ depend critically on the choice of the bandwidth sequence h_n but the choice of the kernel $K(.)$ is less important. It can be shown that if the function f is continuous, then the estimator $f_n(x)$ is a consistent estimator for $f(x)$ if the sequence h_n is chosen so that $nh_n \to \infty$ and $h_n \to 0$ as $n \to \infty$. We will not go into further discussion on this topic.

8.9 Exercises

8.1 Suppose that $X_1, \ldots, X_n$ are i.i.d. Poisson with mean λ. Show that $\bar{X}$ is an unbiased estimator for λ.

8.2 Suppose that $X_1, \ldots, X_n$ are i.i.d. random variables uniform on $(0, \theta)$. Show that the estimators $2\bar{X}$ and $2X_1$ are both unbiased estimators of θ. Which is better and why?

8.3 Suppose there is a random sample of observations $X_i, 1 \leq i \leq n$ from the normal distribution with unknown mean μ and known variance σ^2. Find the MLE of μ.

8.4 Obtain the Cramer-Rao lower bound for the parameter λ of the Poisson distribution with mean λ. Show that the sample mean is the MVUE for λ.

8.5 Suppose that $X_1, \ldots, X_n$ are i.i.d. random variables with the Poisson distribution with mean λ. Show that the MLE of λ is asymptotically normal and asymptotically efficient.

8.6 Suppose X is a random variable with the Pareto distribution function

$$
\begin{aligned}
F(x; \alpha, \beta) &= 1 - \left(\frac{\alpha}{x}\right)^{\beta}, x \geq \alpha \\
&= 0 \qquad \text{otherwise}
\end{aligned}
$$

where $\alpha > 0$ and $\beta > 0$. Derive the maximum likelihood estimators of α and β based on a random sample of size n.

8.7 Let $X_1, X_2, \ldots, X_n$ be a random sample of size n from a population with the double exponential probability density function

$$
f(x, \theta) = \frac{1}{2} e^{-|x-\theta|}, -\infty < x < \infty, \quad -\infty < \theta < \infty.
$$

Estimate the parameter θ by the method of maximum likelihood.

8.8 Suppose that $X_1, X_2, \ldots, X_n$ is a random sample of size $n > 2$ from a distribution with the probability density function

$$
\begin{aligned}
f(x, \theta) &= \theta x^{\theta - 1}, 0 < x < 1 \\
&= 0 \qquad \text{otherwise.}
\end{aligned}
$$

Obtain the Cramer-Rao lower bound for an unbiased estimator of the parameter θ and hence show that $(n-1)/W$ where $W = -\sum_{i=1}^{n} \log X_i$ is an asymptotically efficient estimator of θ.

8.9 Suppose that $X_1, X_2, \ldots, X_n$ is a random sample from the $N(\mu, \sigma^2)$ where μ and σ^2 are unknown. Show that $(\bar{X}, S^2)$ is a sufficient statistic for the parameter $\theta = (\mu, \sigma^2)$ where $\bar{X} = \frac{1}{n}\sum_{i=1}^{n} X_i$ and $S^2 = \frac{1}{n-1}\sum_{i=1}^{n}(X_i - \bar{X})^2$ for $n > 2$.

8.10 Let $X_1, \ldots, X_n$ be a random sample from a population with the distribution function $F(x)$. Show that the empirical distribution function $F_n(x)$ is a consistent estimator of $F(x)$. Find the asymptotic distribution of $\sqrt{n}(F_n(x) - F(x))$ as $n \to \infty$. (Hint: Apply the Central limit theorem.)

8.11 (Jensen's inequality) A function $g(.)$ is said to be convex on an interval $[a, b]$ if for every x_1 and x_2 in $[a, b]$,

$$g(\lambda x_1 + (1 - \lambda)x_2) \leq \lambda\ g(x_1) + (1 - \lambda)\ g(x_2)$$

for all $0 \leq \lambda \leq 1$. Suppose X is a random variable taking values in the interval $[a, b]$ such that $E(X) < \infty$. Then show that

$$g(EX) \leq E(g(X))$$

with strict inequality unless g is a linear function. In particular, show that the function $g(x) = -\log x$ is convex on the interval $(0, \infty)$ and hence prove that for any positive valued random variable with $0 < EX < \infty$,

$$E(\log X) \leq \log E(X),$$

with equality if and only if X is a degenerate random variable.

8.12 Suppose X is a random variable with mean μ and finite variance σ^2. Let $f(a) = E(X - a)^2, -\infty < a < \infty$. Prove that the function is minimum when $a = \mu$.

8.13 Suppose X is a random variable with $E|X| < \infty$ and unique median m. Let $f(a) = E|X - a|, -\infty < a < \infty$. Prove that the function is minimum when $a = m$.

8.14 Suppose $X_i, 1 \leq i \leq n$ are i.i.d. with Poisson distribution with mean λ. Suppose the parameter λ has a prior Gamma probability density function with parameters α and β. Under the squared error loss function, find the Bayes estimator of λ.

8.15 Suppose $X_i, 1 \leq i \leq n$ are i.i.d. normal random variables with mean θ and known variance σ^2. Suppose the parameter θ has a normal prior with mean θ and variance τ^2. Under the squared error loss function, find the Bayes estimator of θ.

Chapter 9

INTERVAL ESTIMATION AND TESTING OF HYPOTHESES

9.1 Introduction

We studied some methods of estimation of parameters involved in probability distributions and some properties associated with these estimators. Estimators of this type are referred to as *point estimators.* In other words, a single value is suggested as an estimator based on the data. Alternately, one might be interested in proposing an interval or a set in which the parameter is likely to be a member, the interval or the set depending, of course, on the observed data. Such an estimation problem is termed as the problem of interval estimation or the problem of obtaining confidence interval or a confidence set for the parameter. We will explain this concept in detail later in this chapter. Another type of problem that is discussed here is the testing of hypotheses. We might be interested, for instance, to know whether a scalar parameter $\theta \leq \theta_0$ or $\theta > \theta_0$ where θ_0 is a preassigned value. Based on the data on hand, we have to choose one of the two actions. Examples of such problems occur in practice. For instance, suppose a manufacturer of automobiles wants to purchase tyres from a vendor. Suppose the vendor supplies the material in lots of size, say N, some of which are possibly defective. It is obvious that it is impossible to check each and every tyre to determine whether it is defective or not as this process is expensive and time consuming for the manufacturer and the vendor in turn might not allow such an inspection. The manufacturer might be willing to purchase the lot if the percentage defective p of the lot is less than or equal to, say, 5%. Since the exact percentage defective in the lot supplied by the vendor is unknown, the manufacturer has to take a decision, whether to accept or reject the lot supplied by the vendor, on the basis of a sample of size, say, n of tyres from the lot consisting of N tyres. This sort of decision making involves errors. The topic of testing of hypotheses discusses different methods of dealing with such problems.

9.2 Interval Estimation (Confidence Interval)

In order to illustrate how to obtain a confidence interval for a parameter of a probability distribution, we will first consider the case of the mean of a normal distribution.

Confidence interval for the mean of a normal distribution

Suppose we are interested in obtaining the information about the unknown parameter θ of a normal population with mean θ and known variance σ^2. Let $X_1, \ldots, X_n$ be a random sample from such a distribution. We have seen that the sample mean $\bar{X}$ is an efficient estimator of θ in the sense that its variance attains the Cramer-Rao lower bound. Hence it is a minimum variance unbiased estimator of θ. It is also a maximum likelihood estimator of θ. Further more

$$Z = \frac{\bar{X} - \theta}{\sigma/\sqrt{n}}$$

has the standard normal distribution when θ is the true parameter. Given $0 < \alpha < 1$, we choose a and b such that

$$P[a \leq Z \leq b] = 1 - \alpha.$$

Since the standard normal distribution is symmetric around 0, let us choose the value $Z_{\alpha/2}$ (see Figure 9.1) such that

$$P[-Z_{\alpha/2} \leq Z \leq Z_{\alpha/2}] = 1 - \alpha.$$

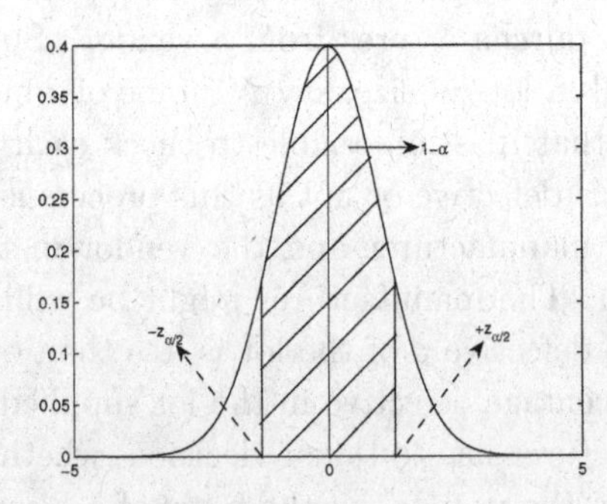

Fig. 9.1

The value $Z_{\alpha/2}$ can be found from the Table C.1 for the standard normal distribution. The above equation can also be written in the form

$$P[-Z_{\alpha/2} \leq \frac{\bar{X}-\theta}{\sigma/\sqrt{n}} \leq Z_{\alpha/2}] = 1-\alpha$$

when θ is the true parameter . This can again be written in the form

$$P[\bar{X} - Z_{\alpha/2}\frac{\sigma}{\sqrt{n}} \leq \theta \leq \bar{X} + Z_{\alpha/2}\frac{\sigma}{\sqrt{n}}] = 1-\alpha.$$

Note that the quantity θ is a parameter but not a random variable and the interval $[\bar{X} - Z_{\alpha/2}\frac{\sigma}{\sqrt{n}}, \bar{X} + Z_{\alpha/2}\frac{\sigma}{\sqrt{n}}]$ has random end points. The end points depend on the observations. How do we interpret such a statement? We are not saying that the conditional probability that the parameter θ belongs to the interval $[\bar{X} - Z_{\alpha/2}\frac{\sigma}{\sqrt{n}}, \bar{X} + Z_{\alpha/2}\frac{\sigma}{\sqrt{n}}]$ is $1-\alpha$ given $\bar{X}$. Since the parameter θ is a constant, the probability that it belongs to a specific interval is zero or one since it either belongs to the interval or it does not. What we are saying is that the probability that the random interval contains the parameter θ is $1-\alpha$. In other words, if the problem of interval estimation is repeated a large number of times, then the relative frequency of the number of intervals, calculated by the above method, containing the parameter θ, approaches the value $1-\alpha$ in the frequentist interpretation of probability. The random interval

$$[\bar{X} - Z_{\alpha/2}\frac{\sigma}{\sqrt{n}}, \bar{X} + Z_{\alpha/2}\frac{\sigma}{\sqrt{n}}] \tag{9.1}$$

is called a $100(1-\alpha)\%$ *confidence interval* or *interval estimate* for the parameter θ and the quantity $1-\alpha$ is called the *confidence coefficient* of the confidence interval. The word confidence comes from the fact that the statistician believes that the random interval is likely to contain the parameter θ about $100(1-\alpha)$ times in repetitions of computing the intervals from 100 i.i.d. samples $\{X_i, 1 \leq i \leq n\}$. Let us now give a formal definition.

Definition. Let $0 \leq \alpha \leq 1$. A random interval $[r_1(\boldsymbol{X}), r_2(\boldsymbol{X})]$ is called a *confidence interval* for a scalar parameter θ with the *confidence coefficient* $(1-\alpha)$ based on the observations $\boldsymbol{X}$ if

$$P_\theta\{\theta \in [r_1(\boldsymbol{X}), r_2(\boldsymbol{X})]\} = 1-\alpha$$

for all θ in the parameter space Θ. The interval $[r_1(\boldsymbol{X}), r_2(\boldsymbol{X})]$ is also called an *interval estimator* for the parameter θ with the confidence coefficient $(1-\alpha)$.

Let us now again consider the normal distribution with an unknown mean θ and an unknown variance σ^2. Suppose we are still interested in obtaining a confidence interval for θ with the confidence coefficient $100(1-\alpha)\%$ based on a random sample $X_1, \ldots, X_n$ from the $N(\theta, \sigma^2)$. Obviously, we cannot use the confidence interval derived earlier for θ as the end points

$$\bar{X} \pm Z_{\alpha/2}\frac{\sigma}{\sqrt{n}}$$

for the confidence interval given in (9.1) depend on the unknown parameter σ. We are interested here in information about the mean θ but not about the variance σ^2. The parameter σ is termed as a *nuisance parameter* in this context. Our problem here is to find a confidence interval for θ with a given confidence coefficient in the presence of the nuisance parameter σ^2. From our discussion in the Chapter 5, we know that the statistic t defined by

$$t = \frac{\sqrt{n}(\bar{X} - \theta)}{S}$$

where

$$S^2 = \frac{1}{n-1}\sum_{i=1}^{n}(X_i - \bar{X})^2$$

has the Student's t-distribution with $(n-1)$ degrees of freedom. Recall that S^2 is the sample variance. Note that the statistic t does not depend on the nuisance parameter σ but only on the parameter θ and its distribution function does not depend on either θ or σ. Hence one can use the t-statistic for constructing a confidence interval for the parameter θ even when σ is unknown. Given a constant $0 < \alpha < 1$, we can find values $t_{\alpha/2}$ such that

$$P[-t_{\alpha/2} \leq t \leq t_{\alpha/2}] = 1 - \alpha$$

using the Table C.2 for the t-distribution with $(n-1)$ degrees of freedom. Rewriting the above equation in terms of the parameter θ, we have

$$P[-t_{\alpha/2} \leq \frac{\sqrt{n}(\bar{X} - \theta)}{S} \leq t_{\alpha/2}] = 1 - \alpha,$$

that is

$$P[\bar{X} - t_{\alpha/2}\frac{S}{\sqrt{n}} \leq \theta \leq \bar{X} + t_{\alpha/2}\frac{S}{\sqrt{n}}] = 1 - \alpha.$$

when θ is the true parameter. Hence the random interval

$$[\bar{X} - t_{\alpha/2}\frac{S}{\sqrt{n}}, \bar{X} + t_{\alpha/2}\frac{S}{\sqrt{n}}] \tag{9.2}$$

is a confidence interval for the parameter θ with the confidence coefficient $100(1-\alpha)\%$. Let us compare the confidence intervals obtained in (9.1) and in (9.2). They are essentially similar in form.

Since the t-distribution can be approximated by the standard normal distribution for large n, one can use the interval (9.2) with $t_{\alpha/2}$ replaced by $Z_{\alpha/2}$ for n large. From empirical observations, it is suggested that the sample size n should be at least 30 in order to use the normal approximation.

Let us again consider the confidence interval

$$[\bar{X} - Z_{\alpha/2}\frac{\sigma}{\sqrt{n}}, \bar{X} + Z_{\alpha/2}\frac{\sigma}{\sqrt{n}}]$$

for the parameter θ of the normal distribution with mean θ and known variance σ^2 with confidence coefficient $100(1-\alpha)\%$. Notice that the length of the interval is

$$2Z_{\alpha/2}\frac{\sigma}{\sqrt{n}}.$$

If we want a confidence interval of prescribed length d, then the relation

$$2Z_{\alpha/2}\frac{\sigma}{\sqrt{n}} = d$$

should hold or equivalently

$$n = \left(\frac{d}{2Z_{\alpha/2}\sigma}\right)^2. \tag{9.3}$$

This formula gives an approximation to the sample size n needed in order to get a confidence interval of prescribed length d with the prescribed confidence coefficient $100(1-\alpha)\%$. Here σ is assumed to be known. In practice the value of n obtained from equation (9.3) need not be an integer. One chooses the sample size to be the smallest integer greater than or equal to n defined by (9.3).

If the variance σ^2 is unknown but a prior estimate of it is known, then we can use the above relation to get an approximate value of the sample size required. However if σ^2 is unknown, then the length of the confidence interval defined by (9.2) is

$$2t_{\sigma/2}\frac{S}{\sqrt{n}}.$$

It is not possible to determine the constant n from the relation

$$2t_{\alpha/2}\frac{S}{\sqrt{n}} = d$$

since S, the sample standard deviation, depends on the data. It is known that there is no fixed sample bounded length confidence interval for the mean of a normal distribution when the variance σ^2 is unknown. It is obvious that $P[\theta \in (-\infty, \infty)] = 1$ for any scalar parameter θ and hence the entire real line is always a confidence interval with the confidence coefficient 100%. But this is not a useful result as the interval is too big. The main aim is to find a shortest confidence interval with the largest confidence coefficient based on a given sample of observations. It turns out that the confidence intervals discussed above for the mean of a normal distribution are the 'best' in this sense. We will not go into the reasons here.

Confidence interval for the difference of means of two normal distributions

Suppose $X_1, \ldots, X_n$ is a random sample from $N(\theta_1, \sigma^2)$ and $Y_1, \ldots, Y_m$ is another random sample from $N(\theta_2, \sigma^2)$ where θ_1, θ_2 and σ^2 are all unknown. Note that both the normal distributions are assumed to have the *common variance.* The problem is to obtain a confidence interval for the difference of the two population means, namely, $\theta_1 - \theta_2$. From the properties discussed in the Chapter 5, it follows that

$$Z = \frac{\bar{X} - \bar{Y} - (\theta_1 - \theta_2)}{\sqrt{\frac{\sigma^2}{n} + \frac{\sigma^2}{m}}}$$

is a standard normal random variable. It would have been possible to construct a confidence interval of a given confidence coefficient for $\theta_1 - \theta_2$ using the distribution of the random variable Z if σ^2 was known. Since the parameter σ^2 is unknown, we can estimate it either by using the first sample or the second sample or both the samples. An estimator for σ^2 based on both the samples is

$$S_p^2 = \frac{1}{n+m-2}\left\{\sum_{i=1}^{n}(X_i - \bar{X})^2 + \sum_{j=1}^{m}(Y_j - \bar{Y})^2\right\}.$$

The statistic S_p^2 is called the *pooled sample variance.* It again follows that

$$\frac{(n+m-2)S_p^2}{\sigma^2}$$

has the Chi-square distribution with $(n+m-2)$ degrees of freedom from the independence of the normal samples $X_i, 1 \leq i \leq n$ and $Y_j, 1 \leq j \leq m$. Hence the statistic

$$t = \frac{\bar{X} - \bar{Y} - (\theta_1 - \theta_2)}{S_p\sqrt{\frac{1}{m} + \frac{1}{n}}}$$

has the t-distribution with $(n+m-2)$ degrees of freedom. It now follows that a confidence interval for $(\theta_1-\theta_2)$ with the confidence coefficient $100(1-\alpha)\%$ is given by

$$\left[\bar{X}-\bar{Y}-t_{\alpha/2}S_p\sqrt{\frac{1}{n}+\frac{1}{m}},\bar{X}-\bar{Y}+t_{\alpha/2}S_p\sqrt{\frac{1}{n}+\frac{1}{m}}\right]$$

where the value $t_{\alpha/2}$ is such that

$$P[|t|\leq t_{\alpha/2}]=1-\alpha$$

and the statistic t has the t-distribution with $(n+m-2)$ degrees of freedom. If the two normal populations do not have a common variance but the ratio of the variances is known, then we can modify the above t-statistic and get a confidence interval with a given confidence coefficient (Check?). However if both the variances are unknown and not equal, then it is not possible to construct a confidence interval with a prescribed confidence coefficient for the difference in the means of normal distributions. We do not discuss the reasons here.

Confidence interval for the variance of a normal distribution

Suppose we are interested in obtaining the information about the unknown variance σ^2 of a normal population with an unknown mean μ. Let $X_1,\ldots,X_n$ be a random sample from such a distribution. Let

$$S^2=\frac{1}{n-1}\sum_{i=1}^{n}(X_i-\bar{X})^2$$

be the sample variance. We have seen earlier that the random variable

$$\chi^2=\frac{(n-1)S^2}{\sigma^2}$$

has the Chi-square distribution with $(n-1)$ degrees of freedom. Hence given any $0<\alpha<1$, it is possible to find constants a and b such that $P(\chi^2<a)=\frac{\alpha}{2}$ and $P(\chi^2>b)=\frac{\alpha}{2}$ where χ^2 is a random variable with the Chi-square distribution with $(n-1)$ degrees of freedom. Hence

$$P(a\leq\frac{(n-1)S^2}{\sigma^2}\leq b)=1-\alpha$$

or equivalently

$$P(\frac{(n-1)S^2}{b}\leq\sigma^2\leq\frac{(n-1)S^2}{a})=1-\alpha.$$

This shows that the interval

$$[\frac{(n-1)S^2}{b},\frac{(n-1)S^2}{a}]$$

is a confidence interval with $100(1-\alpha)\%$ confidence coefficient for the variance σ^2 of the normal distribution.

Confidence interval for the ratio of variances of two normal distributions

Suppose we are interested in obtaining the information about the ratio $\frac{\sigma_1^2}{\sigma_2^2}$ of the unknown variances σ_1^2 and σ_2^2 of two normal distributions with unknown means μ_1 and μ_2. Let $X_1, \ldots, X_n$ be a random sample from the normal distribution $N(\mu_1, \sigma_1^2)$ and $Y_1, \ldots, Y_m$ be another independent random sample from the normal distribution $N(\mu_2, \sigma_2^2)$. Let

$$S_X^2 = \frac{1}{n-1}\sum_{i=1}^{n}(X_i - \bar{X})^2$$

be the sample variance corresponding to the first sample and

$$S_Y^2 = \frac{1}{m-1}\sum_{j=1}^{m}(Y_j - \bar{Y})^2$$

be the sample variance corresponding to the second sample. We have seen earlier that the random variable

$$\frac{(n-1)S_X^2}{\sigma_1^2}$$

has the Chi-square distribution with $(n-1)$ degrees of freedom and the random variable

$$\frac{(m-1)S_Y^2}{\sigma_2^2}$$

has the Chi-square distribution with $(m-1)$ degrees of freedom and these random variables are independent. Hence the ratio

$$\frac{\frac{1}{(n-1)}\frac{(n-1)S_X^2}{\sigma_1^2}}{\frac{1}{(m-1)}\frac{(m-1)S_Y^2}{\sigma_2^2}} = \frac{S_X^2\sigma_2^2}{S_Y^2\sigma_1^2}$$

has the F-distribution with $(n-1)$ and $(m-1)$ degrees of freedom. Therefore, given any $0 < \alpha < 1$, it is possible to find constants a and b such that

$$P(a \leq \frac{S_X^2\sigma_2^2}{S_Y^2\sigma_1^2} \leq b) = 1 - \alpha$$

or equivalently

$$P(a\frac{S_Y^2}{S_X^2} \leq \frac{\sigma_2^2}{\sigma_1^2} \leq b\frac{S_Y^2}{S_X^2}) = 1 - \alpha.$$

This shows that the interval

$$[a\frac{S_Y^2}{S_X^2}, b\frac{S_Y^2}{S_X^2}]$$

is a confidence interval with $100(1-\alpha)\%$ confidence coefficient for the ratio $\frac{\sigma_2^2}{\sigma_1^2}$ of the variances σ_1^2 and σ_2^2 of the two normal distributions.

Remarks. The confidence intervals derived above for the mean of a normal distribution are exact and are valid even for small samples. Even if the original observations are not from a normal population but from a population with finite mean μ and finite positive variance σ^2, the confidence interval

$$[\bar{X} - Z_{\alpha/2}\frac{\sigma}{\sqrt{n}}, \bar{X} + Z_{\alpha/2}\frac{\sigma}{\sqrt{n}}]$$

when σ^2 is known and the confidence interval

$$[\bar{X} - Z_{\alpha/2}\frac{S}{\sqrt{n}}, \bar{X} + Z_{\alpha/2}\frac{S}{\sqrt{n}}]$$

when σ^2 is unknown can still be used as an approximate $100(1-\alpha)\%$ confidence interval for the mean μ of the distribution in view of the fact

$$\frac{\bar{X}-\mu}{\sigma/\sqrt{n}} \xrightarrow{\mathcal{L}} N(0,1) \text{ as } n \to \infty$$

by the Central limit theorem. It is empirically observed that this approximation is good if the sample size is at least 30. If the sample size is large, then the t-distribution is close to the standard normal distribution and the value of $t_{\alpha/2}$ is approximately equal to $Z_{\alpha/2}$ with $P(Z \geq Z_{\alpha/2}) = \alpha/2$ and Z has the standard normal distribution function.

9.3 Testing of Hypotheses

In order to elucidate the concept of testing of hypotheses, let us start with an example. A manufacturer of big items like a car or a refrigerator generally does not make all the components at his or her firm but buys some of the components from individual vendors. He or she has to make sure that the components supplied by the vendor are as per the specifications prescribed by the manufacturer as otherwise the unit produced after using the defective component supplied by the vendor will turn out to be defective. Since it is practically impossible to test each and every component supplied by a particular vendor for several reasons like the time and the cost involved for inspection, the manufacturer has to make a decision whether to accept or to reject the lot supplied by the vendor based on a sample of items supplied by the vendor. It is obvious that the percentage defective in the sample need not be the same as the percentage defective in the total lot supplied by the vendor. In other words, by taking a decision to accept or reject the lot based on the sample, the manufacturer might possibly commit two types of errors; either the lot is accepted when it should not be accepted or the lot is rejected when it should not have been rejected. One of the two decisions, acceptance or rejection of the lot, is made on the basis of a sample. The problem is to devise test procedures by which the probability of these two types of errors can be made as small as possible.

We now discuss an example, which might not be of interest in practical situations, but will explain the notion of testing of hypotheses in a more lucid manner. Consider a coin with the probability p for a *head* to appear in a single toss of the coin. Suppose p is unknown and we have to get information about the parameter p. We conduct a random experiment consists of tossing the coin independently (say) n times. Suppose we are interested to know whether the coin is unbiased ($p = \frac{1}{2}$) or it is biased ($p \neq \frac{1}{2}$). We set up two hypotheses that the coin is unbiased as H_0 and the coin is biased as H_1. The hypothesis H_0 is called the null hypothesis and the hypothesis H_1 is called the alternate hypothesis. Suppose we have observed r heads in n tosses. The fraction $\frac{r}{n} = \hat{p}$ is an estimator for p. Intuitively, it is natural to decide that the hypothesis H_0 is true (H_0 is accepted) if $\hat{p}$ is close to $\frac{1}{2}$ and to decide that H_1 is true (H_0 is rejected) if $\hat{p}$ is away from $\frac{1}{2}$ on either side of the value $\frac{1}{2}$. Since the estimator $\hat{p}$ is based on a sample, both types of errors of decision as mentioned earlier are possible and the problem is to device test procedures to minimize these errors.

Let us now formulate the problem more precisely. Suppose X is a random variable with the probability distribution P_θ where θ is an unknown parameter. Let Θ denote the set of possible values for θ. We have to decide, based of course on some data observed, whether $\theta \in \Theta_0$ or $\theta \in \Theta_1$ where Θ_0 and Θ_1 are disjoint subsets of the parameter space Θ. Let us postulate the hypotheses $H_i : \theta \in \Theta_i$ for $i = 0, 1$. The hypothesis H_0 is called the *null hypothesis* and the hypothesis H_1 is called the *alternate hypothesis.* Historically H_0 is called the null hypothesis to indicate the possible state of the nature at present and the hypothesis H_1 denotes the likely change in the state. In any case, for our purposes, it is not important at the moment to determine which should be called null hypothesis and which one the alternate hypothesis. If the subset Θ_0 or the subset Θ_1 contain just one point, then the corresponding hypothesis is called *simple*, otherwise it is called *composite.* As mentioned earlier, the problem for the statistician is to decide whether the hypothesis H_0 is true or the hypothesis H_1 is true on the basis of an observed sample through some test procedure designed to minimize the two types of errors. Suppose $X_1, \ldots, X_n$ is a random sample of observations from a population with distribution P_θ. Let $\mathcal{X}^n$ denote the sample space corresponding to the observation vector $(X_1, \ldots, X_n)$, and choose a subset $C \subset \mathcal{X}^n$. We will call C a *critical region.* The test procedure for testing the null hypothesis H_0 against the alternate hypothesis H_1 is as follows. Suppose $(x_1, \ldots, x_n)$ is the observed sample: if $(x_1, \ldots, x_n) \in C$, then we reject the hypothesis H_0 and if $(x_1, \ldots, x_n) \notin C$, then we accept the hypothesis H_0. Such a procedure is called a *non-randomized test* for testing the hypothesis H_0 against the alternate H_1. Some caution should be observed when the second action is taken. Acceptance of the hypothesis H_0 does not necessarily mean that we conclude for definiteness that $\theta \in \Theta_0$. What we mean is that the data or the sample of observations we have on hand does not indicate anything contrary not to support the hypothesis that $\theta \in \Theta_0$. At times, we might have to take more data to get stronger evidence either for or against H_0. From the test procedure described above, it is clear that two types of error are possible in such a decision making process. The test procedure might lead to the rejection of the hypothesis H_0 when in reality the hypothesis H_0 is true or it might to lead to acceptance of the hypothesis H_0 when in fact the hypothesis H_0 is not true or equivalently the hypothesis H_1 is true. This can be tabulated in the following manner.

		Action taken	
		H_0 Accepted	H_0 Rejected
State of nature	H_0 true	Correct	Type *I* error
	H_0 false	Type *II* error	Correct

The problem is to devise some test procedures to control probabilities of both the type of errors. It would be best if the test procedures can be formulated so that α, the probability of Type I error and β, the probability of Type II error are as small as possible, preferably zero. However, this is not possible. Let

$$\alpha(\theta) = P_\theta[\text{Reject } H_0]$$

and

$$\beta(\theta) = P_\theta[\text{Accept } H_0]$$

when θ is the true parameter. The function $\alpha(\theta)$ denotes the probability of Type I error when $\theta \in \Theta_0$ and the function $\beta(\theta)$ is the probability of Type II error when $\theta \in \Theta_1$. Let $\gamma(\theta) = 1 - \beta(\theta)$. The function $\gamma(\theta)$ is called the *power* of the test at $\theta \in \Theta_1$. Given $0 \leq \alpha \leq 1$, our interest is to construct a test procedure for which

$$\alpha(\theta) \leq \alpha \text{ for all } \theta \in \Theta_0$$

and

$$\beta(\theta) \text{ is as small as possible for all } \theta \in \Theta_1$$

or equivalently

$$\gamma(\theta) = 1 - \beta(\theta) \text{ is as high as possible for all } \theta \in \Theta_1.$$

The value α is called the *size* or the *significance level* of the test. In practice, the value α is chosen a priori. The choice of α depends on the maximum possible probability of Type I error that the the decision maker is prepared to tolerate.

Test for the mean of a normal distribution

Let us consider the problem of testing for the normal mean μ with known variance σ^2. Let X be a random variable with probability distribution $N(\mu, \sigma^2)$ where σ^2 is known. Suppose the problem is to test the hypothesis $H_0 : \mu = 0$ against $H_1 : \mu = 1$. Here H_0 and H_1 are both simple hypotheses. Let us draw a random sample of size n from $N(\mu, \sigma^2)$, that is, $X_1, \ldots, X_n$ are random variables independent and identically distributed $N(\mu, \sigma^2)$. Since the sample mean $\bar{X}$ is an estimator for μ, it is natural to decide that the hypothesis H_1 is true if $\bar{X}$ is large and H_0 is true if $\bar{X}$ is small. In other words, we should choose a value k such that

$$\text{if } \bar{X} > k, \text{ then reject } H_0 \text{ (or equivalently accept } H_1)$$

and

$$\text{if } \bar{X} \leq k, \text{ then accept } H_0 \text{ (or equivalently reject } H_1).$$

How does one choose k? Let us now compute the probabilities of the Type I error and the Type II error for this test procedure. Here

$$\alpha(\mu) = P_\mu[\bar{X} > k]$$

where P_μ denotes the normal distribution when μ and σ^2 are the parameters. If we want a test with the significance level α, then k should be chosen so that

$$P_0[\bar{X} > k] = \alpha$$

since the mean $\mu = 0$ under the hypothesis H_0. From results studied in the Chapter 6, it follows that the random variable

$$\frac{\bar{X} - 0}{\sigma/\sqrt{n}}$$

has the standard normal distribution when the hypothesis H_0 holds since $X_i, 1 \le i \le n$ are i.i.d. $N(0, \sigma^2)$. Hence

$$P_0\left[\frac{\bar{X} - 0}{\sigma/\sqrt{n}} > \frac{k}{\sigma/\sqrt{n}}\right] = \alpha.$$

Suppose we have chosen $\alpha = .05$. From a more extensive table than Table C.1 for the standard normal distribution function, it can be checked that

$$\frac{k}{\sigma/\sqrt{n}} = 1.645$$

or

$$k = 1.645\, \frac{\sigma}{\sqrt{n}}.$$

The test procedure can be written in the form:

$$\text{reject } H_0 \text{ if } \bar{X} > 1.645 \frac{\sigma}{\sqrt{n}}$$

and

$$\text{accept } H_0 \text{ if } \bar{X} \le 1.645 \frac{\sigma}{\sqrt{n}}.$$

The probability β of Type II error for this test is given by

$$\beta = P_1\left(\bar{X} \le 1.645 \frac{\sigma}{\sqrt{n}}\right)$$

or the power γ of the test is

$$
\begin{aligned}
1 - \beta &= 1 - P_1\left(\bar{X} \leq 1.645\frac{\sigma}{\sqrt{n}}\right) \\
&= P_1\left(\bar{X} > 1.645\frac{\sigma}{\sqrt{n}}\right) \\
&= P_1\left(\frac{\bar{X} - 1}{\sigma/\sqrt{n}} > \frac{1.645\frac{\sigma}{\sqrt{n}} - 1}{\sigma/\sqrt{n}}\right) \\
&= P\left(Z > 1.645 - \frac{\sqrt{n}}{\sigma}\right)
\end{aligned}
$$

where Z is a standard normal random variable. We can explicitly compute this probability given n and σ using the tables for the standard normal distribution. Let us note that the constant k determining the threshold point for acceptance or rejection does not depend on the alternate hypothesis $H_1 : \mu = 1$ in this case. The test obtained above can be used for testing the hypothesis

$$H_0 : \mu = 0 \text{ against } H_1' : \mu > 0$$

and the power function of this test at any value $\mu > 0$ is given by

$$
\begin{aligned}
\gamma(\mu) &= P_\mu\left(\bar{X} > 1.645\frac{\sigma}{\sqrt{n}}\right) \\
&= P\left(Z > 1.645 - \frac{\mu\sqrt{n}}{\sigma}\right) \qquad (9.4)
\end{aligned}
$$

where Z is a standard normal random variable. Note that as μ increases to $+\infty$, the function $\gamma(\mu)$ increases to 1. Further more $\gamma(0) = \alpha$. Define $\gamma(\mu)$ for all μ using the equation (9.4). Then the function $\gamma(\mu)$ decreases to zero as μ decrease to $-\infty$. See the graph in Figure 9.2.

It is easy to see from the definitions of $\gamma(.)$ and $\alpha(.)$ that $\gamma(\mu) = \alpha(\mu)$ for $\mu \leq 0$ and $\gamma(\mu)$ is nondecreasing for all μ. In particular, it follows that $\alpha(\mu) \leq \alpha(0) = P_0(\bar{X} > k] = .05$ for all $\mu \leq 0$. In other words, the test we derived above is also a test of significance level $\alpha = .05$ for testing

$$H_0' : \mu \leq 0 \text{ against } H_1' : \mu > 0.$$

The question that arises now: did we obtain the *best* test, best in the sense of maximum power? It turns out that the test derived above is the best for reasons to be explained later in this section.

Remarks. Suppose we want to test the hypothesis H_0 that the mean $\mu = 0$ for a probability distribution, that is not necessarily normal, with

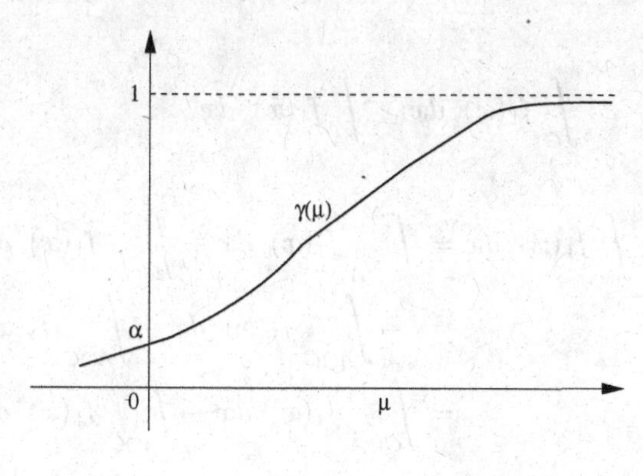

Fig. 9.2

mean μ and known variance σ^2. The test of significance level α, derived above for testing the mean μ of a normal distribution, can still be used as a test with an approximate significance level α whenever the sample size n is large since

$$\frac{\bar{X}-\mu}{\sigma/\sqrt{n}} \xrightarrow{\mathcal{L}} N(0,1) \text{ as } n \to \infty$$

by the Central limit theorem.

Neyman-Pearson Lemma

Suppose we are interested in testing the null hypothesis $H_0 : f(\boldsymbol{x}) = f_0(\boldsymbol{x})$ against the alternate hypothesis $H_1 : f(\boldsymbol{x}) = f_1(\boldsymbol{x})$ were f_0 and f_1 are completely specified probability density functions (or probability functions) and f is the true but unknown probability density function (or probability function). Let $\boldsymbol{X}$ be a random vector with probability density function $f(\boldsymbol{x})$.

Theorem 9.1. Let $k > 0$ and $C = \{\boldsymbol{x} : \frac{f_1(\boldsymbol{x})}{f_0(\boldsymbol{x})} \geq k\}$ and α be the probability of the set C under the hypothesis H_0. Then the set C gives the most powerful critical region of size α for testing the simple null hypothesis $H_0 : f(\boldsymbol{x}) = f_0(\boldsymbol{x})$ against the simple alternate hypothesis $H_1 : f(\boldsymbol{x}) = f_1(\boldsymbol{x})$.

Proof. We will prove the result for the case when f_0 and f_1 are probability density functions. Suppose A is another critical region of size α. In other words

$$\int_C f_0(\boldsymbol{x})\ d\boldsymbol{x} = \alpha = \int_A f_0(\boldsymbol{x})\ d\boldsymbol{x}.$$

We will prove that

$$\int_C f_1(\boldsymbol{x})\ d\boldsymbol{x} \geq \int_A f_1(\boldsymbol{x})\ d\boldsymbol{x}.$$

Observe that

$$\begin{aligned}\int_C f_1(\boldsymbol{x})\ d\boldsymbol{x} - \int_A f_1(\boldsymbol{x})\ d\boldsymbol{x} &= \int_{C\cap A} f_1(\boldsymbol{x})\ d\boldsymbol{x} + \int_{C\cap \bar{A}} f_1(\boldsymbol{x})\ d\boldsymbol{x}\\ &\quad - \int_{A\cap C} f_1(\boldsymbol{x})\ d\boldsymbol{x} - \int_{A\cap \bar{C}} f_1(\boldsymbol{x})\ d\boldsymbol{x}\\ &= \int_{C\cap \bar{A}} f_1(\boldsymbol{x})\ d\boldsymbol{x} - \int_{A\cap \bar{C}} f_1(\boldsymbol{x})\ d\boldsymbol{x}. \quad (9.5)\end{aligned}$$

Note that for every $\boldsymbol{x}$ in the set C and hence in the set $C \cap \bar{A}$,

$$f_0(\boldsymbol{x}) \leq \frac{1}{k} f_1(\boldsymbol{x})$$

and hence

$$\int_{C\cap \bar{A}} f_1(\boldsymbol{x})\ d\boldsymbol{x} \geq k \int_{C\cap \bar{A}} f_0(\boldsymbol{x})\ d\boldsymbol{x}.$$

Similarly for every $\boldsymbol{x}$ in the set $\bar{C}$ and hence in the set $A \cap \bar{C}$,

$$f_0(\boldsymbol{x}) > \frac{1}{k} f_1(\boldsymbol{x})$$

which implies that

$$\int_{A\cap \bar{C}} f_1(\boldsymbol{x})\ d\boldsymbol{x} \leq k \int_{A\cap \bar{C}} f_0(\boldsymbol{x})\ d\boldsymbol{x}.$$

Combining the above two inequalities, we get that

$$\int_{C\cap \bar{A}} f_1(\boldsymbol{x})\ d\boldsymbol{x} - \int_{A\cap \bar{C}} f_1(\boldsymbol{x})\ d\boldsymbol{x} \geq k \int_{C\cap \bar{A}} f_0(\boldsymbol{x})\ d\boldsymbol{x} - k \int_{A\cap \bar{C}} f_0(\boldsymbol{x})\ d\boldsymbol{x}.$$

From the equation (9.5), it follows that

$$\int_C f_1(\boldsymbol{x})\ d\boldsymbol{x} - \int_A f_1(\boldsymbol{x})\ d\boldsymbol{x} \geq k\left[\int_{C\cap \bar{A}} f_0(\boldsymbol{x})\ d\boldsymbol{x} - \int_{A\cap \bar{C}} f_0(\boldsymbol{x})\ d\boldsymbol{x}\right] (9.6)$$

But

$$\begin{aligned}\int_{C\cap \bar{A}} f_0(\boldsymbol{x})\ d\boldsymbol{x} - \int_{A\cap \bar{C}} f_0(\boldsymbol{x})\ d\boldsymbol{x} &= \int_{C\cap A} f_0(\boldsymbol{x})\ d\boldsymbol{x} + \int_{C\cap \bar{A}} f_0(\boldsymbol{x})\ d\boldsymbol{x}\\ &\quad - \int_{A\cap C} f_0(\boldsymbol{x})\ d\boldsymbol{x} - \int_{A\cap \bar{C}} f_0(\boldsymbol{x})\ d\boldsymbol{x}\\ &= \int_C f_0(\boldsymbol{x})\ d\boldsymbol{x} - \int_A f_0(\boldsymbol{x})\ d\boldsymbol{x}\\ &= \alpha - \alpha = 0.\end{aligned}$$

Combining this result with the inequality (9.6), we get that

$$\int_C f_1(\boldsymbol{x})\ d\boldsymbol{x} - \int_A f_1(\boldsymbol{x})\ d\boldsymbol{x} \geq 0$$

which proves the theorem. The case when the functions f_0 and f_1 are probability functions goes through on similar lines by replacing the integrals by the sums in the above proof.

Remarks. Theorem 9.1 gives sufficient conditions for a critical region C of size α to be the best in the sense that it gives rise to a most powerful test of size α. However these conditions are also necessary. A slightly more general version of the result is as follows.

Theorem 9.2. **(Neyman-Pearson Lemma)** The most powerful test for testing a simple hypothesis $H_0 : f(\boldsymbol{x}) = f_0(\boldsymbol{x})$ against a simple alternate hypothesis $H_1 : f(\boldsymbol{x}) = f_1(\boldsymbol{x})$ at a level of significance $0 \leq \alpha \leq 1$ is of the form

$$\psi(\boldsymbol{x}) = \begin{cases} 1 \text{ if } \frac{f_1(\boldsymbol{x})}{f_0(\boldsymbol{x})} > k \\ \gamma \text{ if } \frac{f_1(\boldsymbol{x})}{f_0(\boldsymbol{x})} = k \\ 0 \text{ if } \frac{f_1(\boldsymbol{x})}{f_0(\boldsymbol{x})} < k \end{cases}$$

where $\psi(\boldsymbol{x})$ denotes the probability of rejecting the hypothesis H_0 when $\boldsymbol{x}$ is the observed sample and k, γ are chosen so that $k \geq 0, 0 \leq \gamma \leq 1$ such that

$$E_0[\psi(\boldsymbol{X})] = \alpha,$$

that is,

$$P_0\left[\frac{f_1(\boldsymbol{X})}{f_0(\boldsymbol{X})} > k\right] + \gamma\ P_0\left[\frac{f_1(\boldsymbol{X})}{f_0(\boldsymbol{X})} = k\right] = \alpha.$$

Here $E_0(.)$ and $P_0(.)$ denote the expectation and the probability computed respectively under the hypothesis H_0.

We will not give a detailed proof of this version of the Neyman-Pearson lemma. It is a slight variation of the proof of Theorem 9.1. However we will discuss some implications of Theorem 9.2. The test procedure described above is a *randomized test* as opposed to a non-randomized test described in Theorem 9.1. A *randomized test* is described via a *test function* $\psi(\boldsymbol{x})$ which gives the probability of rejecting the null hypothesis when $\boldsymbol{x}$ is the observed value. For instance, the function ψ given by Theorem 9.2 gives a randomized test. If $[f_1(\boldsymbol{x})/f_0(\boldsymbol{x})] > k$, then $\psi(\boldsymbol{x}) = 1$, that is, the probability of rejecting H_0 is one. If $[f_1(\boldsymbol{x})/f_0(\boldsymbol{x})] < k$, then $\psi(\boldsymbol{x}) = 0$, that is the probability of rejecting H_0 is zero. If $[f_1(\boldsymbol{x})/f_0(\boldsymbol{x})] = k$, then the probability of rejecting H_0 is γ where $0 \leq \gamma \leq 1$. Note that there is a probability of $1 - \gamma$ of accepting H_0 in this case. In the last case, we conduct an independent random experiment with two types of outcomes, for instance, call them success and failure with probabilities γ and $1 - \gamma$ respectively. If the outcome of this experiment is a success, we reject H_0; if the outcome is a failure, then we accept H_0. This type of randomized test leads to an exact level of significance α for the test. Since the most powerful test derived in Theorem 9.2 involves additional randomization, we consider the non-randomized test

$$\begin{aligned} \psi^*(x) &= 1 \text{ if } \frac{f_1(\boldsymbol{x})}{f_0(\boldsymbol{x})} > k \\ &= 0 \text{ if } \frac{f_1(\boldsymbol{x})}{f_0(\boldsymbol{x})} \leq k \end{aligned}$$

as an approximation. In case the ratio $[f_1(\boldsymbol{X})/f_0(\boldsymbol{X})]$ has a distribution under H_0 with $P_0([f_1(\boldsymbol{X})/f_0(\boldsymbol{X})] = k) = 0$, then the test functions $\psi^*(\boldsymbol{x})$ and $\psi(\boldsymbol{x})$ are equivalent and both the tests lead to the same conclusions for a given α. However, if $P_0([f_1(\boldsymbol{X})/f_0(\boldsymbol{X})] = k) > 0$, then the significance level of the test ψ^* is approximately equal to α. However the test ψ^* is easy to implement in practice. Hereafter we consider non-randomized tests only in this book. We observe that, if the ratio $[f_1(\boldsymbol{X})/f_0(\boldsymbol{X})]$ has a continuous distribution when $f = f_0$, then the (randomized) test function $\psi(.)$ and the (non-randomized) test function $\psi^*(.)$ are equal with probability one. The test ψ given by Theorem 9.2 is the most powerful test in the sense that if η is any other test function for testing H_0 versus H_1 with $E_0[\eta] \leq \alpha$, then $E_1[\eta] \leq E_1[\psi]$. The ratio $[f_1(\boldsymbol{x})/f_0(\boldsymbol{x})]$ is called the *likelihood ratio*.

Let us again consider the following example discussed earlier.

Suppose $X_1, \ldots, X_n$ are i.i.d. random variables with the probability distribution $N(\mu, \sigma^2)$ where σ^2 is known and we are interested in testing the simple null hypothesis $H_0 : \mu = \mu_0$ against the simple alternate hypothesis $H_1 : \mu = \mu_1$ at the significance level α where $\mu_0 < \mu_1$. An application of the Neyman-Pearson lemma shows that the most powerful test is given by

$$\psi(\boldsymbol{x}) = \begin{cases} 1 \text{ if } [f_1(\boldsymbol{x})/f_0(\boldsymbol{x})] > k \\ \gamma \text{ if } [f_1(\boldsymbol{x})/f_0(\boldsymbol{x})] = k \\ 0 \text{ if } [f_1(\boldsymbol{x})/f_0(\boldsymbol{x})] < k \end{cases}$$

where $k \geq 0$ and $0 \leq \gamma \leq 1$ are chosen so that

$$E_0[\psi(\boldsymbol{X})] = \alpha.$$

Here

$$\begin{aligned}
\frac{f_1(\boldsymbol{X})}{f_0(\boldsymbol{X})} &= \frac{\left(\frac{1}{\sqrt{2\pi\sigma^2}}\right)^n \exp\left\{-\frac{1}{2}\sum_{i=1}^n \frac{(X_i-\mu_1)^2}{\sigma^2}\right\}}{\left(\frac{1}{\sqrt{2\pi\sigma^2}}\right)^n \exp\left\{-\frac{1}{2}\sum_{i=1}^n \frac{(X_i-\mu_0)^2}{\sigma^2}\right\}} \\
&= \exp\left\{-\frac{1}{2}\frac{n(\bar{X}-\mu_1)^2}{\sigma^2} + \frac{1}{2}\frac{n(\bar{X}-\mu_0)^2}{\sigma^2}\right\} \\
&= \exp\{-\frac{n}{2\sigma^2}[(\bar{X}-\mu_1)^2 - (\bar{X}-\mu_0)^2]\} \\
&= \exp\{-\frac{n}{2\sigma^2}[-2\bar{X}(\mu_1-\mu_0) + \mu_1^2 - \mu_0^2]\}
\end{aligned}$$

where $\bar{X} = n^{-1}\sum_{i=1}^n X_i$. Hence

$$[f_1(\boldsymbol{X})/f_0(\boldsymbol{X})] > k$$

if and only if

$$-\frac{n}{2\sigma^2}[-2\bar{X}(\mu_1-\mu_0) + \mu_1^2 - \mu_0^2] > \log k$$

or equivalently

$$\bar{X}(\mu_1 - \mu_0) > c$$

where c is a suitable constant. Since $\mu_1 > \mu_0$, the most powerful test can be written in the form

$$\psi(\boldsymbol{x}) = \begin{cases} 1 \text{ if } \bar{X} > d \\ \gamma \text{ if } \bar{X} = d \\ 0 \text{ if } \bar{X} < d \end{cases}$$

where $0 \leq \gamma \leq 1$ and d are suitable constants chosen such that

$$E_0[\psi(\boldsymbol{X})] = \alpha.$$

Since $\bar{X}$ has a continuous distribution under H_0, it follows that $P_0(\bar{X} = d) = 0$. Hence the most powerful test is non-randomized in this case and the test can be written in the form:

$$\text{reject } H_0 \text{ if } \bar{X} > d$$

and

$$\text{accept } H_0 \text{ if } \bar{X} \leq d$$

where d is chosen so that $P_0(\bar{X} > d) = \alpha$. Note that

$$P_0(\bar{X} > d) = P_0\left(\frac{\bar{X} - \mu_0}{\sigma/\sqrt{n}} > \frac{d - \mu_0}{\sigma/\sqrt{n}}\right)$$
$$= P\left(Z > \frac{d - \mu_0}{\sigma/\sqrt{n}}\right)$$

where Z is a standard normal random variable. Depending on the value of α, the constant d can be found out using the tables for the standard normal distribution.

Likelihood ratio tests

We have given a method of finding the most powerful test for testing a simple null hypothesis against simple alternate hypothesis. This is achieved by an application of the Neyman-Pearson lemma and the test statistic is determined by the likelihood ratio. If the null hypothesis H_0 is composite or the alternate hypothesis H_1 is composite, it is not obvious how one can obtain a *best* test for testing a hypothesis H_0 against another hypothesis H_1 at a given significance level. For some special types of hypotheses and special classes of probability density functions or probability functions, tests can be developed which have some optimum properties. We will not discuss these results here. However we give now a general procedure to derive tests for testing a composite null hypothesis against a composite alternate hypothesis. The tests derived by this procedure *may not* be optimum or best in the sense of maximizing the power of a test or in any other sense. However the procedure does give the best test available in several problems.

Let $\{f(x;\theta), \theta \in \Theta\}$ be a family of probability density functions (or probability functions) and the problem is to test the hypothesis $H_0 : \theta \in \Theta_0$ against the alternate $H_1 : \theta \in \Theta_1$ where $\Theta_0 \cap \Theta_1 = \Phi$ and $\Theta_0 \cup \Theta_1 = \Theta$. Suppose $X_1, \ldots, X_n$ are i.i.d. random variables with the probability density function (or probability function) $f(x;\theta)$. Then the joint probability density function (or joint probability function) of $\boldsymbol{X} = (X_1, \ldots, X_n)$ is

$$L_n(\boldsymbol{x};\theta) = \Pi_{i=1}^n f(x_i;\theta).$$

Define

$$\lambda(\boldsymbol{x}) = \frac{\sup\{L_n(\boldsymbol{x};\theta) : \theta \in \Theta_0\}}{\sup\{L_n(\boldsymbol{x};\theta) : \theta \in \Theta\}}.$$

If the random variables X_1 has a discrete probability distribution, then the random variable $\lambda(\boldsymbol{X})$ can be interpreted as follows. The numerator represents the maximum of the probability for the occurrence of the sample $(x_1, \ldots, x_n)$ when $\theta \in \Theta_0$ or equivalently under H_0 and the denominator represents the maximum of the probability for the occurrence of the sample $(x_1, \ldots, x_n)$ over all $\theta \in \Theta$ or equivalently under H_0 as well as H_1. The random variable $\lambda(\boldsymbol{X})$ is called the *likelihood ratio.* It is clear that $0 \leq \lambda(\boldsymbol{x}) \leq 1$ for all $\boldsymbol{x} = (x_1, \ldots, x_n)$. If $\lambda(\boldsymbol{x})$ is small, then we expect the parameter θ to be in the subset Θ_1 rather than in the subset Θ_0 since the numerator is small compared to the denominator. If $\lambda(\boldsymbol{x})$ is large, then the parameter θ is likely to be in the subset Θ_0 since the numerator has almost the same value as the denominator. Motivated by this reasoning, a likelihood ratio test for testing the hypothesis $H_0 : \theta \in \Theta_0$ against the alternate hypothesis $H_1 : \theta \in \Theta_1$ is defined as follows:

$$\text{reject the hypothesis } H_0 \text{ if } \lambda(\boldsymbol{x}) \leq k$$

and

$$\text{accept the hypothesis } H_0 \text{ if } \lambda(\boldsymbol{x}) > k.$$

The critical region here is of the form $\{\boldsymbol{x} \in \mathcal{X}^n : \lambda(\boldsymbol{x}) \leq k\}$ where $\mathcal{X}^n$ is the sample space and k is a constant chosen suitably so that the level of significance of the test is equal to α. We will now give an example to illustrate an application of the likelihood ratio test.

Example 9.1. Suppose $X_1, \ldots, X_n$ are i.i.d. random variables with the uniform distribution over $[0, \theta]$ where $\theta > 0$ is unknown. The problem is to test the hypothesis $H_0 : \theta = \theta_0$ against $H_1 : \theta \neq \theta_0$ where $\theta_0 > 0$ is given. Here the parameter space $\Theta = \{\theta : 0 < \theta < \infty\}$ and $\Theta_0 = \{\theta_0\}$. The likelihood function of the sample $(x_1, \ldots, x_n)$ can be written in the form

$$L_n(\boldsymbol{x}; \theta) = \Pi_{i=1}^n f(x_i, \theta)$$

where

$$\begin{aligned} f(x, \theta) &= \frac{1}{\theta} \text{ if } 0 \leq x_i \leq \theta \\ &= 0 \text{ otherwise.} \end{aligned}$$

Hence

$$\begin{aligned} L_n(\boldsymbol{x}; \theta) &= \frac{1}{\theta^n} \text{ if } 0 \leq x_{(n)} \equiv \max\{x_i, 1 \leq i \leq n\} \leq \theta \\ &= 0 \text{ otherwise.} \end{aligned}$$

We have seen earlier that the likelihood function $L_n(\boldsymbol{x}; \theta)$ is maximum over the parameter space$\Theta = (0, \infty)$ when $\theta = \max\{x_i, 1 \leq i \leq n\}$. It is obvious that $\sup\{L_n(\boldsymbol{x}; \theta) : \theta \in \Theta_0\} = L_n(\boldsymbol{x}; \theta_0) = \frac{1}{\theta_0^n}$ for $0 \leq x_{(n)} \leq \theta_0$. Hence

$$\lambda(\boldsymbol{x}) = \frac{(1/\theta_0)^n}{(1/x_{(n)})^n} = \left(\frac{x_{(n)}}{\theta_0}\right)^n \text{ for } 0 \leq x_{(n)} \leq \theta_0$$

and the likelihood ratio test for testing the hypothesis H_0 against the hypothesis H_1 is of the form:

$$\text{reject } H_0 \text{ if } \lambda(\boldsymbol{x}) = \left(\frac{x_{(n)}}{\theta_0}\right)^n \leq k,$$

and

$$\text{accept } H_0 \text{ if } \lambda(\boldsymbol{x}) = \left(\frac{x_{(n)}}{\theta_0}\right)^n > k$$

where k is chosen so that

$$P_{\theta_0}\left[\left(\frac{X_{(n)}}{\theta_0}\right)^n \leq k\right] = \alpha$$

or equivalently

$$P_{\theta_0}\left[\frac{X_{(n)}}{\theta_0} \leq k^{1/n}\right] = \alpha.$$

Check that the constant k is equal to α.

Remarks. Under some regularity conditions on the probability density function (or the probability function) $f(x, \theta)$ (similar to those which ensure the asymptotic normality of the maximum likelihood estimator as discussed in the Chapter 8), it can be shown the distribution of the random variable $-2\log \lambda(\boldsymbol{X})$ converges to the Chi-square distribution as $n \to \infty$ with the degrees of freedom equal to the difference of the number of independent parameters in Θ and the number of independent parameters in Θ_0 under H_0.

9.4 Chi-square Tests

Tests for the goodness-of-fit

Suppose that we are able to classify n independent observations from a population into k different groups $G_i, 1 \leq i \leq k$ with an observed frequency of n_i in the group G_i for $1 \leq i \leq k$ or suppose the observed data itself consists of the frequencies n_i in the groups G_i for $1 \leq i \leq k$. It is of interest and of importance to check whether these observations are from a pre-specified distribution such as the normal distribution or the Poisson distribution. Such a problem can be formulated as a particular case of testing for the probabilities of different outcomes in a Multinomial distribution. Tests of this type are called *tests for goodness-of-fit.*

Let a random vector $(n_1, \ldots, n_k)$ have a Multinomial distribution with n_i denoting the number of outcomes of type i, p_i the probability of the i-th outcome and n the number of independent trials. It is clear that $n_1 + \cdots + n_k = n$. The expected number of outcomes of type i is np_i. We are interested in testing the null hypothesis of the form

$$H_0 : p_i = p_{i0}, 1 \leq i \leq k$$

where $p_{i0} > 0, 1 \leq i \leq k$ are fully specified and $\sum_{i=1}^{k} p_{i0} = 1$. Note that there are only $k-1$ independent p_i since $p_1 + \cdots + p_k = 1$. If the hypothesis H_0 is true, then we expect the observed frequency n_i of the group G_i to be close to np_{i0} for $1 \leq i \leq k$. Define

$$\chi^2 = \sum_{i=1}^{k} \frac{(n_i - np_{i0})^2}{np_{i0}}.$$

It is expected that the value of χ^2 computed from the data should be small if the hypothesis H_0 is true and should be large otherwise. On the basis of this observation, we can suggest a test of the form

$$\text{reject } H_0 \text{ if } \chi^2 > c$$

and

$$\text{accept } H_0 \text{ if } \chi^2 \leq c$$

where c is suitable value to be chosen so that the test has a preassigned significance level α. We have to find c such that

$$P(\chi^2 > c) = \alpha$$

when the hypothesis H_0 holds. The exact distribution of the χ^2 statistic is difficult to compute. It is known however that the distribution of the statistic χ^2 converges to the Chi-square distribution with $(k-1)$ degrees of freedom as the number of observations $n \to \infty$ under H_0. Note that the number of degrees of freedom for the limiting Chi-square distribution is the number of groups less one. One can use the Table C.3 for the Chi-square distribution with $(k-1)$ degrees of freedom to find an approximate value for the threshold constant c. See Figure 9.3 for the graphs of Chi-square distribution with n degrees of freedom for $n = 1, 2, 3, 4$.

Empirical studies show that the approximation of the distribution of the χ^2 statistic by a Chi-square distribution is good if the expected frequencies np_{i0} are greater than or equal to five in every group and the observed frequencies n_i are not too small. If the expected frequency in any particular

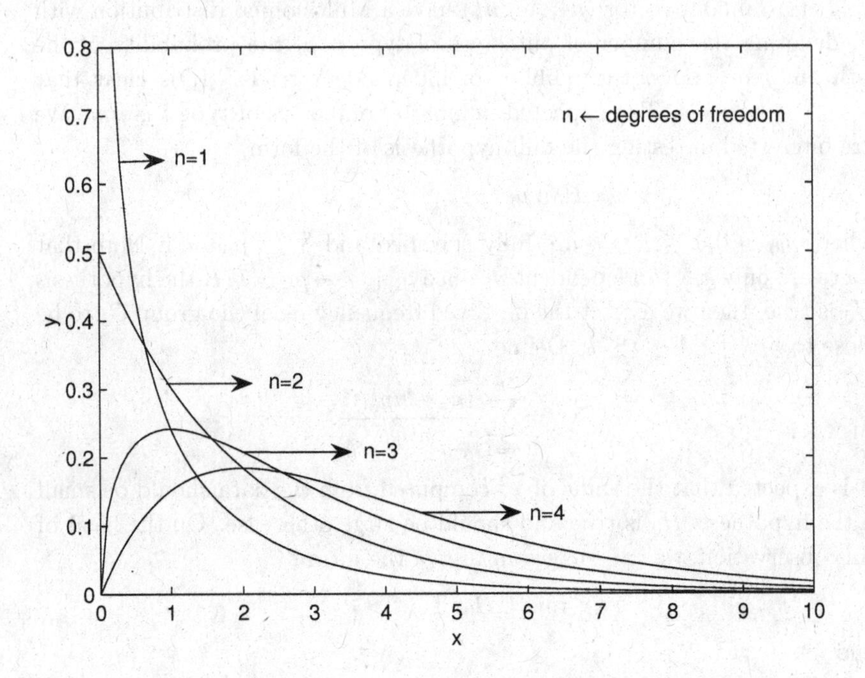

Fig. 9.3 Chi-Square Probability Density Functions

group G_i is less than 5 under H_0, it is suggested that the group G_i may be pooled with another group so that the expected frequency in the combined groups is at least 5. The choice of the second group is arbitrary but the pooling of the groups should be minimum as far as possible as the number of degrees of freedom for the limiting Chi-square statistic depends on the final number of groups after pooling. The degrees of freedom associated with the limiting Chi-square distribution is the number of groups after pooling less one. Recall that, if we conclude, after testing, that we should accept the hypothesis H_0, it only indicates that the data is not strong enough to conclude that the model is not correct.

Example 9.2. A city has four roads in each direction . It was decided to investigate whether the drivers of automobiles preferred any particular road to others for traveling. Observed frequencies of the usage of the four different roads by 1000 automobiles on a particular day are recorded.

Road	Frequency
1	294
2	276
3	238
4	192

Let p_i be the probability that a driver prefers to travel by the i-th road. If no road is preferred to others, then $p_1 = p_2 = p_3 = p_4$. The problem is to test the hypothesis

$$H_0 : p_1 = p_2 = p_3 = p_4.$$

Since $p_1 + p_2 + p_3 + p_4 = 1$, it follows that the hypothesis H_0 can also be specified in the form

$$H_0 : p_i = \frac{1}{4}.$$

We can use the Chi-square test developed above for this problem. Let n_i be the observed frequency of usage for the i-th road. Under the hypothesis H_0, the expected frequency of usage is $np_i = 1000(1/4) = 250$ for the i-th road. The observed value of χ^2

$$\begin{aligned}\chi^2 &= \sum_{i=1}^{4} \frac{(n_i - 250)^2}{250} \\ &= \frac{(294-250)^2}{250} + \frac{(276-250)^2}{250} \\ &\quad + \frac{(238-250)^2}{250} + \frac{(192-250)^2}{250} \\ &= 24.48.\end{aligned}$$

Since the number of groups is four, the limiting Chi-square distribution has three degrees of freedom in this case. If we choose the significance level $\alpha = .05$, it can be checked that the cut off value is $c = 7.81$ from the tables for the Chi-square distribution with three degrees of freedom. Since the observed value 24.48 of the test statistic exceeds 7.81, we reject the hypothesis H_0. Hence there is evidence that the drivers prefer some roads to others.

We have just now given a method for testing whether the probability vector $\boldsymbol{p} = (p_1, \ldots, p_k)$, of the k types of outcomes of a Multinomial distribution with k groups, is a completely specified vector $\boldsymbol{p}_0 = (p_{10}, \ldots, p_{k0})$.

The probability vector $\boldsymbol{p} = (p_1, \ldots, p_k)$ might arise from any probability distribution such as the normal or the Poisson distribution with known parameters with pre-specified rules for classification into k groups. Suppose that we would like to test whether a certain data follows a normal distribution or a Poisson distribution or any other distribution where the parameters $\theta = (\theta_1, \ldots, \theta_r)$ involved in the probability distribution are *not specified*. We now describe a modification of the Chi-square test discussed above. It is clear that the probability vector $\boldsymbol{p} = (p_1, \ldots, p_k)$ of the Multinomial distribution depends on the unknown parameter vector $\theta = (\theta_1, \ldots, \theta_r)$ and the underlying probability distribution. The hypothesis H_0 is no longer a simple hypothesis. Let $p_i(\theta)$ be the probability that an observation belongs to the group G_i for $1 \leq i \leq k$. Suppose that we would like to test the hypothesis

$$H_0 : p_i = p_i(\theta), 1 \leq i \leq k.$$

Here $p_i(\theta)$ is computed from the underlying probability distribution under H_0 when θ is the true parameter vector. Let n_i be the number of observations in the group G_i. If the hypothesis H_0 is true, then the expected number in the group G_i is $n\, p_i(\theta)$. Let

$$Q(\theta) = \sum_{i=1}^{k} \frac{(n_i - n\, p_i(\theta))^2}{n\, p_i(\theta)}.$$

If the parameter θ is known, then the statistic $Q(\theta)$ can be computed and it will have an approximate Chi-square distribution with $k-1$ degrees of freedom for large n under H_0. Since the parameter θ is unknown, we have to estimate the parameter θ from the data. Let $\hat{\theta}_n$ be a maximum likelihood estimator of the parameter θ. Consider

$$Q(\hat{\theta}_n) = \sum_{i=1}^{k} \frac{(n_i - n\, p_i(\hat{\theta}_n))^2}{n\, p_i(\hat{\theta}_n)}.$$

Under certain regularity conditions (similar to those as discussed in the Chapter 8 under which the asymptotic normality of a maximum likelihood estimator holds), it can be shown that the statistic $Q(\hat{\theta}_n)$ will have the Chi-square distribution with $k-1-r$ degrees of freedom for large n under the hypothesis H_0. Note that the degrees of freedom were reduced by r due to the estimation of the r-dimensional parameter θ. It should be noted that the limiting distribution of the statistic $Q(\hat{\theta}_n)$ need not be a Chi-square distribution if the estimator used for the unknown parameter θ is an estimator obtained by the method of moments or the method of minimum

Chi-square or any other method unless these estimators have asymptotic properties similar to those of the maximum likelihood estimator. We can now specify the test for the hypothesis

$$H_0 : p_i = p_i(\theta), 1 \leq i \leq k.$$

The test is given by

$$\text{reject } H_0 \text{ if } Q(\hat{\theta}_n) > c$$

and

$$\text{accept } H_0 \text{ if } Q(\hat{\theta}_n) \leq c$$

where c is suitable value to be chosen so that the test has a preassigned significance level α. We have to find c such that

$$P(Q(\hat{\theta}_n) > c) = \alpha$$

when the hypothesis H_0 holds. For large n, the threshold constant c can be computed using the fact that, the statistic $Q(\hat{\theta}_n)$ will have an approximate Chi-square distribution with $k-1-r$ degrees of freedom for large n under H_0.

Example 9.3. A survey of $n = 200$ families each with four children revealed the following frequency distribution of the number of girls per family.

Number of girls	0	1	2	3	4
Number of families	5	32	65	75	23

We would like to test whether the data follows a Binomial distribution. Let X denote the number of girls in a family with four children. Let θ be the probability for a girl child. Let p_i be the probability that there are i girls in a family of size $m = 4$. Then

$$\begin{aligned} p_i = P(X = i) &= m_i^C \theta^i (1-\theta)^{m-i} \\ &= 4_i^C \theta^i (1-\theta)^{4-i} \\ &= p_i(\theta) \end{aligned}$$

for $i = 0, 1, 2, 3, 4$. Since the parameter θ is unknown, it has to be estimated from the data. Check that the maximum likelihood estimate $\hat{\theta}_n$ is 0.5987. Let us now compute

$$Q(\hat{\theta}_n) = \sum_{i=0}^{4} \frac{(n_i - np_i(\hat{\theta}_n))^2}{np_i(\hat{\theta}_n)}.$$

Check that

$$p_0(\hat{\theta}_n) = 0.02594; p_1(\hat{\theta}_n) = 0.15476; p_2(\hat{\theta}_n) = 0.34634$$

and

$$p_3(\hat{\theta}_n) = 0.34448; p_4(\hat{\theta}_n) = 0.12848.$$

We can tabulate the observed and the expected frequencies:

j	Observed frequency	Expected frequency
0	5	5
1	32	31
2	65	69
3	75	69
4	23	36

Since the number of groups is 5 and the dimension of the parameter space is one, the number of degrees of freedom for the limiting Chi-square distribution of the statistic $Q(\hat{\theta}_n)$ is 5-1-1=3. Note that the number of observations n is 200 which can be considered as large. Check that

$$Q(\hat{\theta}_n) = 1.132$$

in this problem. Suppose we choose $\alpha = .05$ as the significance level. It can be checked from the tables for the Chi-square distribution with three degrees of freedom that $P(\chi^2 > 7.81) = .05$ when the random variable χ^2 has the Chi-square distribution with 3 degrees of freedom. Since the observed value of the test statistic is smaller than 7.81, we conclude that the data could be from a Binomial distribution.

Test for the comparison of two Multinomial distributions

Let us now consider the problem of comparing two independent Multinomial distributions with parameters $(N_j; p_{ij}, 1 \leq i \leq k), j = 1, 2$. We are interested in testing the hypothesis

$$H_0 : p_{i1} = p_{i2}, 1 \leq i \leq k$$

where the values of $p_{i1} = p_{i2}, 1 \leq i \leq k$ are unspecified. Let $n_{ij}, 1 \leq i \leq k$ be the observed frequencies of the N_j independent observations for $j = 1, 2$ under the two Multinomial distributions. Let

$$Q_j = \sum_{i=1}^{k} \frac{(n_{ij} - N_j p_{ij})^2}{N_j p_{ij}}, j = 1, 2.$$

Suppose that N_1 and N_2 are large and the observations under the two Multinomial distributions are independent of each other. If the probabilities $p_{ij}, 1 \leq i \leq k$ are known, then the random variable Q_j has a limiting Chi-square distribution with $k-1$ degrees of freedom for $j = 1, 2$. Since Q_1 and Q_2 are independent random variables, the limiting distribution of

$$Q = Q_1 + Q_2 = \sum_{j=1}^{2} \sum_{i=1}^{k} \frac{(n_{ij} - N_j p_{ij})^2}{N_j p_{ij}}$$

is a Chi-square distribution with $(k - 1 + k - 1) = 2k - 2$ degrees of freedom. However the random variable Q is not computable as it involves the unknown probabilities $p_{ij}, 1 \leq i \leq k, j = 1, 2..$ However, under the hypothesis H_0, we have $p_{i1} = p_{i2}, 1 \leq i \leq k$, and the maximum likelihood estimator of $p_{i1} = p_{i2}$ based on the observed frequencies $\{n_{ij}, j = 1, 2\}$ is

$$\hat{p}_{i1} = \hat{p}_{i2} = \frac{n_{i1} + n_{i2}}{N_1 + N_2}$$

for $1 \leq i \leq k$. Substituting these estimates for $\{p_{ij}\}$ in Q, we obtain

$$\hat{Q} = \sum_{j=1}^{2} \sum_{i=1}^{k} \frac{(n_{ij} - N_j \hat{p}_{ij})^2}{N_j \hat{p}_{ij}}.$$

It can be checked that the statistic $\hat{Q}$ has a limiting Chi-square distribution with $2k - 2 - (k - 1) = k - 1$ degrees of freedom under H_0. Note that we have estimated the $(k - 1)$ common parameters

$$p_{i1} = p_{i2}, 1 \leq i \leq k$$

of the two Multinomial distributions under H_0 since $\sum_{i=1}^{k} p_{ij} = 1, i = 1, 2.$ The test can be framed in the form:

$$\text{reject } H_0 \text{ if } \hat{Q} > c$$

and

$$\text{accept } H_0 \text{ if } \hat{Q} \leq c$$

where c is a suitable value to be chosen so that the test has the preassigned significance level α.

Remarks. The above test can be generalized for comparison of more than two Multinomial distributions along similar lines and it is some times called the *test for homogeneity.* Suppose there are r Multinomial distributions and we want to test for their homogeneity, that is,

$$H_0 : p_{i1} = p_{i2} = \cdots = p_{ir}, 1 \leq i \leq k.$$

Then the statistic

$$\tilde{Q} = \sum_{j=1}^{r}\sum_{i=1}^{k} \frac{(n_{ij} - N_j\hat{p}_{ij})^2}{N_j\hat{p}_{ij}}.$$

has the limiting Chi-square distribution when $N_j \to \infty$ for $1 \le j \le r$ with $(k-1)(r-1)$ degrees of freedom under H_0. The estimator $\hat{p}_{ij}$ of the common value of $p_{i1} = p_{i2} = \cdots = p_{ir}, 1 \le i \le k$ under H_0 is the maximum likelihood estimator

$$\hat{p}_{ij} = \frac{n_{i1} + n_{i2} + \cdots + n_{ir}}{N_1 + N_2 + \cdots + N_r}$$

and one can use the test statistic $\tilde{Q}$ for testing H_0. Large values of $\tilde{Q}$ indicate that there is no homogeneity. We will leave it to the reader to formulate the test.

Test for independence (Contingency table)

Suppose we are interested in studying the dependence or the independence of two characteristics such as the education (A) and the annual income (B) of individuals in a large population. Suppose the characteristic A can be classified into m mutually exclusive cases $A_1, A_2, \ldots, A_m$ and the characteristic B can be classified into n mutually exclusive cases $B_1, B_2, \ldots, B_n$. Suppose N individuals are randomly selected and their characteristics A and B are recorded. Let n_{ij} denote the observed frequency of individuals who possess the characteristics A_i *and* B_j. Let $p_{ij} = P(A_i \cap B_j)$, that is, the probability that a randomly selected individual has the characteristics A_i and B_j. The random vector $(n_{11}, \ldots, n_{nm})$ has the Mutinomial distribution with parameters $(N; p_{11}, \ldots, p_{nm})$ and the random variable

$$Q = \sum_{j=1}^{n}\sum_{i=1}^{m} \frac{(n_{ij} - Np_{ij})^2}{Np_{ij}}$$

has the limiting Chi-square distribution with $mn-1$ degrees of freedom as $N \to \infty$. Note that

$$P(A_i) = \sum_{j=1}^{n} p_{ij} \equiv p_{i.}$$

and

$$P(B_j) = \sum_{i=1}^{m} p_{ij} \equiv p_{.j}.$$

Suppose we want to test the hypothesis that the characteristics A and B are *independent*, that is, we want to test the hypothesis

$$H_0 : P(A_i \cap B_j) = P(A_i)P(B_j); 1 \leq i \leq m, 1 \leq j \leq n.$$

Note that

$$\sum_{i=1}^{m} p_{i.} = \sum_{j=1}^{n} p_{.j} = \sum_{i=1}^{m}\sum_{j=1}^{n} p_{ij} = 1.$$

The hypothesis H_0 can be reformulated as

$$H_0 : p_{ij} = p_{i.}p_{.j}, 1 \leq i \leq m, 1 \leq j \leq n.$$

If $p_{i.}$ and $p_{.j}$ were known for $1 \leq i \leq m, 1 \leq j \leq n$, then we compute the observed value of Q under H_0. Since these probabilities are unknown, they have to be estimated. We can estimate them by the method of maximum likelihood and the estimators are

$$\hat{p}_{i.} = \frac{n_{i.}}{N}, \quad n_{i.} = \sum_{j=1}^{n} n_{ij}$$

and

$$\hat{p}_{.j} = \frac{n_{.j}}{N}, \quad n_{.j} = \sum_{i=1}^{m} p_{ij}.$$

Since

$$\sum_{i=1}^{m} p_{i.} = \sum_{j=1}^{n} p_{.j} = 1,$$

we have to estimate the $(m-1)+(n-1)$ parameters. If we use the estimators $\hat{p}_{ij} = \hat{p}_{i.}\hat{p}_{.j}$ for p_{ij}, then the resulting statistic

$$\hat{Q} = \sum_{j=1}^{n}\sum_{i=1}^{m} \frac{(n_{ij} - N\hat{p}_{ij})^2}{N\hat{p}_{ij}}$$

has the limiting Chi-square distribution with $mn - 1 - (m-1) - (n-1) = (m-1)(n-1)$ degrees of freedom when $N \to \infty$ under the hypothesis H_0. We reject the hypothesis of independence of the characteristics A and B if the computed value of $\hat{Q}$ from the data exceeds the constant c determined by the Chi-square distribution with $(m-1)(n-1)$ degrees of freedom at the given significance level α.

It is convenient to tabulate the observed frequencies in a tabular form with $m \times n$ cells and such a table is called a *contingency table.*

	B_1	B_2	...	B_n
A_1	n_{11}	n_{12}	...	n_{1n}
A_2	n_{21}	n_{22}	...	n_{2n}
...	...	...	...	...
A_m	n_{m1}	n_{m2}	...	n_{mn}

9.5 Exercises

9.1 Find a confidence interval for the unknown variance σ^2 of a normal distribution with unknown mean θ with a given confidence coefficient $100(1-\alpha)\%$. (Hint: Use the fact $\frac{\sum_{i=1}^{n}(X_i-\bar{X})^2}{\sigma^2}$ has a Chi-square distribution with $(n-1)$ degree of freedom).

9.2 A random sample of size 16 from a normal distribution with mean μ and variance σ^2 is obtained. It was found that $\bar{X} = 3.5$ and $S^2 = 4.41$. Determine a 90% confidence interval for the parameter σ^2.

9.3 Suppose X_1, X_2, X_3, X_4 are i.i.d. random variables with the Poisson distribution with unknown mean λ. Test the hypothesis $H_0 : \lambda = 1$ against $H_1 : \lambda = 2$ using a test of the form: reject H_0 if $\sum_{i=1}^{4} X_i \geq k$, accept otherwise. Find k so that the significance level α is about 0.05. Hence find β, the probability of Type II error.

9.4 Let $X_1, \ldots, X_n$ be independent Bernoulli random variables with $P(X_1 = 1) = p$ and $P(X_1 = 0) = 1 - p$. Derive an approximate confidence interval for the parameter p with 95% confidence coefficient.

9.5 Derive the form of the most powerful test of level α for testing the hypothesis $H_0 : p = \frac{1}{2}$ against $H_1 : p = \frac{3}{4}$ where p is the probability of a success in a sequence of n independent Bernoulli trials.

9.6 Let $X_1, \ldots, X_n$ be a random sample from the probability distribution $N(\mu, \sigma^2)$ where μ is known. Find the most powerful test for testing the hypothesis $H_0 : \sigma^2 = \sigma_0^2$ against the alternate $H_1 : \sigma^2 = \sigma_1^2$ where σ_0^2 and σ_1^2 are specified and $\sigma_1^2 > \sigma_0^2$.

9.7 Suppose that X is a random variable with the probability density function

$$f(x, \theta) = \theta x^{\theta-1}, 0 < x < 1$$
$$= 0 \text{ elsewhere.}$$

In order to test the hypothesis $H_0 : \theta = 2$ against the alternate hypothesis $H_1 : \theta = 3$, the following test is used: reject the hypothesis if $X_1 X_2 \geq \frac{1}{2}$ and accept otherwise based on a sample of size 2. Determine the probabilities of Type I and Type II errors and the power of the test.

9.8 Suppose that X is a random variable with the probability density function

$$f(x, \theta) = \theta^{-1} e^{-x/\theta}, \quad 0 < x < \infty$$
$$= 0 \qquad \text{otherwise.}$$

Obtain the form of the most powerful test for testing $H_0 : \theta = 1$ against $H_1 : \theta = 2$ based on a random sample of size n at a significance level α.

9.9 Suppose that X is a random variable with the probability density function $f(x)$. Obtain the form of the most powerful test for testing the hypothesis $H_0 : f(x) = f_0(x)$ against the alternate hypothesis $H_1 : f(x) = f_1(x)$ on the basis of a sample of size one where

$$f_0(x) = \frac{1}{\sqrt{2\pi}} \exp\{-x^2/2\}$$

and

$$f_1(x) = \frac{1}{2} \exp -|x|.$$

What is the size and power of this test?

9.10 Let $X_1, \ldots, X_m$ and $Y_1, \ldots, Y_n$ be independent random samples from the probability distributions $N(\mu_1, \sigma^2)$ and $N(\mu_2, \sigma^2)$ respectively where σ^2 is unknown. Derive the likelihood ratio test for testing the hypothesis $H_0 : \mu_1 = \mu_2$ against the alternate hypothesis $H_1 : \mu_1 \neq \mu_2$.

9.11 Suppose that $X_i, 1 \leq i \leq n$ is a random sample of size n from $N(\mu, \sigma^2)$ where μ and σ^2 are both unknown. Show that the likelihood ratio test for testing $H_0 : \sigma^2 = \sigma_0^2$ against $H_1 : \sigma^2 \neq \sigma_0^2$ is of the form "reject H_0 if and only if $\sum_{i=1}^{n}(X_i - \bar{X})^2 \leq c_1$ or $\sum_{i=1}^{n}(X_i - \bar{X})^2 \geq c_2$" where $c_1 < c_2$ are specified constants depending on the significance level α.

9.12 Let $X_1, \ldots, X_m$ and $Y_1, \ldots, Y_n$ be independent random samples from probability distributions $N(\mu_1, \sigma_2^2)$ and $N(\mu_1, \sigma_1^2)$ respectively where μ_1 and μ_2 are unknown. Obtain the likelihood ratio statistic for testing the hypothesis $H_0 : \sigma_1^2 = \sigma_2^2$ against the alternate hypothesis $H_1 : \sigma_1^2 \neq \sigma_2^2$ at a significance level α.

9.13 Let $(X_i, Y_i), 1 \leq i \leq n$ be a random sample from a bivariate normal distribution with parameters $\mu_1, \mu_2, \sigma_1^2, \sigma_2^2$ and ρ. Obtain the likelihood ratio test for testing the hypothesis $H_0 : \rho = 0$ against the alternate hypothesis $H_1 : \rho \neq 0$ at a significance level α. 9.14. A genetic model suggests that the probabilities of a certain Multinomial distribution should be of the form $p_1 = \theta^2$, $p_2 = 2\theta(1-\theta)$, and $p_3 = (1-\theta)^2$ where $0 < \theta < 1$. If n_1, n_2 and n_3 are the observed frequencies corresponding to n independent observations, explain how you will check the suitability of the suggested genetic model.

9.14 Consider the following data:

x	Observed frequency
0	20
1	40
2	16
3	18
4	6

Check whether a Poisson distribution is a good-fit for the data at 5% level of significance.

9.15 An examination was given to 500 high school students in each of two cities and the marks obtained by the students are recoded as low, medium and high. The results obtained in the examination are given in the following table:

City	Low	Medium	High
1	103	145	252
2	140	136	224

Test the hypothesis that the performance of the students in both the cities is the same at 5% level of significance.

9.16 In order to study whether the level of education and the level of annual income are independent, a survey of 100 persons was taken in a particular city. The following data was observed.

	Annual Income		
Education level	Low	Medium	High
Undergraduate	10	15	30
Graduate	8	17	20

Test whether the annual income of a person is independent of his or her level of education at the 5% significance level.

Chapter 10

LINEAR REGRESSION AND CORRELATION

10.1 Introduction

In the study of the underlying mechanisms for various phenomena which depend on several characteristics, most often we find that relationships exist between two or more characteristics. For instance, it is known that the weight of an adult male depends on his height, the stopping distance of an automobile depends on the speed at which the automobile is traveling, the pressure of a given mass of gas depends on its temperature and volume, the potency of a drug depends on the amount of time it was under storage and so on. We will now study statistical methods which can be used to obtain the possible relationship between a dependent or response or output of a random variable y based on one or more non-random independent variables or input variables also called regressors or covariates $x_i, 1 \leq i \leq p$. Such a study is called *regression analysis* and our discussion is to develop methods for building regression models between the dependent variable y and the covariates $x_i, 1 \leq i \leq p$. Apart from the fact that such an analysis explains the possible relationship between the response variable y and the independent variables $x_i, 1 \leq i \leq p$, it can also be used for prediction of the response variable y at values close to the range of values of the covariates $x_i, 1 \leq i \leq p$ where it was not observed. As we shall see later, such a prediction or extrapolation has to be made with caution. Suppose there is only one independent variable x and a corresponding dependent variable y. In order to determine a possible relation between the dependent variable y and the independent variable x, a first step is the collection of data based on x and observing the corresponding y. Suppose the number of regressors or covariates is one and the data consists of $(x_j, y_j), 1 \leq j \leq n$. In order to get an idea of the functional relationship between y and x, a visual or graphical method is to form a *scatter diagram*, as shown in Figure 10.1, with the covariate values of x_i on the x-axis and the corresponding responses y_i on the y-axis.

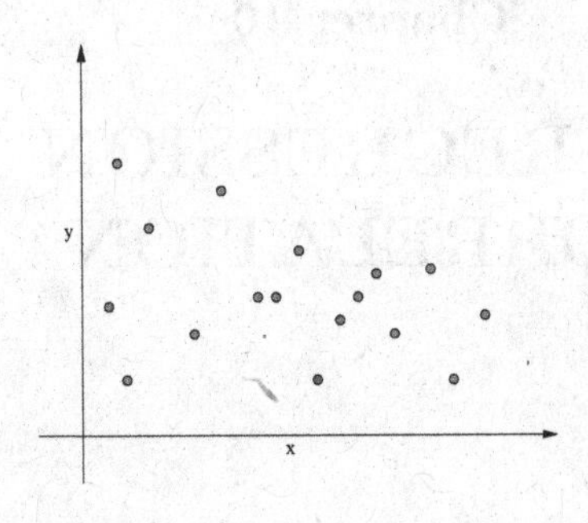

Fig. 10.1

10.2 Simple Linear Regression Model

We first discuss the case when there is only one independent or regressor variable x. Suppose we propose a *simple linear regression model* that y_i and x_i are related by the equation

$$y_i = \beta_0 + \beta_1 x_i + \varepsilon_i, 1 \leq i \leq n \tag{10.1}$$

where ε_i, representing the error component, is a random variables with mean zero and finite variance σ^2. We assume that the errors $\varepsilon_i, 1 \leq i \leq n$ are uncorrelated. It is obvious that

$$E(y|x) = \beta_0 + \beta_1 x \tag{10.2}$$

and

$$Var(y|x) = \sigma^2. \tag{10.3}$$

Observe that the mean response of y given x is linear in x under the proposed model. The constants β_0 and β_1 are called the *regression coefficients.* We will now discuss the method of least squares for obtaining estimates for the regression coefficients β_0 and β_1. These estimates are dependent on the data $(x_i, y_i), 1 \leq i \leq n$. The line

$$E(y|x) = \beta_0 + \beta_1 x$$

is called the *regression line.* The parameter β_0 is the intercept of the regression line and β_1 is the slope of the regression line. The slope β_1 measures the change in the mean response $E(y|x)$ for a unit change in x. If the range of values of the independent variable x includes $x = 0$, then β_0 can be interpreted as the mean response when $x = 0$. We assume that there is no error in the measurement of the regressor variable x. Furthermore, for each value of $x = x_i$, the random variable $y = y_i$ has a probability distribution with mean $\beta_0 + \beta_1 x_i$ and variance σ^2 and the random variables $y_i, 1 \le i \le n$ are uncorrelated.

Estimation of regression coefficients β_0 and β_1

In order to estimate the regression coefficients β_0 and β_1, it is natural to choose their estimators $\hat{\beta}_0$ and $\hat{\beta}_1$ so that y_i is close to $\beta_0 + \beta_1 x_i, 1 \le i \le n$. A natural measure of such closeness is

$$\sum_{i=1}^{n} |y_i - \beta_0 - \beta_1 x_i|^r$$

for some $r > 0$. It is convenient to choose $r = 2$ which gives rise to the method of least squares. Let

$$S(\beta_0, \beta_1) = \sum_{i=1}^{n} (y_i - \beta_0 - \beta_1 x_i)^2. \tag{10.4}$$

We choose $\hat{\beta}_0$ and $\hat{\beta}_1$ so that $S(\hat{\beta}_0, \hat{\beta}_1)$ is minimum. Hence the partial derivatives of $S(\beta_0, \beta_1)$ with respect to β_0 and β_1 should be zero at the point $(\hat{\beta}_0, \hat{\beta}_1)$. Differentiating $S(\beta_0, \beta_1)$ with respect to β_0 and β_1 respectively, we obtain the equations

$$\sum_{i=1}^{n} (y_i - \hat{\beta}_0 - \hat{\beta}_1 x_i) = 0, \tag{10.5}$$

and

$$\sum_{i=1}^{n} (y_i - \hat{\beta}_0 - \hat{\beta}_1 x_i) x_i = 0. \tag{10.6}$$

These equations are called the *normal equations*. Let $\bar{x} = n^{-1} \sum_{i=1}^{n} x_i$ and $\bar{y} = n^{-1} \sum_{i=1}^{n} y_i$. Define

$$S_{xy} = \sum_{i=1}^{n} (x_i - \bar{x})(y_i - \bar{y}), \tag{10.7}$$

and

$$S_{xx} = \sum_{i=1}^{n} (x_i - \bar{x})^2; S_{yy} = \sum_{i=1}^{n} (y_i - \bar{y})^2. \tag{10.8}$$

Solving the system of equations given by (10.5) and (10.6), it can be checked that

$$\hat{\beta}_1 = \frac{S_{xy}}{S_{xx}} \tag{10.9}$$

and

$$\hat{\beta}_0 = \bar{y} - \hat{\beta}_1 \bar{x}. \tag{10.10}$$

It is clear that these estimators are well defined only when $S_{xx} = \sum_{i=1}^{n}(x_i - \bar{x})^2 \neq 0$. It is obvious that $S_{xx} = 0$ if and only if $x_1 = \cdots = x_n$. Hence we assume that at least two of the values $x_i, 1 \leq i \leq n$ are not equal. The estimators $\hat{\beta}_0$ and $\hat{\beta}_1$ of β_0 and β_1 respectively are called the *least squares estimators* (LSE). Define

$$\hat{y}_i = \hat{\beta}_0 + \hat{\beta}_1 x_i. \tag{10.11}$$

The quantity $\hat{y}_i$ is called the *fitted value* at $x = x_i$. If the proposed simple linear regression model is good or correct, then we expect the fitted value $\hat{y}_i$ to be close to the observed value y_i when the value of the regressor variable is x_i. The difference between the observed value y_i and the fitted value $\hat{y}_i$ is called the *residual* and we denote it by e_i. Hence

$$e_i = y_i - \hat{y}_i, 1 \leq i \leq n. \tag{10.12}$$

Observe that

$$y_i = \hat{y}_i + e_i = \hat{\beta}_0 + \hat{\beta}_1 x_i + e_i \tag{10.13}$$

from the fitted values and

$$y_i = \beta_0 + \beta_1 x_i + \varepsilon_i \tag{10.14}$$

from the proposed simple linear regression model. The above two equations indicate that the residuals $e_i, 1 \leq i \leq n$ can in some sense be viewed as a realization of the model errors $\varepsilon_i, 1 \leq i \leq n$. However the residuals $e_i, 1 \leq i \leq n$ are not independent random variables even if we assume that $\varepsilon_i, 1 \leq i \leq n$ are independent random variables. After obtaining the least squares fit, one has to check the following.

(i) How well does the proposed model fit the data?
(ii) Is the model useful for prediction?
(iii) Are any of the basic assumptions such as the constancy of the variance and uncorrelatedness of errors violated and if yes, how serious is it?

Let us now study some properties of the least square estimators derived above for a simple linear regression model.

Properties of least square estimators

(1) The least square estimators $\hat{\beta}_0$ and $\hat{\beta}_1$ are linear functions of the observations $y_i, 1 \leq i \leq n$. For instance

$$\hat{\beta}_1 = \frac{S_{xy}}{S_{xx}} = \sum_{i=1}^{n} c_i y_i$$

where

$$c_i = \frac{x_i - \bar{x}}{S_{xx}}.$$

(2) The least square estimators $\hat{\beta}_0$ and $\hat{\beta}_1$ are unbiased estimators of β_0 and β_1 respectively. For example,

$$\begin{aligned} E(\hat{\beta}_1) &= E(\sum_{i=1}^{n} c_i y_i) \qquad (10.15)\\ &= \sum_{i=1}^{n} c_i(\beta_0 + \beta_1 x_i) \\ &= \beta_1 \end{aligned}$$

since $\sum_{i=1}^{n} c_i = 0$ and $\sum_{i=1}^{n} c_i x_i = 1$. Further more it is easy to check, from the representation of $\hat{\beta}_1$ given above, that

(i) $Var(\hat{\beta}_1) = \frac{\sigma^2}{S_{xx}}$,
(ii) $Var(\hat{\beta}_0) = \sigma^2(\frac{1}{n} + \frac{\bar{x}^2}{S_{xx}})$, and
(iii) $Cov(\bar{y}, \hat{\beta}_1) = 0$.

(3) **Gauss-Markov Theorem:** The least square estimators $\hat{\beta}_0$ and $\hat{\beta}_1$ are unbiased estimators for β_0 and β_1 respectively and have the minimum variance among all the unbiased estimators of β_0 and β_1 which are linear functions of $y_i, 1 \leq i \leq n$.

We will not prove the Gauss-Markov Theorem. For the proof, see Chapter 4 of Rao (1974). In view of the result stated in (3) above, the least square estimators $\hat{\beta}_0$ and $\hat{\beta}_1$ are also called the *Best Linear Unbiased Estimators* or BLUE for short.

(4) The least squares regression line $y = \hat{\beta}_0 + \hat{\beta}_1 x$ passes through the centroid $(\bar{y}, \bar{x})$ of the data.

(5) Furthermore, the residuals $e_i, 1 \leq i \leq n$ satisfy the properties (a) $\sum_{i=1}^{n} e_i = 0$, (b) $\sum_{i=1}^{n} x_i e_i = 0$ and (c) $\sum_{i=1}^{n} \hat{y}_i e_i = 0$.

Estimation of error variance σ^2

The sum of squares of the residuals $e_i, 1 \leq i \leq n$ is called he *residual sum of squares* and is denoted by SS_{Res}. If the errors $\varepsilon_i, 1 \leq i \leq n$ are assumed to be independent and identically distributed (i.i.d.) random variables with the normal distribution with mean zero and variance σ^2, then it can be shown that

$$E(SS_{Res}) = (n-2)\sigma^2$$

since SS_{Res}/σ^2 has the Chi-square distribution with $(n-2)$ degrees of freedom. Let the *residual mean square* be defined by $MS_{Res} = \frac{SS_{Res}}{n-2}$. Then

$$E(MS_{Res}) = \sigma^2$$

and hence

$$\hat{\sigma}^2 = MS_{Res} = \frac{SS_{Res}}{n-2}$$

is an unbiased estimator of the error variance σ^2 under the assumption that the errors are i.i.d. normal with mean zero and variance σ^2.

Testing for the slope β_1

We now develop the tests of hypotheses for the slope β_1 and the intercept β_0. Assume that the model errors $\varepsilon_i, 1 \leq i \leq n$ are i.i.d. $N(0, \sigma^2)$ where σ^2 is unknown. Before one uses these tests in practice, one has to check whether such an assumption is reasonable for the problem under consideration. It is possible to develop some diagnostic checks for checking the validity of such model assumptions (cf. Montgomery et al. (2001)).

Suppose we want to test the hypothesis

$$H_0 : \beta_1 = \beta_{10} \text{ against } H_1 : \beta_1 \neq \beta_{10}$$

where β_{10} is pre-specified. In case $\beta_{10} = 0$, then the test can be thought of as a test of significance of linear regression of the response y on the independent variable or the covariate x. If we do not reject the hypothesis in this case, it indicates the lack of linear dependence between x and y. In other words, either there is no dependence between x and y or the dependence is possibly nonlinear. Since the errors $\varepsilon_i, 1 \leq i \leq n$ are assumed to be i.i.d. $N(0, \sigma^2)$, it follows that $y_i, 1 \leq i \leq n$ are independent random variables and y_i has the normal distribution with mean $\beta_0 + \beta_1 x_i$ and variance σ^2. Hence the linear function $\hat{\beta}_1 = \sum_{i=1}^{n} c_i y_i$ of $y_i, 1 \leq i \leq n$ has a normal distribution. Check that $\hat{\beta}_1$ has $N(\beta_1, \frac{\sigma^2}{S_{xx}})$. In particular, it follows that the statistic

$$\frac{\hat{\beta}_1 - \beta_{10}}{\sigma^2 / S_{xx}}$$

will have the standard normal distribution $N(0,1)$ under the hypothesis H_0. Since σ^2 is not known, we can use the estimator $\hat{\sigma}^2 = MS_{Res}$ as an estimator for σ^2. It follows from general results using the Cochran's theorem (see Chapter 6) that

$$\frac{(n-2)MS_{Res}}{\sigma^2}$$

has the Chi-square distribution with $(n-2)$ degrees of freedom. Furthermore the statistics MS_{Res} and $\hat{\beta}_1$ are statistically independent. Hence the statistic

$$t_0 = \frac{\hat{\beta}_1 - \beta_{10}}{\sqrt{MS_{Res}/S_{xx}}}$$

has the Student $t-$*distribution* with $(n-2)$ degrees of freedom under the hypothesis H_0. Suppose we choose α as the level of significance for testing the null hypothesis H_0 against the alternate hypothesis H_1. Let $t_{\alpha/2,n-2}$ be the critical value such that

$$P(|t| > t_{\alpha/2,n-2}) = \alpha.$$

The test procedure consists of rejection of hypothesis H_0 at the significance level α if the observed value t satisfies the condition $|t| > t_{\alpha/2,n-2}$.

Testing for the intercept β_0

We leave it to the reader to develop a test, similar to the test discussed above, for testing the hypothesis $H_0 : \beta_0 = \beta_{00}$ against $H_1 : \beta_0 \neq \beta_{00}$.

Test procedure via the Analysis of Variance (ANOVA)

We now describe another approach to tests of hypotheses which can be generalized to multiple linear regression models which involve more than one covariate. Note that

$$y_i - \bar{y} = y_i - \hat{y}_i + \hat{y}_i - \bar{y}$$

and hence

$$\sum_{i=1}^{n}(y_i - \bar{y})^2 = \sum_{i=1}^{n}(y_i - \hat{y}_i)^2 + \sum_{i=1}^{n}(\hat{y}_i - \bar{y})^2 + 2\sum_{i=1}^{n}(y_i - \hat{y}_i)(\hat{y}_i - \bar{y}).$$

But

$$\begin{aligned}\sum_{i=1}^{n}(y_i-\hat{y}_i)(\hat{y}_i-\bar{y}) &= \sum_{i=1}^{n}\hat{y}_i(y_i-\hat{y}_i)-\bar{y}\sum_{i=1}^{n}(y_i-\hat{y}_i)\\ &= \sum_{i=1}^{n}\hat{y}_ie_i-\bar{y}\sum_{i=1}^{n}e_i\\ &= 0.\end{aligned}$$

The last equality follows from the properties of the residuals. Hence

$$\sum_{i=1}^{n}(y_i-\bar{y})^2=\sum_{i=1}^{n}(y_i-\hat{y}_i)^2+\sum_{i=1}^{n}(\hat{y}_i-\bar{y})^2.$$

Let

$$SS_T=\sum_{i=1}^{n}(y_i-\bar{y})^2$$

and

$$SS_R=\sum_{i=1}^{n}(\hat{y}_i-\bar{y})^2.$$

The sum of squares denoted above by SS_T is called the *Corrected or Total sum of squares.* It measures the sum of squares of deviations of the observations y_i from their average value $\bar{y}$ and does not depend on the regression model. The sum of squares SS_R measures the sum of squares of deviations between the fitted values $\hat{y}_i$ based on the regression model and $\bar{y}$ and hence it is a measure how good the model explains the data under consideration with respect to the average value $\bar{y}$. It is called the *Regression sum of squares* or *model sum of squares.* As we have seen earlier, the term SS_{Res} indicates the *residual sum of squares* or *error sum of squares.* It measures the sum of squares of deviations between fitted values based on the model and the observed values. If the model is a good fit, then SS_T should be close to SS_R and SS_{Res} should be small. This can be seen from the relation

$$SS_T=SS_R+SS_{Res}$$

which follows from the above equations. It is convenient to denote SS_T by S_{yy} following the quantities S_{xx} and S_{xy} introduced earlier. Assuming that the errors $\varepsilon_i, 1\le i\le n$ are i.i.d. with probability distribution $N(0,\sigma^2)$, it can be shown that

$$\frac{SS_{Res}}{\sigma^2}\sim\chi^2_{n-2}, \tag{10.16}$$

and

$$\frac{SS_R}{\sigma^2} \sim \chi_1^2 \tag{10.17}$$

and that the random variables SS_R and SS_{Res} are independent. Here we write $Z \sim \chi_r^2$ if the random variable Z has the Chi-square distribution with r degrees of freedom. All the observations made above follow from an application of the Cochran's theorem discussed in the Chapter 6. We denote $\frac{SS_{Res}}{n-2}$ as MS_{res} and $\frac{SS_R}{1}$ as MS_R. The denominators in both the terms are the respective degrees of freedom of the corresponding Chi-square distributions. Furthermore it can be checked that

$$E(MS_{Res}) = \sigma^2$$

and

$$E(MS_R) = \sigma^2 + \beta_1^2 S_{xx}.$$

Observe that $E(MS_{Res}) < E(MS_R)$ if $\beta_1 \neq 0$ and they are equal if $\beta_1 = 0$. Hence the quantity MS_R is likely to be large compared to MS_{Res} on the average if $\beta_1 \neq 0$.

We now consider again the problem of testing the hypothesis

$$H_0 : \beta_1 = 0 \text{ against } H_1 : \beta_1 \neq 0.$$

Define

$$F_0 = \frac{SS_R/1}{SS_{Res}/(n-2)} = \frac{MS_R}{MS_{Res}}. \tag{10.18}$$

The test statistic F_0 follows the F-distribution with 1 and $(n-2)$ degrees of freedom under the null hypothesis H_0 and it has the noncentral F-distribution with 1 and $(n-2)$ degrees of freedom under the alternate hypothesis H_1 with the non-centrality parameter $\lambda = \frac{\beta_1^2 S_{xx}}{\sigma^2}$. We have not discussed properties of noncentral distributions such as noncentral chi-square and noncentral F in this book. Let α be the level of significance and $F_{\alpha;1,n-2}$ denote the critical value such that

$$P(F_0 > F_{\alpha;1,n-2}) = \alpha$$

under H_0. The test procedure consists in rejecting the hypothesis H_0 if the observed

$$F_0 > F_{\alpha;1,n-2}.$$

The above calculations can be compactly displayed in a tabular form known as the *Analysis of variance* or ANOVA in short.

ANOVA

Source of variation	Sum of squares	Degrees of freedom	Mean sum of squares	F_0
Regression	$SS_R = \hat{\beta}_1 S_{xy}$	1	MS_R	$\frac{MS_R}{MS_{Res}}$
(Residual) Error	$SS_{Res} = SS_T - \hat{\beta}_1 S_{xy}$	$n-2$	MS_{Res}	
Total	SS_T	$n-1$		

Check that $t_0^2 = F_0$ and both the F-test and the t-test described earlier lead to the same conclusions for testing the null hypothesis $H_0 : \beta_1 = 0$ against the alternate hypothesis $H_1 : \beta_1 \neq 0$ at a given level of significance α.

Confidence intervals for β_0, β_1 and σ^2

Suppose that the errors $\varepsilon_i, 1 \leq i \leq n$ are i.i.d. as $N(0, \sigma^2)$. We will now construct a $100(1-\alpha)\%$ confidence interval for the intercept β_1 based on the data $(x_i, y_i), 1 \leq i \leq n$. We have seen earlier that

$$\frac{\hat{\beta}_1 - \beta_1}{\sqrt{\hat{\sigma}^2 / S_{xx}}}$$

has a t-distribution with $(n-2)$ degrees of freedom. Using this information, it is possible to construct a $100(1-\alpha)\%$ confidence interval for β_1. Similarly it is possible to construct a confidence interval for β_0 with a given confidence coefficient using the fact that

$$\frac{\hat{\beta}_0 - \beta_0}{\sqrt{\hat{\sigma}^2 (\frac{1}{n} + \frac{\bar{x}^2}{S_{xx}})}}$$

has a t-distribution with $(n-2)$ degrees of freedom. Observing that

$$\frac{(n-2)MS_{Res}}{\sigma^2} \sim \chi^2_{n-2},$$

One can construct a confidence interval for the error variance σ^2 with a prescribed confidence coefficient.

Confidence interval for mean response

Apart from being useful for predicting the response y for values of x within a valid range, a regression model is also useful in estimating the mean response $E(y)$ at a given value of x. From the simple linear regression model connecting y and x, we observe that

$$E(y|x) = \beta_0 + \beta_1 x.$$

Hence an estimate for the mean response $E(y|x)$ at a value $x = x_0$ is

$$\hat{E}(y|x_0) = \hat{\beta}_0 + \hat{\beta}_1 x_0$$

and the variance of this estimator is given by

$$\begin{aligned} Var(\hat{\beta}_0 + \hat{\beta}_1 x_0) &= Var(\hat{\beta}_1 x_0 + \bar{y} - \hat{\beta}_1 \bar{x}) \\ &= Var(\bar{y} + \hat{\beta}_1 (x_0 - \bar{x})) \\ &= \frac{\sigma^2}{n} + \frac{\sigma^2 (x_0 - \bar{x})^2}{S_{xx}} \\ &= \sigma^2 (\frac{1}{n} + \frac{(x_0 - \bar{x})^2}{S_{xx}}). \end{aligned}$$

Here we have used the result that $Cov(\bar{y}, \hat{\beta}_1) = 0$. An estimate for the variance of the estimator can be obtained by estimating σ^2 using the residual mean sum of squares MS_{Res} as was done earlier. If we assume that the errors $\varepsilon_i, 1 \leq i \leq n$ are i.i.d. $N(0, \sigma^2)$, then check that

$$\frac{\hat{E}(y|x_0) - E(y|x_0)}{\sqrt{MS_{Res}(\frac{1}{n} + \frac{(x_0 - \bar{x})^2}{S_{xx}})}} \sim t_{n-2}.$$

Using this observation, one can construct a confidence interval for the mean response $E(y|x_0)$ of y at $x = x_0$. The length of such a $100(1-\alpha)\%$ confidence interval for $E(y|x_0)$ is

$$2t_{\alpha/2, n-2}\sqrt{MS_{Res}(\frac{1}{n} + \frac{(x_0 - \bar{x})^2}{S_{xx}})}$$

and it depends on the value x_0. It is smallest when $x_0 = \bar{x}$ and increases as $|x_0 - \bar{x}|$ increases. Hence it is possible to get a best estimate of the mean response at values of x near the mean or the centre of the range of values of x and the quality of the estimate decreases as we go away from the centre towards the boundary values in the range of x.

Prediction of new observation

Suppose the errors $\varepsilon_i, 1 \leq i \leq n$ are i.i.d. $N(0, \sigma^2)$. One of the important uses of a regression model is for prediction of the response y at a specified value x_0 of the covariate x. If $x = x_0$, then

$$\hat{y} = \hat{\beta}_0 + \hat{\beta}_1 x_0$$

can be chosen as the prediction for the observation y for $x = x_0$. This should not be confused with an estimator for $E(y|x_0)$ as the latter is an estimator for the mean response at $x = x_0$. Let y_0 denote the observed response at $x = x_0$. Check that $y_0 - \hat{y}_0$ has a normal distribution with mean zero and variance

$$Var(y_0 - \hat{y}_0) = \sigma^2(1 + \frac{1}{n} + \frac{(x_0 - \bar{x})^2}{S_{xx}})$$

since the observation y_0 is independent of the predicted value $\hat{y}_0$. Hence

$$\frac{y_0 - \hat{y}_0}{\sqrt{MS_{Res}(1 + \frac{1}{n} + \frac{(x_0 - \bar{x})^2}{S_{xx}})}} \sim t_{n-2}$$

and a $100(1-\alpha)\%$ prediction interval for the observation y_0 at $x = x_0$ is

$$[\hat{y}_0 - t_{\alpha/2,n-2}\jmath_n, \hat{y}_0 + t_{\alpha/2,n-2}\jmath_n]$$

where

$$\jmath_n = \sqrt{MS_{Res}(1 + \frac{1}{n} + \frac{(x_0 - \bar{x})^2}{S_{xx}})}.$$

The prediction interval is of shortest length at $x_0 = \bar{x}$ and the size increases as $|x_0 - \bar{x}|$ increases. Observe that the $100(1-\alpha)\%$ prediction interval for y_0 at $x = x_0$ is larger than the $100(1-\alpha)\%$ confidence interval for the mean response $E(y|x_0)$ at $x = x_0$ since the prediction depends on the errors from the model fitted earlier and the error connected with the future observation.

Coefficient of determination R^2

In order to know whether the variation in the response or dependent $y_i, 1 \leq i \leq n$ is explained by the regressor or independent $x_i, 1 \leq i \leq n$, we define

$$R^2 = \frac{\sum_{i=1}^n (\hat{y}_i - \bar{y})^2}{\sum_{i=1}^n (y_i - \bar{y})^2} = \frac{SS_R}{SS_T}$$
$$= 1 - \frac{SS_{Res}}{SS_T}.$$

This quantity R^2 is called the *Coefficient of determination.* Note that SS_T is a measure of the variability in y not taking into account the effect of the regressor variable x and SS_R is a measure of the variability in y after the variable x has been considered through the regression model. Hence the quantity R^2 measures the proportion of variation explained by the regressor x. Check that $0 \leq R^2 \leq 1$. If R^2 is close to one, then it will imply that most of the variability in y is explained by the proposed regression model.

Remarks. (1) The interpretation of the value of R^2 should be made with caution. It is always possible to increase R^2 by adding more terms to the model. For instance, one can always construct a polynomial of degree $n-1$ passing through the n points $(x_i, y_i), 1 \leq i \leq n$ irrespective of any relation between x and y and it always gives a perfect fit to the data leading to $R^2 = 1$. The magnitude of R^2 depends on the variability in the regressor variable x. We should caution that although R^2 cannot decrease if we add another regressor variable to the model, it does not mean that the new model is better than the earlier model. Increasing the number of regressor variables might decrease the SS_{Res} but may increase the MS_{Res} which is an estimate for the error variance σ^2.

(2) Regression models can be used for interpolation over the range of the regressor variables. However they should be used for extrapolation with caution, that is to predict at values which are outside the range of the regressor variables. Outliers or bad values in the data, as shown in the Figure 10.2 and the Figure 10.3, can give bad estimators for β_0 and β_1 leading to a bad estimator for σ^2 and hence for a bad least squares fit. It is necessary to detect the outliers if any and check whether these values are really bad in the sense that they are there because of wrong recording of the data for instance or these outliers are part and parcel of the observed data indicating specific changes in the phenomenon under study. Detection of outliers is of primary importance as the least squares estimators are significantly affected by their presence in the data.

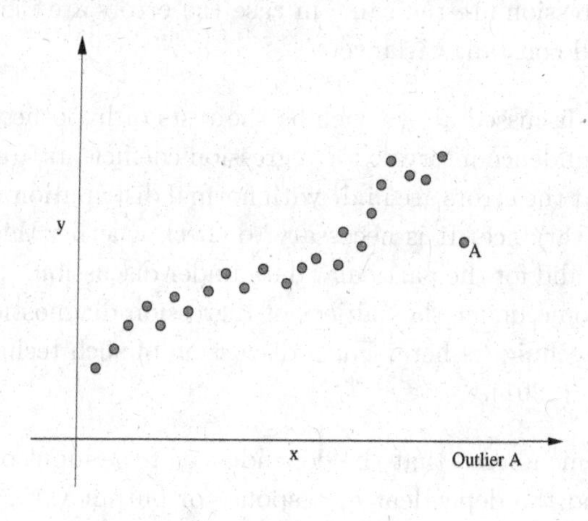

Fig. 10.2

(3) The basic assumption behind the use of the simple linear regression model studied above is that the errors are independent and normally distributed with mean zero and constant variance. As we have seen above, validity of the test procedures, confidence intervals and prediction intervals discussed earlier hinges on this assumption. If one believes that the distributions of the errors is not normal but known, then the regression coefficient

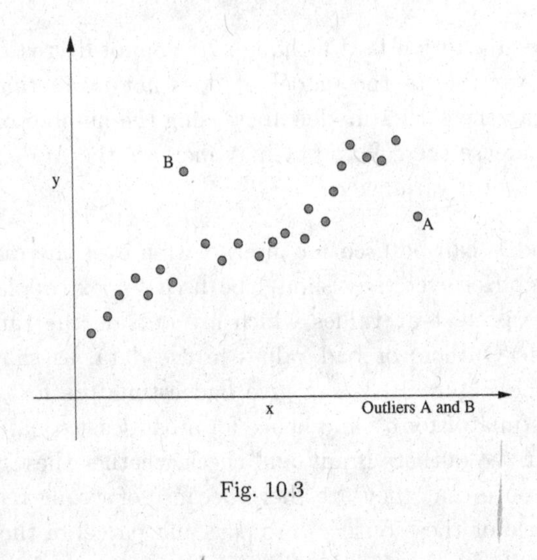

Fig. 10.3

can be estimated by the maximum likelihood method. However it can be shown that the maximum likelihood estimators and the least squares estimators for the regression are the same in case the errors are i.i.d. normal with mean zero and constant variance.

(4) All the results discussed above such as the tests of hypotheses or construction of the confidence intervals for regression coefficients are based on the assumption that the errors are i.i.d. with normal distribution with mean zero and constant variance. It is necessary to check whether this underlying assumption is valid for the particular data under discussion. Techniques for such a study come under the subject of regression diagnostics. We do not go into those techniques here. For a discussion of such techniques, see Montgomery et al. (2001).

(5) It should be remembered that the dependence of the input or independent variable x and the dependent or response or output variable y via a regression model does not necessarily indicate causal relationship between y and x..

10.3 Multiple Linear Regression Model

A regression model that involves more than one regressor variable or covariate is called a *multiple linear regression model.* An example of such a model is

$$y = \beta_0 + \beta_1 x_1 + \beta_2 x_2 + \varepsilon.$$

For example, y might be the output observed in a chemical process for a given amount of raw material x_1 at the temperature level x_2. Note that the word linear here as well as in the previous section refers to the linearity in the regression coefficients β_0, β_1, and β_2 and not necessarily in the covariates x_1 and x_2. For instance the covariate x_1 could be x and the covariate x_2 could be x^2 and the proposed regression model might be

$$y = \beta_0 + \beta_1 x + \beta_2 x^2 + \varepsilon.$$

Observe that this model leads to a second order or quadratic model in the covariate x but linear in the regression coefficients. Another model such as

$$y = \beta_0 + \beta_1 x_1 + \beta_2 x_2 + \beta_{12} x_1 x_2 + \varepsilon$$

is also a special case of a multiple linear regression model. There is a component in this model representing the *interaction effect* between the covariates or regressors x_1 and x_2. This model can be considered as a multiple linear regression model by defining the regressor or covariate $x_3 = x_1 x_2$. We do not go into more details. For a comprehensive discussion of the multiple linear regression models, estimation of the parameters and the corresponding regression diagnostics , see Montgomery et al. (2001).

10.4 Correlation

We have discussed regression method for investigating the relationship between a dependent or response variable y which is possibly random and independent variable or covariate x which is not random. There are situations when the input variable or the regressor x itself is random. Suppose that x and y are jointly distributed but the form of the joint distribution is unknown. All the results described earlier continue to hold if the following conditions hold:

(i) the conditional distribution of y_i given x_i is normal with the conditional expectation $\beta_0 + \beta_1 x_i$ and and conditional variance σ^2 is independent of x_i for $1 \leq i \leq n$ and

(ii) the $x_i, 1 \leq i \leq n$ are independent random variables with probability distributions *not* depending on β_0, β_1 and σ^2.

Under the above assumptions, all the results obtained earlier for developing different test procedures and for constructing the confidence intervals, remain valid. However the confidence coefficients of the confidence intervals and the errors of estimators have to be interpreted in a different way. When the regressor x is a random variable, the confidence coefficients and the error variances are based on repeated sampling of the values of (x_i, y_i) and *not* on repeated sampling of y_i at a fixed value of x_i.

Correlation model

When the regressor is random, a basic model that is used for studying the relationship between the variable x and the variable y is the *correlation model.* We assume that the pair (y, x) has the bivariate normal distribution with $E(y) = \mu_1, E(x) = \mu_2, Var(y) = \sigma_1^2, Var(x) = \sigma_2^2$ and $Corr(y, x) = \rho$. Here after we denote this bivariate normal distribution by $N(\mu_1, \mu_2; \sigma_1^2, \sigma_2^2; \rho)$. We assume that $\sigma_i^2 > 0, i = 1, 2$. From the general results on the bivariate normal distributions, it follows that the conditional expectation of y given x is linear and the conditional variance y given x is constant, some times known as the property of as *homoscedasticity*, that is,

$$E(y|x) = \beta_0 + \beta_1 x$$

and

$$Var(y|x) = \sigma_{1.2}^2$$

where

$$\beta_0 = \mu_1 - \mu_2 \rho \frac{\sigma_1}{\sigma_2}, \beta_1 = \rho \frac{\sigma_1}{\sigma_2}$$

and

$$\sigma_{1.2}^2 = \sigma_1^2(1 - \rho^2).$$

We observe that $\beta_1 = 0$ if and only if $\rho = 0$ and in such a case the knowledge of x does not assist in predicting the value of y. Applying the maximum likelihood method, check that the maximum likelihood estimators of β_0 and β_1 are

$$\hat{\beta}_0 = \bar{y} - \hat{\beta}_1 x$$

and

$$\hat{\beta}_1 = \frac{S_{xy}}{S_{xx}}$$

which are the least squares estimators when x was considered as the regressor variable. An estimator for the correlation coefficient ρ is the *Sample correlation coefficient* r given by

$$r = \frac{S_{xy}}{\sqrt{S_{xx}S_{yy}}}.$$

We leave it to the reader to check that $r^2 = R^2$ defined in the earlier section. Hence the coefficient of determination is the same as the square of the sample correlation coefficient under the correlation model.

Remarks. Correlation coefficient measures the possible linear association between the variables x and y and is not of use in prediction of y given x where as the regression methods are useful to obtain relationship between x and y for possible prediction.

In order to test the hypothesis $H_0 : \rho = 0$ against the hypothesis $H_1 : \rho \neq 0$, one can use the test statistic

$$t_0 = \frac{r\sqrt{n-2}}{\sqrt{1-r^2}} \sim t_{n-2}$$

under the hypothesis H_0. The test consists in rejecting the hypothesis H_0 for values of t_0 such that $|t_0| > t_{\alpha/2,n-2}$ where $t_{\alpha/2,\, n-2}$ depends on the level of significance α.

10.5 Exercises

10.1 Determine the regression line which best fits the following points (x, y) : (1,3), (2,3), (4,7), (5,6), (8,12).

10.2 The following data gives the percentage of marks x obtained in the final examinations in the high school and the percentage of marks y obtained in the final examinations of the college for ten students: (40%, 50%), (30%, 40%), (40%, 40%),(90%, 80%), (90%, 90%), (80%, 90%), (90%,70%), (70%,80%), (60%, 50%), (50%, 60%).
Can we conclude that the performance of a student in the high school and in the college are related?

10.3 A life insurance company is interested to find the relationship between the sales experience and the sales volumes for its insurance agents. The following data gives the number of years of experience x and the annual sales y in a particular year in lakhs of Rupees for nine agents:

(1,2), (2,1), (3,3), (4,3), (5,4), (6,5), (7,6), (8,5), (9,7).
Obtain an estimate for the volume of the sales per year for an agent who had 10 years of experience.

10.4 The following data gives the travel time y in minutes and the distance to work x in kilometers for 15 employees of an IT Company:
(3,7), (5,20),(7,20), (8,15), (10,25), (11,17), (12,20), (12,35), (13,26), (15,25), (15,35), (16,32), (18,44), (19, 37), (20, 45).

(i) Fit a linear regression model to the data and estimate the corresponding regression coefficients. Obtain an estimator for the error variance.
(ii) Construct a 95% confidence interval for the mean travel time for all employees who travel 7 kilometers.
(iii) Obtain a 95% prediction interval for the time to travel a distance of 7 kilometers.

10.5 It is generally accepted that the amount of time a student puts in for studies for an examination is reflected in his performance in that examination. The following data gives the number x of hours spent for study and the percentage marks y he or she obtained for 10 students:
(10,51), (6,35), (15, 67), (11, 63), (7, 44), (19, 89), (17, 80), (3,26), (13, 50), (17,85).

(i) Fit a linear regression model to the data and estimate the corresponding regression coefficients. Obtain an estimator for the error variance.
(ii) Construct a 95% confidence interval for the average percentage of marks for all students who spend 12 hours in study for the examination.
(iii) Obtain a 95% prediction interval for the percentage of marks for all students who spend 12 hours in study for the examination.
(iv) Calculate the sample correlation coefficient r.
(v) Test whether $\rho = 0$ at 5% level of significance.

10.6 It is known that the yield of rice per acre depends on the amount of fertilizer used subject of course to a threshold. The following data gives the amount of fertilizer x in kilograms used per acre on a plot and the corresponding yield y of rice in kilograms:
(30, 550), (30, 940), (30, 1420), (40, 650), (40, 600), (40, 1450), (40, 1820), (50, 1230), (50, 1450), (50, 2300), (60, 1830), (60, 2450), (60, 2750).

(i) Can we conclude that the more the fertilizer used the higher is the yield? Test at 5% level of significance.

(ii) Obtain a 95% confidence interval for the mean yield if 40 kilograms of fertilizer is used per acre.

(iii) Construct a 95% prediction interval for the yield when 40 kilograms of fertilizer is used per care.

10.7 An experiment is conducted over 9 years to determine the relationship between the rainfall x in centimetres in a district and the rice yield y in quintals in that district for that year. The following data are obtained:
(1,1), (2,3), (3,2), (4,5), (5,5), (5,4), (6,7), (7,6), (8,9), (9,8).
Obtain an estimator for the rice yield if the rain fall is 10 centimetres.

10.8 The management of a manufacturing company of electronic tools has kept a record of maintenance cost for six machines of the same type of different ages. The company wants to know the relationship between the age x in years of the machine and cost y in hundreds of rupees of its maintenance. The following data was obtained for the six machines:
(2,70), (1,40), (3,100), (2,80), (1,30), (3,100).
What is the estimated maintenance cost for a machine which is 4 years old?

10.9 An advertising company is interested in finding out whether the number of television commercials x are related to the number of sales y of a certain product. The following data was obtained from 12 cities:
(9,29), (11,65), (12, 48), (4,12), (6,11), (6, 23), (5,10), (16, 55), (8,30), (2,11), (12, 75), (15, 90).

(i) Find the sample correlation coefficient r.

(ii) Test the null hypothesis that the correlation coefficient ρ between x and y is zero.

Appendix A

REFERENCES

Montgomery, D.C., Peck, E.A., and Vining, G.G. (2003) Introduction to Linear Regression Analysis, Third Edition, John Wiley and Sons (Asia), Singapore.

Rao, C.R. (1974) Linear Statistical Inference and its Applications, Second Edition, Wiley Eastern, New Delhi.

Appendix B

ANSWERS TO SELECTED EXERCISES

Chapter 2: 2.1: (a)$\frac{4}{5}$ (b)$\frac{1}{5}$; 2.2: $\frac{5}{9}$; 2.3: 0.0396; 2.4: 0.8; 2.7: 0.5; 2.8: $\frac{1}{3}$; 2.9: 0.36; 2.10: $\frac{3}{4}$; 2.14: 0.26.

Chapter 3: 3.1: $\frac{1}{2}$.

Chapter 4: 4.1: (a)$\frac{1}{4}$ (b)0 (c)$\frac{1}{4}$ (d)0; 4.2 (a)0 (b)$\frac{1}{4}$ (c)$\frac{1}{2}$ (d) $\frac{3}{4}$ (e) $\frac{1}{4}$ (f) 0.625; 4.9: $\frac{1}{3}$; 4.10: No; 4.14: $E(U) = 0, Var(U) = \frac{1}{6}$; 4.15: $E(X) = 1$; 4.16: $Var(X) = \frac{5}{6}$; 4.17: $E(X) = 2$; 4.19: $x_0 = 2.$; 4.20: 0; 4.22:$(\frac{1}{4})^{\frac{1}{4}}$; 4.31: $E(X) = 3, Var(X) = 2$; 4.39: (a)0.3078 (b) 0.4803 (c) 0.7711 (d) 0.3030; 4.40: 0.2112; 4.44: 0.9022; 4.51: $\alpha = \frac{4}{7}, \lambda = \frac{2}{7}$; 4.53: $\frac{\alpha-1}{\alpha+\beta-2}$.

Chapter 5: 5.3: $e^{-(500)\lambda}(1 + (500)\lambda)$; 5.8: (b)$\frac{1}{3}$; 5.9: $\frac{1-2e^{-1}+e^{-2}}{1-e^{-2}}$; 5.14: $\frac{5}{81}$; 5.28: $\rho = \frac{4}{5}$.

Chapter 6: 6.12: 0.78; 6.14: n-2; 6.17: 0.046.

Chapter 7: 7.6: 0.0616; 7.7: 0.4725; 7.8: 0.0527; 7.10: 0.7797; 7.12: 0.9876.

Appendix C

TABLES

C.1 Table for Standard Normal Probability Distribution

$$\Phi(x) = \int_{-\infty}^{x} \frac{1}{\sqrt{2\pi}} e^{-y^2/2}\, dy, \quad x > 0$$

x	$1-\Phi(x)$	x	$1-\Phi(x)$	x	$1-\Phi(x)$	x	$1-\Phi(x)$	x	$1-\Phi(x)$
0.00	0.5000	0.41	0.3409	0.82	0.2061	1.23	0.1093	1.64	0.0505
0.01	0.4960	0.42	0.3372	0.83	0.2033	1.24	0.1075	1.65	0.0495
0.02	0.4920	0.43	0.3336	0.84	0.2005	1.25	0.1056	1.66	0.0485
0.03	0.4880	0.44	0.3300	0.85	0.1977	1.26	0.1038	1.67	0.0475
0.04	0.4840	0.45	0.3264	0.86	0.1949	1.27	0.1020	1.68	0.0465
0.05	0.4801	0.46	0.3228	0.87	0.1922	1.28	0.1003	1.69	0.0455
0.06	0.4761	0.47	0.3192	0.88	0.1894	1.29	0.0985	1.70	0.0446
0.07	0.4721	0.48	0.3156	0.89	0.1867	1.30	0.0968	1.71	0.0436
0.08	0.4681	0.49	0.3121	0.90	0.1841	1.31	0.0951	1.72	0.0427
0.09	0.4641	0.50	0.3085	0.91	0.1814	1.32	0.0934	1.73	0.0418
0.10	0.4602	0.51	0.3050	0.92	0.1788	1.33	0.0918	1.74	0.0409
0.11	0.4562	0.52	0.3015	0.93	0.1762	1.34	0.0901	1.75	0.0401
0.12	0.4522	0.53	0.2981	0.94	0.1736	1.35	0.0885	1.76	0.0392
0.13	0.4483	0.54	0.2946	0.95	0.1711	1.36	0.0869	1.77	0.0384
0.14	0.4443	0.55	0.2912	0.96	0.1685	1.37	0.0853	1.78	0.0375
0.15	0.4404	0.56	0.2877	0.97	0.1660	1.38	0.0838	1.79	0.0367
0.16	0.4364	0.57	0.2843	0.98	0.1635	1.39	0.0823	1.80	0.0359
0.17	0.4325	0.58	0.2810	0.99	0.1611	1.40	0.0808	1.81	0.0351
0.18	0.4286	0.59	0.2776	1.00	0.1587	1.41	0.0793	1.82	0.0344
0.19	0.4247	0.60	0.2743	1.01	0.1562	1.42	0.0778	1.83	0.0336
0.20	0.4207	0.61	0.2709	1.02	0.1539	1.43	0.0764	1.84	0.0329
0.21	0.4168	0.62	0.2676	1.03	0.1515	1.44	0.0749	1.85	0.0322
0.22	0.4129	0.63	0.2643	1.04	0.1492	1.45	0.0735	1.86	0.0314
0.23	0.4090	0.64	0.2611	1.05	0.1469	1.46	0.0721	1.87	0.0307
0.24	0.4052	0.65	0.2578	1.06	0.1446	1.47	0.0708	1.88	0.0301
0.25	0.4013	0.66	0.2546	1.07	0.1423	1.48	0.0694	1.89	0.0294
0.26	0.3974	0.67	0.2514	1.08	0.1401	1.49	0.0681	1.90	0.0287
0.27	0.3936	0.68	0.2483	1.09	0.1379	1.50	0.0668	1.91	0.0281
0.28	0.3897	0.69	0.2451	1.10	0.1357	1.51	0.0655	1.92	0.0274
0.29	0.3859	0.70	0.2420	1.11	0.1335	1.52	0.0643	1.93	0.0268
0.30	0.3821	0.71	0.2389	1.12	0.1314	1.53	0.0630	1.94	0.0262
0.31	0.3783	0.72	0.2358	1.13	0.1292	1.54	0.0618	1.95	0.0256
0.32	0.3745	0.73	0.2327	1.14	0.1271	1.55	0.0606	1.96	0.0250
0.33	0.3707	0.74	0.2296	1.15	0.1251	1.56	0.0594	1.97	0.0244
0.34	0.3669	0.75	0.2266	1.16	0.1230	1.57	0.0582	1.98	0.0239
0.35	0.3632	0.76	0.2236	1.17	0.1210	1.58	0.0571	1.99	0.0233
0.36	0.3594	0.77	0.2206	1.18	0.1190	1.59	0.0559	2.00	0.0228
0.37	0.3557	0.78	0.2177	1.19	0.1170	1.60	0.0548	2.01	0.0222
0.38	0.3520	0.79	0.2148	1.20	0.1151	1.61	0.0537	2.02	0.0217
0.39	0.3483	0.80	0.2119	1.21	0.1131	1.62	0.0526	2.03	0.0212
0.40	0.3446	0.81	0.2090	1.22	0.1112	1.63	0.0516	2.04	0.0207

x	$1-\Phi(x)$	x	$1-\Phi(x)$	x	$1-\Phi(x)$	x	$1-\Phi(x)$	x	$1-\Phi(x)$
2.05	0.0202	2.46	0.0069	2.87	0.0021	3.28	0.0005	3.69	0.0001
2.06	0.0197	2.47	0.0068	2.88	0.0020	3.29	0.0005	3.70	0.0001
2.07	0.0192	2.48	0.0066	2.89	0.0019	3.30	0.0005	3.71	0.0001
2.08	0.0188	2.49	0.0064	2.90	0.0019	3.31	0.0005	3.72	0.0001
2.09	0.0183	2.50	0.0062	2.91	0.0018	3.32	0.0004	3.73	0.0001
2.10	0.0179	2.51	0.0060	2.92	0.0018	3.33	0.0004	3.74	0.0001
2.11	0.0174	2.52	0.0059	2.93	0.0017	3.34	0.0004	3.75	0.0001
2.12	0.0170	2.53	0.0057	2.94	0.0016	3.35	0.0004	3.76	0.0001
2.13	0.0166	2.54	0.0055	2.95	0.0016	3.36	0.0004	3.77	0.0001
2.14	0.0162	2.55	0.0054	2.96	0.0015	3.37	0.0004	3.78	0.0001
2.15	0.0158	2.56	0.0052	2.97	0.0015	3.38	0.0004	3.79	0.0001
2.16	0.0154	2.57	0.0051	2.98	0.0014	3.39	0.0003	3.80	0.0001
2.17	0.0150	2.58	0.0049	2.99	0.0014	3.40	0.0003	3.81	0.0001
2.18	0.0146	2.59	0.0048	3.00	0.0013	3.41	0.0003	3.82	0.0001
2.19	0.0143	2.60	0.0047	3.01	0.0013	3.42	0.0003	3.83	0.0001
2.20	0.0139	2.61	0.0045	3.02	0.0013	3.43	0.0003	3.84	0.0001
2.21	0.0136	2.62	0.0044	3.03	0.0012	3.44	0.0003	3.85	0.0001
2.22	0.0132	2.63	0.0043	3.04	0.0012	3.45	0.0003	3.86	0.0001
2.23	0.0129	2.64	0.0041	3.05	0.0011	3.46	0.0003	3.87	0.0000
2.24	0.0125	2.65	0.0040	3.06	0.0011	3.47	0.0003	3.88	0.0000
2.25	0.0122	2.66	0.0039	3.07	0.0011	3.48	0.0002	3.89	0.0000
2.26	0.0119	2.67	0.0038	3.08	0.0010	3.49	0.0002	3.90	0.0000
2.27	0.0116	2.68	0.0037	3.09	0.0010	3.50	0.0002	3.91	0.0000
2.28	0.0113	2.69	0.0036	3.10	0.0010	3.51	0.0002	3.92	0.0000
2.29	0.0110	2.70	0.0035	3.11	0.0009	3.52	0.0002	3.93	0.0000
2.30	0.0107	2.71	0.0034	3.12	0.0009	3.53	0.0002	3.94	0.0000
2.31	0.0104	2.72	0.0033	3.13	0.0009	3.54	0.0002	3.95	0.0000
2.32	0.0102	2.73	0.0032	3.14	0.0008	3.55	0.0002	3.96	0.0000
2.33	0.0099	2.74	0.0031	3.15	0.0008	3.56	0.0002	3.97	0.0000
2.34	0.0096	2.75	0.0030	3.16	0.0008	3.57	0.0002	3.98	0.0000
2.35	0.0094	2.76	0.0029	3.17	0.0008	3.58	0.0002	3.99	0.0000
2.36	0.0091	2.77	0.0028	3.18	0.0007	3.59	0.0002	4.00	0.0000
2.37	0.0089	2.78	0.0027	3.19	0.0007	3.60	0.0002	4.01	0.0000
2.38	0.0087	2.79	0.0026	3.20	0.0007	3.61	0.0001	4.02	0.0000
2.39	0.0084	2.80	0.0026	3.21	0.0007	3.62	0.0001	4.03	0.0000
2.40	0.0082	2.81	0.0025	3.22	0.0006	3.63	0.0001	4.04	0.0000
2.41	0.0080	2.82	0.0024	3.23	0.0006	3.64	0.0001	4.05	0.0000
2.42	0.0078	2.83	0.0023	3.24	0.0006	3.65	0.0001	4.06	0.0000
2.43	0.0075	2.84	0.0023	3.25	0.0006	3.66	0.0001	4.07	0.0000
2.44	0.0073	2.85	0.0022	3.26	0.0006	3.67	0.0001	4.08	0.0000
2.45	0.0071	2.86	0.0021	3.27	0.0005	3.68	0.0001	4.09	0.0000

C.2 Table for t-Distribution

$$P(t \le t_0) = \int_{-\infty}^{t_0} \frac{\Gamma[(n+1)/2]}{\sqrt{n\pi}\,\Gamma(n/2)(1+(u^2/n))^{(n+1)/2}}\,du$$

The following table gives the values of t_0 such that $P(t \le t_0) = p$ for some values of p when the number of degrees of freedom is n.

n	p=0.90	p=0.95	p=0.975	p=0.99	p=0.995
1	3.0777	6.3138	12.7062	31.8205	63.6567
2	1.8856	2.9200	4.3027	6.9646	9.9248
3	1.6377	2.3534	3.1824	4.5407	5.8409
4	1.5332	2.1318	2.7764	3.7469	4.6041
5	1.4759	2.0150	2.5706	3.3649	4.0321
6	1.4398	1.9432	2.4469	3.1427	3.7074
7	1.4149	1.8946	2.3646	2.9980	3.4995
8	1.3968	1.8595	2.3060	2.8965	3.3554
9	1.3830	1.8331	2.2622	2.8214	3.2498
10	1.3722	1.8125	2.2281	2.7638	3.1693
11	1.3634	1.7959	2.2010	2.7181	3.1058
12	1.3562	1.7823	2.1788	2.6810	3.0545
13	1.3502	1.7709	2.1604	2.6503	3.0123
14	1.3450	1.7613	2.1448	2.6245	2.9768
15	1.3406	1.7531	2.1314	2.6025	2.9467
16	1.3368	1.7459	2.1199	2.5835	2.9208
17	1.3334	1.7396	2.1098	2.5669	2.8982
18	1.3304	1.7341	2.1009	2.5524	2.8784
19	1.3277	1.7291	2.0930	2.5395	2.8609
20	1.3253	1.7247	2.0860	2.5280	2.8453
21	1.3232	1.7207	2.0796	2.5176	2.8314
22	1.3212	1.7171	2.0739	2.5083	2.8188
23	1.3195	1.7139	2.0687	2.4999	2.8073
24	1,3178	1.7109	2.0639	2.4922	2.7969
25	1.3163	1.7081	2.0595	2.4851	2.7874
26	1.3150	1.7056	2.0555	2.4786	2.7787
27	1.3137	1.7033	2.0518	2.4727	2.7707
28	1.3125	1.7011	2.0484	2.4671	2.7633
29	1.3114	1.6991	2.0452	2.4620	2.7564
30	1.3104	1.6973	2.0423	2.4573	2.7500

C.3 Table for Chi-square Distribution

$$P(\chi^2 \leq u) = \int_0^u \frac{v^{(n/2)-1}e^{-u/2}}{\Gamma(n/2)2^{n/2}}\,dv$$

The following table gives the values of u such that $P(\chi^2 \leq u) = p$ for some values of p when the number of degrees of freedom is n.

N	p=0.005	p=0.01	p=0.025	p=0.975	p=0.99	p=0.995
1	0.0000	0.0002	0.0010	5.0239	6.6349	7.8794
2	0.0100	0.0201	0.0506	7.3778	9.2103	10.5966
3	0.0717	0.1148	0.2158	9.3484	11.3449	12.8382
4	0.2070	0.2971	0.4844	11.1433	13.2767	14.8603
5	0.4117	0.5543	0.8312	12.8325	15.0863	16.7496
6	0.6757	0.8721	1.2373	14.4494	16.8119	18.5476
7	0.9893	1.2390	1.6899	16.0128	18.4753	20.2777
8	1.3444	1.6465	2.1797	17.5345	20.0902	21.9550
9	1.7349	2.0879	2.7004	19.0228	21.6660	23.5894
10	2.1559	2.5582	3.2470	20.4832	23.2093	25.1882
11	2.6032	3.0535	3.8157	21.9200	24.7250	26.7568
12	3.0738	3.5706	4.4038	23.3367	26.2170	28.2995
13	3.5650	4.1069	5.0088	24.7356	27.6882	29.8195
14	4.0747	4.6604	5.6287	26.1189	29.1412	31.3193
15	4.6009	5.2293	6.2621	27.4884	30.5779	32.8013
16	5.1422	5.8122	6.9077	28.8454	31.9999	34.2672
17	5.6972	6.4078	7.5642	30.1910	33.4087	35.7185
18	6.2648	7.0149	8.2307	31.5264	34.8053	37.1565
19	6.8440	7.6327	8.9065	32.8523	36.1909	38.5823
20	7.4338	8.2604	9.5908	34.1696	37.5662	39.9968
21	8.0337	8.8972	10.2829	35.4789	38.9322	41.4011
22	8.6427	9.5425	10.9823	36.7807	40.2894	42.7957
23	9.2604	10.1957	11.6886	38.0756	41.6384	44.1813
24	9.8862	10.8564	12.4012	39.3641	42.9798	45.5585
25	10.5197	11.5240	13.1197	40.6465	44.3141	46.9279
26	11.1602	12.1981	13.8439	41.9232	45.6417	48.2899
27	11.8076	12.8785	14.5734	43.1945	46.9629	49.6449
28	12.4613	13.5647	15.3079	44.4608	48.2782	50.9934
29	13.1211	14.2565	16.0471	45.7223	49.5879	52.3356
30	13.7867	14.9535	16.7908	46.9792	50.8922	53.6720

C.4 Table for F-Distribution

$$P(F \leq u) = \int_0^u \frac{\Gamma[(n_1+n_2)/2](n_1/n_2)^{n_1/2} v^{(n_1/2)-1}}{\Gamma(n_1/2)\Gamma(n_2/2)(1+(n_1/n_2)v)^{(n_1+n_2)/2}} dv$$

The following tables give the values of u such that $P(F \leq u) = p$ for $p = 0.95$ and $p = 0.975$ when the number of degrees of freedom is n_1=1 to 15 and n_2=1 to 15.

p = 0.95	n_2=1	n_2=2	n_2=3	n_2=4	n_2=5	n_2=6	n_2=7	n_2=8
n_1=1	161.4476	18.5128	10.1280	7.7086	6.6079	5.9874	5.5914	5.3177
n_1=2	199.5000	19.0000	9.5521	6.9443	5.7861	5.1433	4.7374	4.4590
n_1=3	215.7073	19.1643	9.2766	6.5914	5.4095	4.7571	4.3468	4.0662
n_1=4	224.5832	19.2468	9.1172	6.3882	5.1922	4.5337	4.1203	3.8379
n_1=5	230.1619	19.2964	9.0135	6.2561	5.0503	4.3874	3.9715	3.6875
n_1=6	233.9860	19.3295	8.9406	6.1631	4.9503	4.2839	3.8660	3.5806
n_1=7	236.7684	19.3532	8.8867	6.0942	4.8759	4.2067	3.7870	3.5005
n_1=8	238.8827	19.3710	8.8452	6.0410	4.8183	4.1468	3.7257	3.4381
n_1=9	240.5433	19.3848	8.8123	5.9988	4.7725	4.0990	3.6767	3.3881
n_1=10	241.8817	19.3959	8.7855	5.9644	4.7351	4.0600	3.6365	3.3472
n_1=11	242.9835	19.4050	8.7633	5.9358	4.7040	4.0274	3.6030	3.3130
n_1=12	243.9060	19.4125	8.7446	5.9117	4.6777	3.9999	3.5747	3.2839
n_1=13	244.6898	19.4189	8.7287	5.8911	4.6552	3.9764	3.5503	3.2590
n_1=14	245.3640	19.4244	8.7149	5.8733	4.6358	3.9559	3.5292	3.2374
n_1=15	245.9499	19.4291	8.7029	5.8578	4.6188	3.9381	3.5107	3.2184

p = 0.95	$n_2 = 9$	$n_2 = 10$	$n_2 = 11$	$n_2 = 12$	$n_2 = 13$	$n_2 = 14$	$n_2 = 15$
$n_1 = 1$	5.1174	4.9646	4.8443	4.7472	4.6672	4.6001	4.5431
$n_1 = 2$	4.2565	4.1028	3.9823	3.8853	3.8056	3.7389	3.6823
$n_1 = 3$	3.8625	3.7083	3.5874	3.4903	3.4105	3.3439	3.2874
$n_1 = 4$	3.6331	3.4780	3.3567	3.2592	3.1791	3.1122	3.0556
$n_1 = 5$	3.4817	3.3258	3.2039	3.1059	3.0254	2.9582	2.9013
$n_1 = 6$	3.3738	3.2172	3.0946	2.9961	2.9153	2.8477	2.7905
$n_1 = 7$	3.2927	3.1355	3.0123	2.9134	2.8321	2.7642	2.7066
$n_1 = 8$	3.2296	3.0717	2.9480	2.8486	2.7669	2.6987	2.6408
$n_1 = 9$	3.1789	3.0204	2.8962	2.7964	2.7144	2.6458	2.5876
$n_1 = 10$	3.1373	2.9782	2.8536	2.7534	2.6710	2.6022	2.5437
$n_1 = 11$	3.1025	2.9430	2.8179	2.7173	2.6347	2.5655	2.5068
$n_1 = 12$	3.0729	2.9130	2.7876	2.6866	2.6037	2.5342	2.4753
$n_1 = 13$	3.0475	2.8872	2.7614	2.6602	2.5769	2.5073	2.4481
$n_1 = 14$	3.0255	2.8647	2.7386	2.6371	2.5536	2.4837	2.4244
$n_1 = 15$	3.0061	2.8450	2.7186	2.6169	2.5331	2.4630	2.4034

$\mathbf{p = 0.975}$	$n_2 = 1$	$n_2 = 2$	$n_2 = 3$	$n_2 = 4$	$n_2 = 5$	$n_2 = 6$	$n_2 = 7$	$n_2 = 8$
$n_1 = 1$	647.7890	38.5063	17.4434	12.2179	10.0070	8.8131	8.0727	7.5709
$n_1 = 2$	799.5000	39.0000	16.0441	10.6491	8.4336	7.2599	6.5415	6.0595
$n_1 = 3$	864.1630	39.1655	15.4392	9.9792	7.7636	6.5988	5.8898	5.4160
$n_1 = 4$	899.5833	39.2484	15.1010	9.6045	7.3879	6.2272	5.5226	5.0526
$n_1 = 5$	921.8479	39.2982	14.8848	9.3645	7.1464	5.9876	5.2852	4.8173
$n_1 = 6$	937.1111	39.3315	14.7347	9.1973	6.9777	5.8198	5.1186	4.6517
$n_1 = 7$	948.2169	39.3552	14.6244	9.0741	6.8531	5.6955	4.9949	4.5286
$n_1 = 8$	956.6562	39.3730	14.5399	8.9796	6.7572	5.5996	4.8993	4.4333
$n_1 = 9$	963.2846	39.3869	14.4731	8.9047	6.6811	5.5234	4.8232	4.3572
$n_1 = 10$	968.6274	39.3980	14.4189	8.8439	6.6192	5.4613	4.7611	4.2951
$n_1 = 11$	973.0252	39.4071	14.3742	8.7935	6.5678	5.4098	4.7095	4.2434
$n_1 = 12$	976.7079	39.4146	14.3366	8.7512	6.5245	5.3662	4.6658	4.1997
$n_1 = 13$	979.8368	39.4210	14.3045	8.7150	6.4876	5.3290	4.6285	4.1622
$n_1 = 14$	982.5278	39.4265	14.2768	8.6838	6.4556	5.2968	4.5961	4.1297
$n_1 = 15$	984.8668	39.4313	14.2527	8.6565	6.4277	5.2687	4.5678	4.1012

$\mathbf{p = 0.975}$	$n_2=9$	$n_2=10$	$n_2=11$	$n_2=12$	$n_2=13$	$n_2=14$	$n_2=15$
$n_1=1$	7.2093	6.9367	6.7241	6.5538	6.4143	6.2979	6.1995
$n_1=2$	5.7147	5.4564	5.2559	5.0959	4.9653	4.8567	4.7650
$n_1=3$	5.0781	4.8256	4.6300	4.4742	4.3472	4.2417	4.1528
$n_1=4$	4.7181	4.4683	4.2751	4.1212	3.9959	3.8919	3.8043
$n_1=5$	4.4844	4.2361	4.0440	3.8911	3.7667	3.6634	3.5764
$n_1=6$	4.3197	4.0721	3.8807	3.7283	3.6043	3.5014	3.4147
$n_1=7$	4.1970	3.9498	3.7586	3.6065	3.4827	3.3799	3.2934
$n_1=8$	4.1020	3.8549	3.6638	3.5118	3.3880	3.2853	3.1987
$n_1=9$	4.0260	3.7790	3.5879	3.4358	3.3120	3.2093	3.1227
$n_1=10$	3.9639	3.7168	3.5257	3.3736	3.2497	3.1469	3.0602
$n_1=11$	3.9121	3.6649	3.4737	3.3215	3.1975	3.0946	3.0078
$n_1=12$	3.8682	3.6209	3.4296	3.2773	3.1532	3.0502	2.9633
$n_1=13$	3.8306	3.5832	3.3917	3.2393	3.1150	3.0119	2.9249
$n_1=14$	3.7980	3.5504	3.3588	3.2062	3.0819	2.9786	2.8915
$n_1=15$	3.7694	3.5217	3.3299	3.1772	3.0527	2.9493	2.8621

Index